MW01680709

A TEXTBOOK
of
GENETICS AND MOLECULAR BIOLOGY

A TEXTBOOK
of
GENETICS AND MOLECULAR BIOLOGY

Dr Sabyasachi Raychoudhuri
Former Head of the Dept. of Zoology (UG & PG)
Suri Vidyasagar College
Suri, Birbhum

New Central Book Agency (P) Ltd
LONDON
HYDERABAD ERNAKULAM BHUBANESWAR
DELHI KOLKATA PUNE

NCBA

REGD OFFICE

8/1 Chintamoni Das Lane, Kolkata 700 009, India
email: ncbapvtltd@eth.net

OVERSEAS

NCBA (UK) Ltd, 149 Park Avenue North, Northampton, NN3 2HY, UK

EXPORT

NCBA Exports Pvt. Ltd, 212 Shahpur Jat, New Delhi 110 049
email: ncbaexp@ncbapvtltd.com

BRANCHES

208B Peacock Lane, Shahpur Jat
New Delhi 110 049
email: ncbadel@ncbapvtltd.com

House No. 3-1-315, 1st Floor
Nimboliadda, Kachiguda, **Hyderabad** 500 027
email: ncbahydb@ncbapvtltd.com

Shop Nos. 3 & 4, Vinayak Towers
681/B Budhwar Peth, Appa Balwant Chowk
Pune 411 002
email: ncbapune@ncbapvtltd.com

GSS Shopping Complex, 1st Floor
Opposite College Post Office, Convent Road
Ernakulam 682 035
email: ncbaernk@ncbapvtltd.com

Shop No. 15, Bharati Towers, Ground Floor
Forest Park, **Bhubaneswar** 751 001
email: ncbabub@ncbapvtltd.com

A TEXTBOOK OF GENETICS AND MOLECULAR BIOLOGY

First Published: 2012

PUBLISHER AND TYPESETTER
New Central Book Agency (P) Ltd
8/1 Chintamoni Das Lane, Kolkata 700 009

PRINTER
New Central Book Agency (P) Ltd
Web-Offset Division, Dhulagarh, Sankrail, Howrah

COVER DESIGNER
Soumen Paul

COVER PRINTER
Biswanath Studio, Kolkata

PROJECT TEAM
Swapan Sarkar, Kakoli Banerjee, Koushik Samanta, and Prabhat Jas

ISBN: 978-81-7381-661-1

Price: ₹ 285.00

To
my respected teachers
and
beloved students

CONTENTS

PREFACE

During my teaching period for about forty years in Suri Vidyasagar College I came across various types of academic problems from hundreds of my students. Most of them were related to inadequacy in the knowledge of basic Genetics. Some were related to wrong or misconception of the subject. Since then I had a desire to write a textbook in Genetics in such a way that would serve the need of the degree students.

But Genetics is a vast subject and it is expanding its filed everyday, it is changing its mode of application every moment, so it became very difficult for me to choose the topics. I had to keep in mind that the topics of the subject should be such that they serve the purpose of basic understanding as well as modern application; they cover the syllabi of most of the Indian universities as well as recent researches in various fields. It was not a simple job.

In this book, I tried to explain the topics in easy language and lucid manner as far as possible. I intended to introduce the subject in such a way that a beginner can understand it without difficulty. I have tried to encourage the students to study the subject step by step so that they can find an interlinked relationship between the chapters. I desired to give them impetus to study and learn more of the subject. I hope they will bear a fair knowledge of genetics following this book.

Lastly, I would like to thank my friends and colleagues for encouraging me to write this book. I shall be glad if any suggestion comes to me for the betterment of the book. Further, I am thankful to Mr. Amitabha Sen, Director of New Central Book Agency (P) Ltd for giving me scope to write this book and I am indebted to Mr Swapan Sarkar, Ms. Kakoli Banerjee and Mr Prabhat Jas of New Central Book Agency (P) Ltd, without whose sincere help this work could not have been done. No thanks are sufficient for them.

Finally, I shall be glad if the students are benefited from this book.

15 August, 2011
Suri

Author

CHAPTER 1

INTRODUCTION

Genetics is the science of heredity and variation. It is inseparably related to the process of reproduction of organisms. The characteristics of individuals pass from one generation to the next through reproduction, sexual or asexual, and are maintained in the individuals through cellular reproduction, i.e., cell-division. The transmission of characteristics or traits through generations is called *heredity.* The other meaning of heredity is 'resemblance among individuals related by descent'. This is why the members of a family descended from the same parents have an overall resemblance. Yet there are differences among them too. The difference among individuals is called *variation.* Thus, *heredity and variation* are inversely related exhibiting both *similarities and differences* among offspring of the same parent. The study of heredity and variation is called *genetics,* the most extraordinary and expanding discipline of biology. It has undergone a phenomenal expansion during the last part of the last century and is still going on meaningful expansion. Meaningful in the sense that *genetics* is helping us understand the role of genes in every aspect of our lives, the process of continuity of life on earth, the process of evolution and formation of new species; meaningful in the sense that it is trying to overcome the problem of food-crisis through *hybridization, transgenesis* and *genetic selection* of plants and animals creating 'green revolution' and plenty of animal-food-products respectively; meaningful in the sense that it is trying to understand the causes of diseases like cancers and AIDS and also to overcome hundreds of hereditary hazards applying the principles of *gene therapy* and *genetic engineering;* meaningful in the sense that it has even invaded the field of politics and law having impact on the administration of drug, the patent law and the criminal law through regulation of *mutagenic chemicals* and *radioactive substances,* by raising questions on *patentability of living organisms* and by using *DNA-fingerprints* respectively. Thus genetics and life are inseparable.

The word 'gene' was coined by *Johannsen* in 1877 and the word 'genetics' (a Greek word, meaning "to generate") was coined by William Bateson in 1905, although the science of genetics itself began its journey as early as in 1865 when *Gregor Johann Mendel* read his paper to the Brünn Society for Natural History, after extensive experiments on inheritance in garden pea-plants for several years (1856-65) postulating *'principles of heredity'.* Even before

Mendel there were innumerable, though fruitless, sporadic attempts to find out the mystery of heredity. After many long years, in 1900, without having the knowledge of Mendel's work and discovery of Laws of heredity, *Hugo de Vries* of Holland, *Carl Correns* of Germany and *Erich Tschermak-Seysenegg* of Austria worked in this field separately and came to the same conclusion what Mendel had advocated in 1865. Thus, they rediscovered Mendel's work and brought him in the light as *'the father of genetics'*. Rediscovery of Mendelism led to many other discoveries in the field of genetics. *William Bateson* translated Mendel's paper in English and confirmed that the 'laws of heredity' as were postulated by Mendel in pea plant are also applicable to animals. Series of experiments and discoveries were going on since then. The *chromosome theory of heredity* (Sutton, 1902), phenomenon of *linkage* (Bateson and Punnett 1906), *inheritance pattern* of *metabolic disorder* in human (Garrod, 1908), the *genotype-phenotype concept* (Johannsen, 1909), the *Drosophila genetics* (Morgan, Bridges, Muller, Sturtevant, 1910-46), *radiation and mutations* (Muller, 1927), *cytoplasmic inheritance* (Jenkins, 1924), *transforming phenomenon of bacteria* (Griffith, 1928), *One-gene one-enzyme hypothesis* (Beadle and Tatum, 1941), *chemical nature of hereditary material* (Avery, MacLeod and McCarty, 1944), *Discovery of DNA-structure* (Watson, Crick, 1953), and hundreds like these came into the light through the study of genetics.

The period before the discovery of the molecular structure of DNA (1953) is known as the period of ***classical genetics***. In this period, the 'gene' remained as an abstract and invisible unit of heredity to the scientists. After the discovery of the chemical and molecular structure of DNA, the mystery of heredity became unveiled and the genetical study during this period is referred to as the study of ***molecular genetics***.

Then after the functions of gene became understandable. Through the investigations of **Benzer** (1955), it was possible to define 'gene' in terms of function, recombination and mutation; he called them *cistron, recon* and *muton* respectively. The genes were found to be little parts of DNA consisting linearly arranged nucleotides. The evidence of gene regulation in bacteria, called *operon* was discovered by Jacob and Monod (1961). Isolation of *DNA polymerase* enzyme from *Escherichia coli* bacteria (**Gaulian** and **Kornberg**), discovery of *base-sequences in t-RNA* (**Holley**), discovery of *Triplet Genetic Code* (**Crick**), synthesis of *RNA molecules* based on the work of genetic code (**Nirenberg**, **Mathaei**, **Leder** and **Khorana**), work and demonstration of *mobile* or *jumping gene* in bacteria by **Hedges** and **Jacob** (originally discovered by McClintock in 1956), discovery of synthetic bacteria that feed on oil in marine environment and make it pollution-free are some important milestones in the field of genetics.

A recent stream of genetics is the **recombinant technology** or the **genetic engineering**. In this younger science, the scientists "can cut apart and paste together DNA molecules almost at will, regardless of the source of the molecules." This has been possible by the discovery and isolation of *restriction endonucleases* and *DNA ligases* for "cutting and pasting" DNA molecules, respectively. The impact of this science has already been immense. The most important contribution of recombinant DNA technology is in the expansion of our basic knowledge of biology; it allowed scientists to isolate and characterize the structure and function of eukaryotic genes. Besides, the recombinant-DNA-technology or the genetic engineering led to the discovery and practice of ***gene therapy*** and the development of the process of ***cloning***. The potential commercial applications of genetic engineering have been strongly advocated too. The last but not the least milestone so far in the field of genetics is the mapping and sequencing *human genome* under the **Human Genome Project**.

Thus genetics, the science of heredity and variations, has reached its culmination in every corner of human life and civilization and in every sphere of biodiversity. The future of this science is of immense magnitude and importance to mankind as it has already laid the foundation of **eugenics** and **euphenics**. Eugenics is the modern branch of science that applies the principles of genetics to the improvement of humankind. The euphenics, on the other hand, is the symptomatic treatment of genetic diseases of human.

To conclude and summarize, there are three great milestones in genetics—**the work of Gregor Mendel** postulating the existence of 'genes' (factors) to explain how traits are inherited; **the work of Watson and Crick** elucidating the structure of DNA, and the researches in **Human Genome Project** determining the sequence of nucleotides in the DNA of human genome.

Is man trying to be immortal? Perhaps not. But surely he wants to live a long life without disease and distress in a pollution free world. The genetics will help.

CHAPTER 2

MENDELISM AND ITS EXTENSION

Gregor Johann Mendel, Mendel's experiments, Terminology and Symbols, Dominance and Recessiveness, Principles of Segregation and Independent Assortment, Methods for prediction of outcome of crosses, Back cross and Test cross, Dominance and its variation, Incomplete dominance, Codominance, Multiple alleles, Gene Interaction, Epistasis and Variations, Lethals, Pleiotropy, Trihybrid cross and Ratio, Probability in Mendelian Inheritance, Chi-square (X^2).

FIGURE 2.1

Gregor Johann Mendel.

2.1 Gregor Johann Mendel

The work of **Gregor Johann Mendel** (1822-84) gave 'birth' to **Genetics** in its truest sense. As such, he is honoured as the "father of genetics". So we begin with Mendel. The work (experiments, observations and inferences) of Mendel is known as ***Mendelism***.

Mendel was born in July 22, (actually on 20th, baptised on 22nd), 1822 in Heinzendorf, Silesia, Austria. In 1843, he joined the Augustinian Monastery at Brunn, Moravia, Austria. In 1847, he was ordained a priest adopting the clerical name, Gregor. In 1949, he became a school teacher in the Imperial Gymnasium, and from 1851 to 1853 he studied Physics, Chemistry, Mathematics and Natural Science in the University of Vienna. In 1854, he became a teacher in physics and Natural Science in a school in Brunn. From 1856 he conducted experiments on garden peas (*Pisum sativum*) and delivered his first lecture on his experiments and findings at the Brunn Natural History

Society in 1865. His works (Experiments on plant hybridization) were published in 1866. His conclusions on the principles of heredity in this paper became the foundation of today's genetics, after a long gap of unfortunate 34 years; he was unfortunate enough to be remained unnoticed and unappreciated in the science-world until 1900, only when he was rediscovered by Hugo de Vries in Holland, Carl Correns in Germany and Eric Von Tschermak-Seysenegg in Austria. In the meantime, in 1884, he died completely unknown and unhonoured. He became posthumously famous from the beginning of the 20th century.

2.1.1 Mendel's Experiments

Mendel performed his experiments with the garden pea plant, *Pisum sativum*. This selection of subject was one of the reasons of his success. Pea plant has the following advantageous characteristic features:

1. **It is a short-living annual plant and can be grown easily.**
2. **The flower is bisexual and self-fertilization is a natural phenomenon.**
3. The floral structure prevents contamination of foreign pollen, either by wind or by insect.
4. **The flowers are large and are so constructed that artificial cross-fertilization is simple to perform.** (Mendel had only to open a flower bud and remove the stamens before any pollen had been shed in order to prevent self-pollination. He could then dust the stigma with foreign pollen to have the fertilization he had desired).
5. **The plant has various contrasting characteristics among its varieties—**
   i. Stems are *tall* or *dwarf*.
   ii. Seeds are *yellow* or *green*.
   iii. Seeds are *round* or *wrinkled*.
   iv. The colour of pods is *green* or *yellow*.
   v. The ripe pods are *inflated* or *constricted*.
   vi. Flowers grow *axially* or *terminally*, and
   vii. The colour of flowers is *violet* or *white*.
6. **The hybrid plants are fertile**

Some consider Mendel as fortunate in choosing a *diploid* and *true-breeding* plant. But it should also be kept in mind that he had a very precised scientific aptitude as well as a very laborious effort for his work. He had a very good judgement in making hybridization between parents (pea) that differed in only one trait. When it was not possible he considered only one trait (for example, plant height) at a time during analysis of the progeny. Like a true and sincere laboratory-scientist he kept careful records of his experiments.

2.1.2 Experiment-I

Monohybrid Cross: Dominance, Recessiveness and Segregation

In this experiment, Mendel selected tall pea plants (about 2 meters high) and crossed them with dwarf pea-plants (half meter high). Both were *true-breeding* (i.e. produced same variety in self-crossing). He *emasculated* (removing of anther) one variety and applied pollen (by a brush, manually !) from the other variety to the stigma. The resultant seeds of this cross-fertilization were sown next year. Surprisingly, Mendel found that the hybrid plants from these seeds were only **tall** plants ! No intermediate height was obtained and the 'dwarf' had disappeared ! Same results were obtained in reciprocal crosses too (i.e., male tall x female dwarf or male dwarf x female tall). Mendel continued his experiments for the next generation and allowed the *hybrid tall* plants to undergo self-fertilization. The second gen-

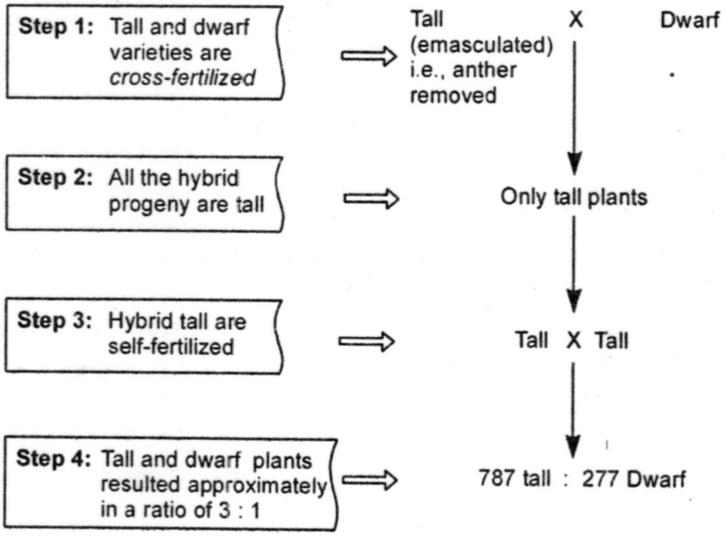

FIGURE 2.2

Mendel's experiment with tall and dwarf varieties of pea-plants.

eration progeny-result was more surprising ! Mendel found that the progeny consisted of *both tall and dwarf*. He counted their numbers and found that in this generation, there were 1064 progeny, of which 787 were tall and 277 were dwarf. Their ratio was approximately 3 tall : 1 dwarf.

Regarding the reappearance of the dwarf characteristics from the cross of two hybrid tall plants, Mendel's analysis was like this : the hybrids that were produced by crossing tall and dwarf varieties had the 'potency' to produce dwarf progeny though they themselves were tall in appearance. He inferred that this potency (factor for dwarfness) remained latent or hidden in the hybrids as it was masked by the expression of the factor for tallness. He explained that the case of expression of factor in hybrid is a phenomenon of *dominance* and the case of dormancy of factor in the hybrid is a phenomenon of *recessiveness*. He named the factors *dominant* and *recessive* respectively. According to this explanation the hybrid tall plants contained both the dominant and recessive factors but only the dominant factor (tallness) was expressed. Not only that, he explained how the dwarf reappeared in the next generation. He inferred that the dominant and recessive factors separated from each other when the hybrid talls were self-fertilized (Fig. 2.2).

TABLE 2.1

Cross	Traits
1. Round seed × Wrinkled seed	(Shape of seed)
2. Violet flower × White flower	(Colour of the flower)
3. Yellow seed × Green seed	(Colour of the seed)
4. Inflated pods × Constricted pods	(Shape of the pod)
5. Green pods × Yellow pods	(Colour of the pod)
6. Axial flower × Terminal flower	(Position of the flower)
7. Tall plant × Dwarf plant	(Height of the plant)

The above cross is called a *monohybrid cross* because a single trait (in this case, the height) was considered in the hybrids.

Mendel performed many other monohybrid crosses by crossing plants with the following traits (Table. 2.1). In each case he found that only *one* of the two contrasting traits appeared in the hybrids; when the hybrids were self-fertilized they produced two types of progeny in 3 : 1 ratio. The experiments revealed that each of the traits is controlled by a heritable factor; and the factor has two forms, one dominant and the other recessive. These factors are now called *genes*. The dominant and recessive forms of the genes are called *alleles* (alternate form of genes) or *allelomorphs*. Mendel also inferred that the factors (genes) in the parental strain (Step 1) remain in pairs (in terms of today's interpretation, *diploid*), and they are two identical copies of a gene (in modern terminology, *homozygous*). In the garden peas the gene for height has two alleles, one for tallness and one for dwarfness. The allele for tall is *dominant* and that for dwarf is *recessive*. What happens at the time of hybridization is that the alleles for each gene *separate* and enter into different reproductive cells (gametes) of the male and female parent. During fertilization those gametes fuse and give rise to progeny, and as a result, different types of offspring are produced. This *separation* of alleles and distribution of them in different gametes is called *segregation*. This concept of segregation is the most significant inference by Mendel from the monohybrid-cross experiment.

Terminology and Symbols

Mendel used some symbols and terminology to describe and explain his experiments. Such as, he used capital letter of the alphabet for dominant allele and the same letter of small script for recessive allele. He used the term 'Unit character' for allele or 'Splitting of hybrids' for segregation, or 'factor' for gene etc. The modern terminology used today to explain Mendelism are as follows:

Allele or Allelomorph:
: One of a pair of alternative forms of a gene that occurs at a given locus in a chromosome.

Back cross:
: The cross of an individual with one of its parents.

Codominance:
: Equal expression of both dominant and recessive alleles in the hybrid.

Dihybrid:
: An individual that is heterozygous for two pairs of alleles.

Diploid:
: An organism or a cell with two sets of chromosomes (2n) or two genomes.

Dominance:
: A condition in which one of the allelic pair is expressed suppressing the other (recessive).

Dominant allele:
: The allele that expresses itself even in heterozygous condition.

Epistasis:
: Interaction between products of non-allelic genes.

Filial generations:
: Successive generation of progeny in a controlled series of crosses starting from the generation next to the parental one.

F_1 The first filial generation. The first resultant generation of a given mating.

F_2 The second filial generation produced by self-fertilization of the F_1.

Gamete:
: A male or female reproductive cell.

Gene:
: The hereditary unit of a specific biological function. The unit of inheritance located at a fixed place in a chromosome.

Haploid:
: An organism or cell having only one complete set (n) of chromosome or genome.

Heredity:
Transmission of traits from parents to offspring.

Hybrid:
An offspring of a cross between unrelated (differing in one or more genes) strains.

Hybridization:
Interbreeding of two species or races or varieties of organisms. The process of forming hybrid.

Incomplete dominance:
Expression of 2 alleles in a heterozygous organism that makes the heterozygote distinct from either of its homozygous parents.

Independent assortment:
The separation and random distribution of alleles to the gametes that occur when genes are located in different chromosomes.

Monohybrid:
An offspring or hybrid of two homozygous parents that differ from one another by the alleles present at only one gene locus.

Monohybrid cross:
A cross between parents in which only one trait is considered.

Multiple allele:
A condition in which a particular gene occurs in 3 or more allelic forms.

Overdominance:
A condition in which heterozygotes are superior to either of the associated homozygotes.

P Symbol of the parental generation in an experiment of heredity.

Phenotype:
The exhibition of characteristics by an organism.

Recessive:
Referred to one member of an allelic pair lacking the ability to express itself when the other (the dominant) member is present.

Reciprocal cross:
Cross between strains with the sexes reversed; for example, Male A × Female B and Female A × Male B.

Segregation:
1. The separation of paternal and maternal chromosomes from each other at the time of meiosis. 2. The separation of alleles from each other in heterozygotes. 3. Mendel's first principle of inheritance. 4. The production of different phenotypes among offspring resulting from chromosome or allele separation in their heterozygous parents.

Self-fertilization:
Fertilization of the pollen and ovule occurring in the same plant.

Test cross:
1. Back cross to the recessive parental type. 2. A cross between genetically unknown individual with a double recessive tester to determine whether the individual in question is homozygous or heterozygous for a certain allele.

Trihybrid:
The offspring of homozygous parents, that differ in 3 pairs of genes.

Variation:
Occurrence of differences among individuals.

Wild type:
The customary phenotype or standard for comparison.

Phenotype and Genotype

In genetics, a small letter *of the recessive trait* is usually chosen to denote the alleles (that is, *d* for the allele of *d*warfism); the allele for *tallness* (which is dominant) is symbolized by corresponding capital letter (that is, *D* for the allele of tallness). Thus a tall pea plant is symbolized as DD and a dwarf as dd. This allelic constitution is called *genotype* and the physical appearance of

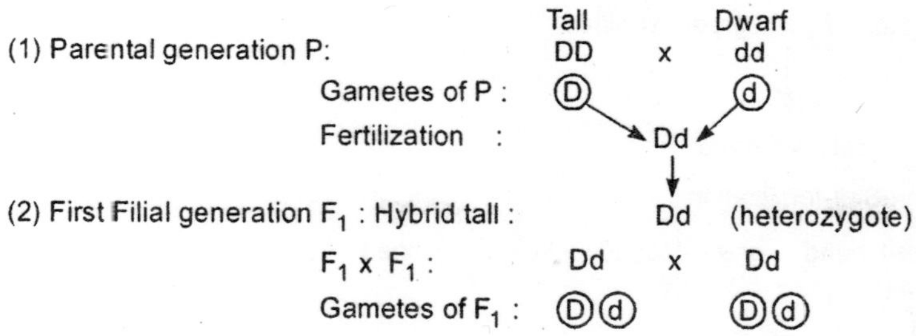

Self-fertilization of F_1 heterozygotes

♀ \ ♂	D	d
D	DD	Dd
d	Dd	dd

(3) Second Filial Generation F_2 :
Produce tall and dwarf plants in 3 : 1 ratio
1 homozygous tall
2 heterozygous tall
1 homozygous dwarf

Phenotypes	Genotypes	Genotypic ratio	Phenotypic ratio
Tall	DD	1	3
	Dd	2	
Dwarf	dd	1	1

(4) Ratio of the F_2 offspring:
Phenotype 3 tall : 1 dwarf, Genotype 1 : 2 : 1

FIGURE 2.3

Monohybrid Cross of Mendel.

each strain (i.e., tall or dwarf) is called *phenotype*.

Using symbols and terminology, the monohybrid cross of Mendel can be charted (Fig. 2.3). Here the symbols stand for alleles (so, doubly written).

Thus Monohybrid Cross of Mendel helped him to discover two important Principles or Laws of heredity:

1. **The Principle of Dominance**: *In a heterozygote, one allele may hide the other.*
2. **The Principle of Segregation**: *In a heterozygote, two different alleles segregate* from each other during the formation of gametes.

2.1.3 Experiment-II

Dihybrid Cross: Independent Assortment

The other experiment of Mendel is called **dihybrid cross** because he considered parental plants that differed in **2** traits. In one of his dihybrid crosses the two parental plants were— (1) plants that produced **yellow** and **round** seeds, and (2) Plants that produced **green** and **wrinkled** seeds. Clearly, there were two traits to be considered in a single plant, i.e., the *colour* and *texture* of the seed. From previous experiments it was known that the *yellow colour* and *round texture* characteristics were dominant. Thus, crossing the parental plants (Yellow, Round × green, wrinkled), he obtained all yellow-round seeds in F_1. He grew plants from these seeds and allowed them to self-fertilize ($F_1 \times F_1$). The F_2 seeds were of 4 types. Mendel classified them and counted them according to their phenotype. The types and number of the seeds were—

1. Yellow-Round 315
2. green-Round 108
3. Yellow-wrinkled 101
4. green-wrinkled 32.

P:	Yellow-Round x green-wrinkled			
	↓			
F_1:	Yellow-Round			
F_1 x F_1:	Self-fertilization			
F_2:	Yellow-Round 315	green-Round 108	Yellow-wrinkled 101	green-wrinkled 32

FIGURE 2.4

Mendel's Experiment considering two traits (Dihybrid Cross).

Out of the 4 types of seeds in F_2, two (Yellow-Round and green-wrinkled) resembled the parental strains while the other two (green-Round, and Yellow-wrinkled) showed a new combination of traits. The ratio of the types of the F_2 plants were approximately 9 Yellow-Round : 3 green-Round : 3 Yellow-wrinkled : 1 green-wrinkled (Fig. 2.4).

By symbol, the dihybrid cross can be illustrated and analyzed as follows: The small letters (*g* and *w*) are used to denote recessive traits (*green* and *wrinkled* respectively) and capital letters (*G* and *W*) are used to denote dominant traits (*Yellow* and *Round* respectively). The P-strains (Parental strains) are true-breeding and doubly homozygous (dominant *GGWW* and recessive *ggww*). The parental cross, formation of their gametes, production of F_1-plants, their self-fertilization and F_2-productions are shown in the (Fig. 2.5).

The *principle of segregation* is well applied here also. The gametes of the dominant plant (Yellow-Round seed) of P generation contain one copy of colour-gene (G) and one copy of texture gene (W) and become **GW**, while those of recessive plants (green-wrinkled seed) contain one copy of colour-gene (g) and one copy of texture-gene (w), being **gw**. The cross-fertilization of these gametes produce plants of F_1 which are hybrids and *heterozygous* for both the traits (doubly heterozygous, Gg Ww).

According to the *principle of segregation* 4 different gametic genotypes are produced from F_1-hybrids : GW, Gw, gW, and gw. Each gene must segregate its alleles *independently* as evidenced by the results that the F_1-hybrids produce 4 types of *gametic genotypes* in equal frequency and 16 types of equally frequent *zygotic genotypes*, by self-fertilization. But due to the *law of dominance*, the *phenotype frequency* in F_2 is : Yellow-Round $\frac{9}{16}$, Yellow-wrinkled $\frac{3}{16}$, green-Round $\frac{3}{16}$, and green-wrinkled $\frac{1}{16}$.

From the experiment it is evident that each gene has segregated its allele *independently*, (that is, there is *no linkage* between two genes, see Chapter VII). From this experiment and result Mendel postulated another principle of heredity, the *Principle of Independent Assortment.*

3. **The Principle of Independent Assortment**: *The alleles of different genes segregate (i.e., assort) independently of each other.*

This principle is more vividly supported by the behaviour of different pairs of chromosomes during *meiosis* (see chapter III) although there occur important exceptions due to *linkage* (see chapter VII).

IMPORTANT NOTES:

1. Mendel experimented on the mode of inheritance of 7 different traits in garden peas (*Pisum sativum*). Each trait was controlled by a different gene.

2. Mendel postulated three principles (laws) of inheritance:
   (a) The alleles of a gene are either dominant or recessive.
   (b) The alleles of a gene segregate from each other during the formation of gametes, and
   (c) The alleles of different genes assort (segregate) independently.

2.3 Methods for Prediction of Outcome of Crosses

There are 3 methods to predict the outcome of crosses. Those are—

1. **Punnett-Square Method**
2. **Forked-line Method** and
3. **Probability Method**.

P : Yellow-Round x green-wrinkled
GGWW ggww

Gametes : (GW) (gw)

F_1 : Yellow-Round (GgWw)

Gametes : (GW) (Gw) (gW) (gw)

F_2 : Self-fertilization :

F_2 : Cross

♀ \ ♂	GW	Gw	gW	gw
GW	GG WW Yellow-Round	GG Ww Yellow-Round	Gg WW Yellow-Round	Gg Ww Yellow-Round
Gw	GG Ww Yellow-Round	GG ww Yellow-wrinkled	Gg WW Yellow-Round	Gg ww Yellow-wrinkled
gW	Gg WW Yellow-Round	Gg Ww Yellow-Round	gg WW green-Round	gg Ww green-Round
gw	Gg Ww Yellow-Round	Gg ww Yellow-wrinkled	gg Ww green-Round	gg ww green-wrinkled

F_2 : Ratio

Phenotypes	Genotypes	Genotypic ratio	Phenotypic ratio
Yellow-Round	GG WW GG Ww Gg WW Gg Ww	1 2 2 4	9
Yellow-wrinkled	GG ww Gg ww	1 2	3
green-Round	gg WW gg Ww	1 2	3
green-wrinkled	gg ww	1	1

FIGURE 2.5
Mendel's Dihybrid Cross: Independent Assortment.

2.3.1 Punnett-Square Method

R.C. Punnett devised a method to predict the outcome of crosses. This method is called **Punnett-Square Method**. One example of the method is given in the (Fig. 2.6). The Punnett-Square contains the genotypes which are obtained by multiplying the frequencies of the two types of gametes produced by the heterozygous parents of the monohybrid cross.

2.3.3 Probability Method

Another method, alternative to the Punnett-Square and Forked-Line, is the **Probability Method** which is useful in complicated crosses. An intercross of heterozygotes with 4 different genes may be calculated as—

1. The fraction of progeny of 4 recessive homozygotes will be $\frac{1}{4} \times \frac{1}{4} \times \frac{1}{4} \times \frac{1}{4} = \frac{1}{256}$.

Cross : Aa × Aa

Female gametes \ Male gametes	A(1/2)	a(1/2)
A(1/2)	AA (1/4)	Aa (1/4)
a(1/2)	Aa (1/4)	aa (1/4)

⟹

Progeny

Genotype	Frequency	Phenotype	Frequency
AA	1/4	Dominant	3/4
Aa	1/2	Dominant	
aa	1/4	Recessive	1/4

FIGURE 2.6

Punnett-Square Method for prediction of outcome of crosses.

2.3.2 Forked-Line Method

The other prediction method is a forked-line method in which branching lines are used to enumerate the progeny. An example for predicting the outcome of an intercross involving *3 independently assorting genes in pea plant* is given in (Fig. 2.7).

2. The fraction of progeny of all 4 homozygotes (both dominant and recessive) will be $\frac{1}{2} \times \frac{1}{2} \times \frac{1}{2} \times \frac{1}{2} = \frac{1}{16}$.

Cross : Dd Gg Ww (Tall-Yellow-Round) × Dd Gg Ww (Tall-Yellow-Round)

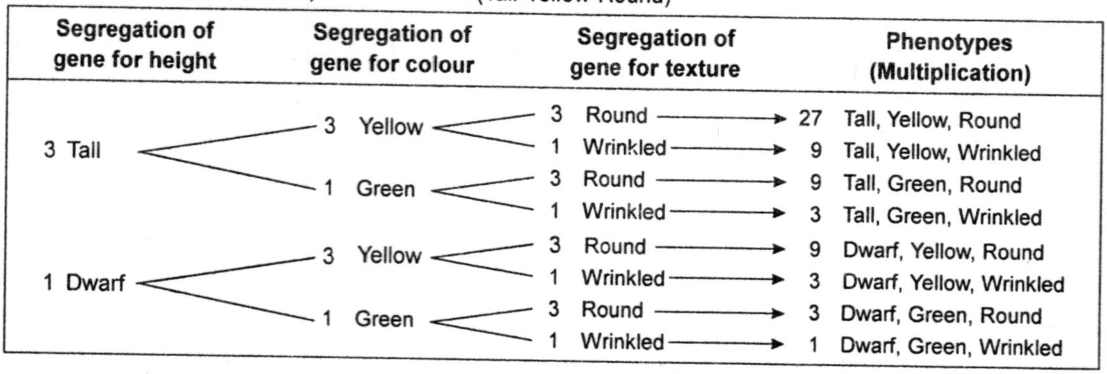

FIGURE 2.7

Forked-line Method.

2.4 Back Cross and Test Cross

When the F_1 individuals of an experiment are crossed *back* with any one of their parental stock (P), it is referred to as **back cross**. The parental stock is of two types—dominant and recessive. When this back cross is done only with the *recessive parents,* it is called **test cross**, because this cross is used to *test* whether an individual in question is *homozygous* or *heterozygous.* Thus, a test cross is also a back cross but it is a *cross of individuals with their recessive parents.* Thus, all test crosses are back crosses but all back crosses are not test crosses.

In a monohybrid experiment (considering single gene for height) of pea plant, the test cross product will be 50% homozygous and 50% heterozygous (i.e., an 1 : 1 ratio) or dwarf and tall will be in 1 : 1 ratio (Fig. 2.8).

- **Examples of Dominance in Animals**

Albinism in Mammals. Albinism means lack of pigmentation. The gene that permits normal pigmentation of the skin or fur is dominant over the gene that causes albinism (absence of pigment). Let us consider a case, where dominant gene be designated by C, and the recessive by c. With these two alleles the monohybrid cross will be as follows (Fig. 2.9).

P : Pigmented CC × Albino cc

Gametes : Ⓒ Ⓒ   ⓒ ⓒ

F_1 :

♀ \ ♂	Ⓒ	Ⓒ
ⓒ	Cc	Cc
ⓒ	Cc	Cc

FIGURE 2.9
Segregation and Dominance.

SUMMARY:

All F_1 individuals are heterozygous pigmented. No albino developed. This supports Mendel's **Law of dominance** and **Law of segregation**.

When two individuals having the genotype of the F_1 are crossed together, the following results are obtained. (For result see the next page).

- **Explanation**

Homologous genes *segregate* during formation of gametes; each gamete carries

P : Homozygous Tall (dominant) DD × Homozygous Dwarf (recessive) dd

↓

F_1 Dd (heterozygous) tall

1. **Back Cross :** F_1 (Tall) Dd × P (Tall) DD

↓

Result : 50% DD (homozygous) 50% Dd (heterozygous) (All Tall)

2. **Test Cross :** F_1 (Tall) Dd × P (Dwarf) dd (Doubly recessive)

↓

Result : 50% Dd, 50% dd
(50% tall, 50% dwarf or, 1 : 1 ratio)

FIGURE 2.8
Back Cross and Test Cross.

F_1 :	Pigmented Cc	x	Pigmented Cc
Gametes :	Ⓒ ⓒ		Ⓒ ⓒ

F_2 :

♀ \ ♂	Ⓒ	ⓒ
Ⓒ	CC Pigmented	Cc Pigmented
ⓒ	Cc Pigmented	cc Albino

Summary of F_2 : Pigmented and albino develop in 3 : 1 ratio.

Phenotypes	Phenotypic ratio	Genotypes	Genotypic ratio
Pigmented	3	CC Cc	1 2
Albino	1	cc	1

one allele for each trait. The F_2 generation contains *albino* which has reappeared ! It indicates that the gene for this trait was not lost but was only *hidden* in the F_1. Segregation of alleles made it possible to *reappear*. The ratio of the individuals is also 3 pigmented : 1 albino, supporting Mendel's *law of dominance* and *law of segregation.*

2.5 Dominance and its Variation

In Mendel's experiment with garden pea-plant the monohybrid and dihybrid crosses showed *complete or full dominance* for all the 7 pairs of alleles he studied (see Table 2.1, Monohybrid Cross, 2.1.2). But in later days some sort of deviation in the expression of dominance was noted in some crosses by some experimenters. Those are—

1. **Incomplete dominance or Semi-dominance**: The white (w) and red (W) flowers in snapdragon (*Antirrhinum majus*) are found in homozygous (ww, WW) condition. But when they (P) are crossed, they produce (F_1) heterozygotes (Ww) with pink flowers *instead of red.* Therefore, the allele for the red colour (W) is said to be **incompletely** or **partially dominant** over the white colour (w). This phenomenon is called **semi-dominance** because the heterozygote's phenotype is *intermediate* between the phenotypes of the two homozygotes (Fig. 2.10). When the F_1-pink heterozygotes are self-fertilized, the F_2-phenotype ratio is 1 red : 2 pink : 1 white (Fig. 2.10).

2. **Codominance:** Another variation from complete dominance is noticed in human **blood groups**. Here, the heterozygote shows characteristics found in each of the associated homozygotes. As for example, the alleles that govern the **M-N blood type** in humans are **codominant**. Allele L^M for M-blood group is *codominant with its allele* L^N for N-blood group. The heterozygote ($L^M L^N$) expresses the characteristics of both M and N antigens (MN-blood type), that is, $L^M L^N$ heterozygotes have RBC with

P	Red (WW) homozygous	×		White (ww) homozygous
F_1		Ww (Pink) heterozygous		
$F_1 \times F_1$		Ww × Ww		
F_2	WW	Ww	Ww	ww
	1 Red	2 Pink		1 White

FIGURE 2.10
Incomplete Dominance in snapdragon flower colour.

M and N antigens on their membranes. Since the two alleles encode for non-identical antigens, a mating between a homozygous M-type person (L^ML^M) and a homozygous N-type person (L^NL^N) would produce all heterozygous (L^ML^N) offspring. Mating between heterozygotes ($L^ML^N \times L^ML^N$) would produce M-type (L^ML^M), MN-type (L^ML^N) and N-type (L^NL^N) in 1 : 2 : 1 ratio. It is an exception to the 3 : 1 ratio of complete dominance. As the alleles L^M and L^N are codominant with each other the alleles are *both expressed* in the heterozygote. This phenomenon is called **codominance** (Fig. 2.11).

Codominance is also exhibited by the alleles determining the coat-colour of the short horn breed of cattle. When a cattle of red-coat (C^RC^R) is crossed with a cattle of white-coat (C^WC^W), the F_1 heterozygotes are produced with **roan-coat** (C^RC^W). In roan, the red and while hairs are found to develop in patches, but no hair of intermediate colour of red and white develop. When these F_1 roan heterozygotes are interbred, the produced F_2 offspring become red, roan and white in 1 : 2 : 1 ratio, indicating deviation from 3 : 1 ratio of simple or complete dominance, and codominance of the alleles for red and white coat-colour (Fig. 2.12).

2.5.1 Some Dominant and Recessive Traits in Human.

Dominant	Recessive	Dominant	Recessive
Brown eyes	Blue eyes	Free ear-lobe	Attached ear-lobe
Long eye-lashes	Short eye-lashes	Normal vision	Colour blind
Dark hair	Blond hair	Normal skin	Albino
Curly hair	Straight hair	Dimpled cheek	Lack of dimple

Genotype	Blood group (antigen present)	*Reactions with anti-sera	
		Anti-M serum	Anti-N serum
L^ML^M	M	+	—
L^ML^N	M and N	+	+
L^NL^N	N	—	+

* + = agglutination, – = nonagglutination.

FIGURE 2.11
Codominance in MN-blood group.

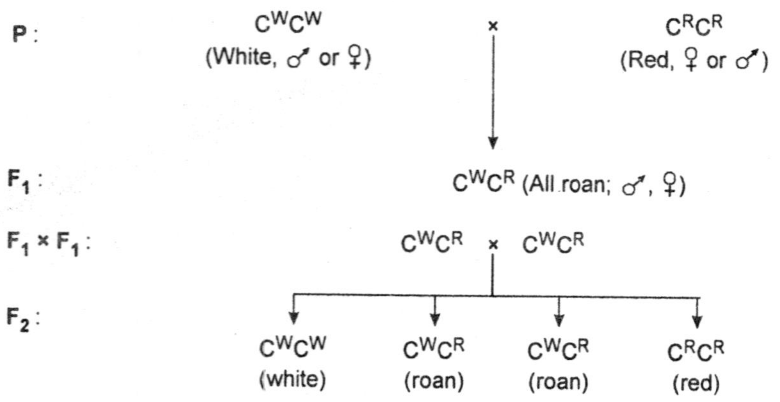

FIGURE 2.12
Codominance in coat colour of shorthorn cattle.

2.6 Multiple Alleles

Mendelian concept of *two alleles per gene* (one dominant the other recessive, or two dominants, or two recessives) was soon found not to be universal. Instead, *genes with three or more alleles* were discovered. These are called genes with **multiple alleles**.

As for example, the coat-colour determining gene of rabbit has *four alleles* : c (albino), c^h (himalayan), c^{ch} (chinchila) and c^+ (wild-type*). When homozygous, cc produces 'colourless' or 'albino' rabbit (white hairs all over the body), c^hc^h produces 'himalayan' rabbit (white hairs everywhere, but black hairs on the extremities), $c^{ch}c^{ch}$ produces 'chinchila' rabbit (white hairs with black tips on the body), c^+c^+ produces 'wild' rabbit (coloured hairs over the entire body).

The human blood-group A, B, AB and O also are governed by *multiple allele.* When only the A antigen is present on the RBC-membrane, the blood-group is A; when only B antigen is present, the blood-group is B; when both antigens are present the blood-group is AB, and when no antigen is present, the group is O (Landsteiner 1900).

The gene responsible for A and B antigens is I. I gene has three alleles— I^A, I^B and i. The I^A produces A antigen, I^B produces B antigen but i does not produce any antigen. Table 2.2 shows the possible 6 genotypes and 4 phenotypes of blood-group.

The inheritance of this system follows Mendelian principles. An individual with blood group O, must be i/i in genotype. Both parents of this person could be O (i/i × i/i),

TABLE 2.2: Genotypes and Phenotypes of ABO blood-group

Genotype	Blood-group	A-antigen present	B-antigen present
I^AI^A or I^Ai	A	+	–
I^BI^B or I^Bi	B	–	+
I^AI^B	AB	+	+
ii	O	–	–

* Because most rabbits in wild populations are homozygous for c^+ allele, it is called the **wild-type.** It is customary to represent wild type alleles by putting a '+' sign after the letter of the gene. In clear context only '+' is used to denote wild. Thus, '++' denotes a wild variety in homozygous condition.

both could be A ($I^A/i \times I^A/i$, to produce $\frac{1}{4}$th i/i progeny), both could be B ($I^B/i \times I^B/i$), or one could be A and one could be B ($I^A/i \times I^B/i$). Each parent must be either homozygous i or heterozygous, with i as one of the two alleles.

2.7 Gene Interaction

Sometimes the expression of an allele of one gene may alter the expression of one or more of the alleles of another (non-allelic) gene.

There are some different comb shapes in chickens. For instance, *Wyandottes* have **rose** comb, *brahmas* have **pea** comb and *Leghorns* have **single** comb. They are all true-breeding. In an experiment of Bateson and Punnett, when wyandottes (rose comb) were crossed with brahmas (pea comb), the F_1 chickens showed a different kind of comb, called **walnut**, which was not present in either parent. When these F_1 chickens (with **walnut** comb) were mated among themselves ($F_1 \times F_1$), an F_2 population was obtained in which about $\frac{9}{16}$ chickens were **walnut**, $\frac{3}{16}$ were **rose**, $\frac{3}{16}$ were **pea** and $\frac{1}{16}$ were with **single** comb. It showed a 9 : 3 : 3 : 1 ratio in F_2, but the phenotypes did not agree with that of the original work of Mendel (See 2.1.3). Neither the **single comb** nor the **walnut** was present in the P generation. These two phenotypes, according to Bateson and Punnett, resulted due to **gene interaction.** The genes R and P are nonallelic but dominant. The **rose** comb developed by the combination of homozygous recessive genes *pp* and homozygous or heterozygous dominant genes *RR* or *Rr*. The **pea** comb was resulted by the combination of homozygous recessive *rr* and homozygous or heterozygous dominant *PP* or *Pp*. The **single** type of comb

P: Rose comb (RR pp) X Pea comb (rr PP)

F_1: Walnut (Rr Pp)

$F_1 \times F_1$:

	Phenotypes	Genotypes	Ratio	Cause
	Walnut	RRPP RRPp RrPP RrPp	9	Interaction of R and p
	Rose	RRpp Rrpp	3	Dominance of R over r
	Pea	rrPP rrPp	3	Dominance of P over p
	Single	rrpp	1	Recessiveness of rr and pp

FIGURE 2.13
Case of Interaction of genes.

was produced by double recessive, *rrpp*. Separately, the **R** and **P** genes determine the shape of **rose** comb and **pea** comb respectively but when they remain together in a single individual (as in the F_1-Rr Pp of this cross), they ***interact*** with each other and produces **walnut**. Again, in the mating of two walnut chickens ($F_1 \times F_1$) the two genes (R and P) show various degrees of ***interaction*** and four types of chickens are produced in F_2 (Fig. 2.13).

2.8 Epistasis

Epistasis is a type of functional interaction of non allelic genes. *When an allele or genotype at one locus "masks" or inhibits the expression of a non-allele or genotype at a distinct locus,* it is called **epistasis.**

In epistatic interaction in dihybrid cross the F_2 phenotypes are produced in less than 4 types (which is a clear deviation from Mendelian result). Their ratio also differ from the Mendelian dihybrid cross, i.e., from 9 : 3 : 3 : 1. In epistatic reaction the F_2 phenotypic ratios may be 12 : 3 : 1, 9 : 3 : 4, 9 : 6 : 1, 9 : 7, 15 : 1 or 13 : 3 as to the degree of interaction between non-allelic genes. The allele of one gene that "masks" the expression of another non-allele, is called **epistatic allele** while the suppressed allele is called **hypostatic allele**.

EXAMPLES:

2.8.1 Dominant Epistasis (12 : 3 : 1)

In dogs, the coat colour depends on the interaction of alleles of two genes, where I* is the dominant epistatic allele. Table 2.3 shows the *epistatic* and *hypostatic* alleles, their *phenotypic expression* and F_2 *phenotypic* ratio, in a dihybrid cross of dog's coat-colour.

2.8.2 Recessive Epistasis (9 : 3 : 4)

When recessive alleles of one gene mask the expression of their non-alleles of other gene, it is called **recessive epistasis.**

In mice, the coat-colour is controlled by the genetic interaction of alleles of two genes. The table 2.4 shows the *epistatic* and *hypostatic* alleles, *phenotypic expression* and F_2*-phenotypic ratio.*

2.8.3 Duplicate Genes and Cumulative Epistasis (9 : 6 : 1)

There are three coat colours—red, sandy and white in the Duroc-Jersey breed of pigs which are expressed by the interaction of dominant alleles of two gene-loci. The

TABLE 2.3: Dominant epistasis in dog's coat-colour

Epistatic alleles	Hypostatic alleles	Phenotypic expression	F_2 Phenotypic ratio
II, Ii	BB, Bb, bb	I (White, nonpigmented)	White = 12
ii	BB, Bb	B (Black)	Black = 3
ii	bb	b (brown)	Brown = 1

TABLE 2.4: Recessive epistasis in the coat-colour of mice

Epistatic alleles	Hypostatic alleles	Phenotypic expression	F_2 Phenotypic ratio
CC, Cc	AA, Aa	A = Agouti	Agouti = 9
CC, Cc	aa	a = Black	Black = 3
cc	AA, Aa, aa	c = Albino	Albino = 4

* I for 'inhibitor'

TABLE 2.5: Duplicate gene and Cumulative epistasis

Epistatic alleles	Hypostatic alleles	Phenotypic expression	F_2-ratio
RR, Rr	SS, Ss	R + S (Cumulative effect) = red	Red = 9
RR, Rr	ss	R } = Sandy	Sandy = 6
rr	SS, Ss	S } = Sandy	
rr	ss	Neither of them (albino)	White = 1

dominant condition, homozygous or heterozygous, produces the same phenotype. Presence of dominant alleles on both epistatic and hypostatic loci produce collective phenotypic expression. Table 2.5 shows the alleles and their interactive expressions.

2.8.4 Duplicate Recessive Genes (9 : 7)

In sweet pea plant (*Lathyrus adoratus*), the colour of flowers are determined by epistatic recessive genes. In an experiment, Bateson and Punnett crossed two different white flowered varieties of sweet peas. The F_1-plants (Fig. 2.14) all had purple flowers. When the F_1 plants were intercrossed, they produced F_2 progeny with purple and white flowers in 9 : 7 ratio.

In this experiment, one variety of plant was ccPP and the other was CCpp. The F_1 hybrids were all CcPp and purple. (Since, C and P are dominant to their alleles c and p, respectively, the F_1 plants could synthesize anthocyanin pigment, and the flowers were purple).

In the F_2 generation, $\frac{9}{16}$ of the progeny carry at least one dominant allele of each gene (C- P-) and bear purple flowers. The remaining $\frac{7}{16}$ are homozygous recessive for

P : White CCpp x White ccPP

F_1 : CcPp (purple)

$F_1 : F_1$ CcPp x CcPp

F_2

	CP	Cp	cP	cp
CP	CCPP Purple	CCPp Purple	CcPP Purple	CcPp Purple
Cp	CCPp Purple	CCpp White	CcPp Purple	Ccpp White
cP	CcPP Purple	CcPp Purple	ccPP White	ccPp White
cp	CcPp Purple	Ccpp White	ccPp White	ccpp White

FIGURE 2.14
Experiment for duplicate recessive gene-epistasis, (9 : 7) ratio.

P : Triangular AABB x Ovoid aabb

↓

F_1 : AaBb (Triangular)

$F_1 \times F_1$ AaBb x AaBb

F_2

	(AB)	(Ab)	(aB)	(ab)
(AB)	AABB Triangular	AABb Triangular	AaBB Triangular	AaBb Triangular
(Ab)	AABb Triangular	AAbb Triangular	AaBb Triangular	Aabb Triangular
(aB)	AaBB Trangular	AaBb Triangular	aaBB Triangular	aaBb Triangular
(ab)	AaBb Triangular	Aabb Triangular	aaBb Triangular	aabb Ovoid

Triangular : Ovoid = 15 : 1

FIGURE 2.15
Experiment for duplicate gene-control (15 : 1 ratio).

one or the other or both of the genes ($\frac{3}{16}$ ccP, $\frac{3}{16}$ C-pp, $\frac{1}{16}$ ccpp) and bear white flowers.

2.8.5 Duplicate Gene Control (15 : 1 Ratio)

If two non-allelic genes code for enzymes that are able to catalyze the same reaction, a 15 : 1 dihybrid phenotype ratio might result. Such ratio has been found to occur in the crosses (Fig. 2.15) involving genes controlling the shape of the seed capsule of *Bursa bursa-pastoris*. The following phenotypes develop in genic interaction of the following genotypes A–B– (triangular), aaB– (triangular), A–bb (triangular) and aabb (ovoid).

2.8.6 Dominant and Recessive Epistasis (13 : 3 Ratio)

A F_2-phenotypic 13 : 3 ratio may be obtained, when the dominant alleles of one gene locus in homozygous and heterozygous condition (AA and Aa respectively) and the homozygous recessive alleles (bb) of another gene locus produce the same phenotype. The white feather-colour of leghorn fowl is caused by dominant genotype CCII, while the white feather colour of plymouth rock

TABLE 2.6: Interaction of dominant and recessive epistasis

Epistatic alleles	Hypostatic alleles	Phenotypic expression	F_2 Phenotypic ratio
II, Ii	CC, Cc, cc	I (dominant inhibition)	White = 12
ii	CC, Cc	C	Coloured = 3
ii	cc	c	White = 1

Summary : White 13 : Coloured 1

fowl is caused by the recessive genotype ccii. When white varieties of both of them are crossed, the F_1 white colour birds (hybrids) are produced. Interbreeding of these ($F_1 \times F_1$) produce white and coloured birds in F_2 in 13 : 3 ratio. The table 2.6 shows the interaction of such type of genes.

2.9 Lethals

Some genes may cause death in some individuals. Death or inviability may result due to impaired biochemical and physical functioning. The responsible genes for inviability are called **lethal genes**. When the lethal genes are *dominant* and immediately expressed, the individuals carrying them die and the gene becomes lost. Some dominant lethal genes may show delayed effect. The *recessive lethal genes*, when present in individuals in heterozygous condition, show no effect. But when mating between carriers of this recessive lethal gene occurs they may cause inviability of the progeny. This may also show a deviation from the 3 : 1 Mendelian ratio. As for example, an aberrant form of chicken is called "creeper" in heterozygous condition

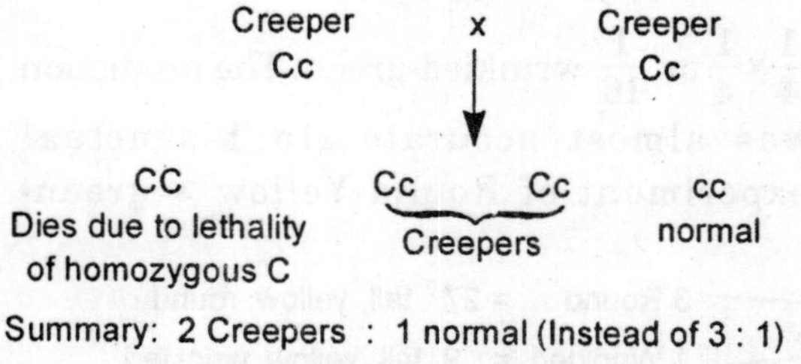

FIGURE 2.16
Cross, showing lethality.

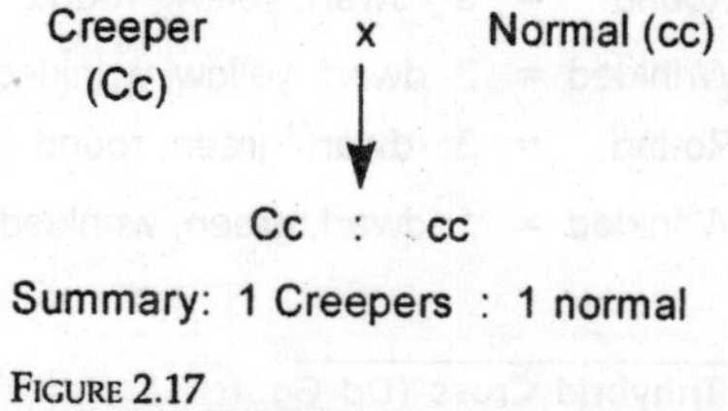

FIGURE 2.17
Cross between a normal and a creeper.

(Cc). Their homozygous genotype (CC) is lethal. When two "creepers" are mated, a ratio of two creepers to one normal (2 : 1) instead of 3 : 1 is produced (Fig. 2.16).

When a creeper is mated to a normal (cc) chicken, the expected 'test cross' result of 1 creeper : 1 normal chicken appears (Fig. 2.17).

Thalassemia is a hemoglobin-disease in human and is nearly a lethal disease. The disease is controlled by a lethal gene c, which produces **thalassemia major** or **Cooley's anaemia** in homozygous (cc) condition and causes death to the patient. Its heterozygous condition (Cc) causes **thalassemia minor** and they are called carriers.

2.10 Pleiotropy

When a gene influences more than one phenotype it is called **pleiotropy**. The gene for PKU (Phenylketoneuria) in man is an example of pleiotropy. Recessive mutation of the gene responsible for PKU causes accumulation of toxic substances in brain, mental impairment, reduction in melanin pigment synthesis etc. This series of phenotypic effects is associated with the pleiotropic gene. The effects result from the interconnection between the biochemical and cellular pathways controlled by the gene.

2.11 Trihybrid Cross and Ratio

Following Mendel's monohybrid and dihybrid crosses, **trihybrid** (differing in 3 traits) **crosses** or crosses considering more traits could be performed and their results, although complex, could be calculated by resolving them into monohybrid crosses.

When a true breeding *tall* (DD) garden pea plant with *yellow* (GG) and *round* (WW) seeds is crossed with a recessive variety of it (*dwarf, green* and *wrinkled* ddggww), the F_1 generation hybrids are all *tall, yellow* and *round* (genotype being DdGg Ww) (Fig. 2.18).

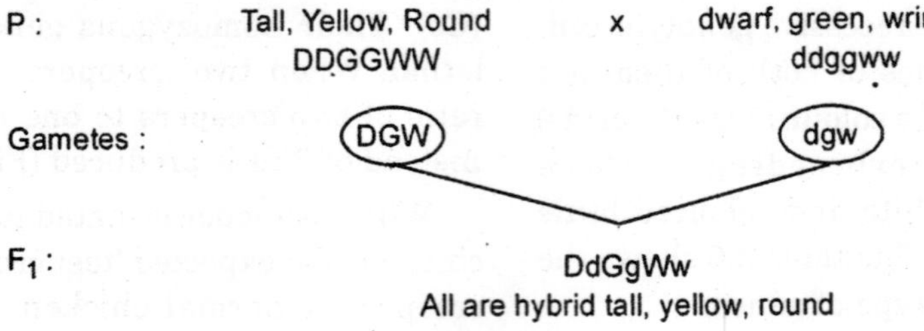

FIGURE 2.18

A trihybrid production.

When the F_1 products are self-crossed (Dd Gg Ww × Dd Gg Ww), the F_2 results are obtained in 27 : 9 : 9 : 3 : 9 : 3 : 3 : 1 phenotypic ratio which can be shown mathematically (resolving into monohybrid results) by forked-line method (Fig. 2.19).

When a *test cross* of these F_1 hybrids (i.e., cross with full recessive type, Dd Gg Ww × dd gg ww), is performed, 8 kinds of plants are produced and their phenotypic and genotypic ratio are 1 : 1 : 1 : 1 : 1 : 1 : 1 : 1. It is explained in the (Fig. 2.20).

2.12 Probability in Mendelian Inheritance

Probability is the chance that may occur as an outcome of an event. The law of probability is applied for the solution of Mendelian problems. It is considered to explain both *independent assortment* and *segregation* principles. Mendel was able to recognize and interpret the dihybrid ratio of 9 : 3 : 3 : 1 as a multiple of 3 : 1 monohybrid ratio. In the explanation the binomial expansion of $(p + q)^n$, can be used. As Mendel's phenotypic ratio in monohybrid cross was 3 : 1, it can be used as $(3 + 1)^2$ in case of a dihybrid cross, or $(3 + 1)^3$ in a trihybrid cross. In dihybrid cross, the result comes $(3 + 1)^2 = 9 + 3 + 3 + 1$ and in trihybrid the result comes $(3 + 1)^3 = 27 + 9 + 9 + 9 + 3 + 3 + 3 + 1$, which satisfy Mendel's experiment and result. On the basis of the law of probability for his dihybrid cross, Mendel predicted that $\frac{3}{4} \times \frac{3}{4}$ or $\frac{9}{16}$ of the F_2 would be round-yellow; $\frac{3}{4} \times \frac{1}{4}$ or $\frac{3}{16}$ round-green; $\frac{1}{4} \times \frac{3}{4}$ or $\frac{3}{16}$ wrinkled-yellow; and $\frac{1}{4} \times \frac{1}{4}$ or $\frac{1}{16}$ wrinkled-green. The prediction was almost accurate (In his actual experiment of Round-Yellow × green-

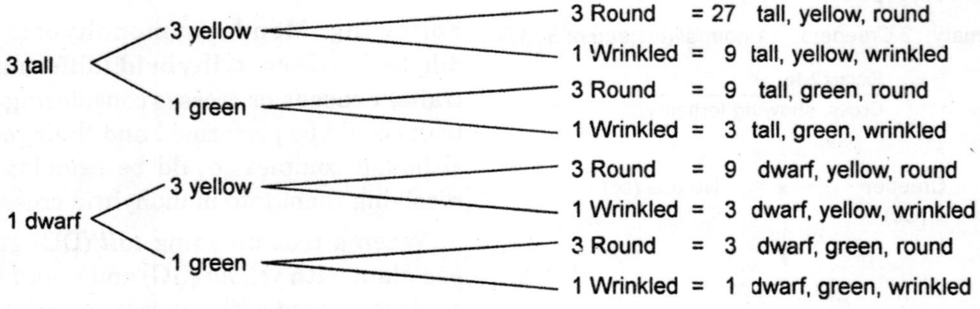

FIGURE 2.19

Forked-line method to predict F_2 result in Trihybrid Cross (Dd Gg Ww × Dd Gg Ww).

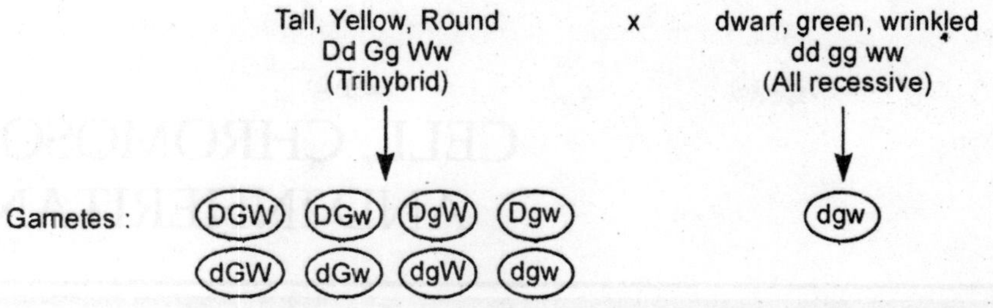

Result :

Phenotype	Genotype	Genotypic ratio	Phenotypic ratio
Tall, Yellow, Round	Dd Gg Ww	1	1
Tall, Yellow, Wrinkled	Dd Gg ww	1	1
Tall, green, Round	Dd gg Ww	1	1
Tall, green, Wrinkled	Dd gg ww	1	1
Dwarf, Yellow, Round	dd Gg Ww	1	1
Dwarf, Yellow, Wrinkled	dd Gg ww	1	1
Dwarf, green, Round	dd gg Ww	1	1
Dwarf, green, Wrinkled	dd gg ww	1	1

FIGURE 2.20
Phenotypic and genotypic ratio of a test cross of trihybrid plants.

wrinkle, the F_2 plants were 315 : 108 : 101 : 32, which was nearly to 9 : 3 : 3 : 1).

In experimental genetics the results are tested by fitting them to hypothesis. In the experimental genetics, **"The goodness of fit"** is a key-concept. With this concept one can testify, how far the experimental results fit good to the hypothesis. Often deviation occurs from the expectation. To know the degree of deviation, and to know that how much deviation is permissible to accept the result, one should follow a statistical method; the **Chi-square.**

2.12.1 Chi-square (X^2)

The Chi-square (X^2)-test is an important tool by which an experimenter can determine the goodness of fit.

Application of Chi-square is based on two considerations—*the size of sample* and the *deviation from the expected ratio*. When the sample consists of two classes, i.e., either 1 : 1 or 3 : 1 ratios, the formula of X^2 is $\frac{(O_1 - E_1)^2}{E_1} + \frac{(O_2 - E_2)^2}{E_2}$, when O is the **observed** value of the first class and E_1 is the **expected** value of the same class derived from the experiment; O_2 is the **observed** number of the second class and E_2 is the **expected** number of it. After getting the deviation, it is *squared* and divided by the expected value. The resultant fractions for each class are *added* and the X^2 is obtained. Now the formula stands as

$X^2 = \sum \frac{(O - E)^2}{E}$, when (O – E) is the deviation between each observed and expected class value and E is the expected value in the respective class.

Thus, **Chi-square test** can be defined as: **The test of significance of overall deviation square in the observed and expected frequencies divided by expected frequencies.**

CHAPTER 3

CELL, CHROMOSOME AND INHERITANCE

Cell, Nucleus, Chromosome, Chromatin, Nucleosome, Solenoid-structure, Cell-cycle, Mitosis, Meiosis, Gametes, Spermatogenesis, Oogenesis, Spore-formation, Comparison between Mitosis and Meiosis, Meiosis and Mendel's Principles, Giant chromosomes—Polytene, Lampbrush.

CELL

Life begins with cell. Cells are units of life. Cells are capable of self-reproduction and are continued to the next generations. Cells divide and multiply. All of these phenomena are concerned with *inheritance*. Cells are the structural basis of living body and are functional machines of life. All eukaryotic cells possess a ***nucleus***. Inside the nucleus reside the **chromosomes**. Chromosomes carry **genes** and genes carry and express the **hereditary characteristics** in individuals.

3.1 Nucleus and Chromatins

In eukaryotic cells, the **nucleus** controls the biosynthetic events. It is an essential organelle and its chromosome stores the **genetic information**. The nucleus is composed of five ultrastructures—**nuclear membrane** or **envelope**, **nuclear sap** or **nucleoplasm**, **chromocenters**, **chromatin fibres** and **nucleolus**.

The **Chromatin fibres** are twisted threads observed only in the interphase nucleus under a powerful microscope. The 'threads' become thick and specific in number before cell-division (mitosis and meiosis). These are called **chromosomes**. The **chromocenters** are condensed region of **heterochromatin** (see 3.1.3.1) which are the parts of chromosomes that remain condensed at interphase nucleus. These are usually found near the **nuclear envelope** and remain attached to the **nucleolus**. All the materials remain embedded in the **nucleoplasm**.

3.1.1 Chromatins and C-value

All **diploid** cells in any organism contain the same amount of DNA-content (2C). Gametes are **haploid** and, therefore, have half the amount of DNA-content (1C).

Each species is characteristic in its DNA-content, which is *constant* in all individuals of that species. It is called **C-value** of the species. Most of the DNA in eukaryotic genomes is of non-coding in nature (that is, they do not synthesize protein). The large amount of DNA remains tightly packed and coiled within the nucleus forming chromatin structures (See 3.1.3) and eventually the chromosomes. The packaging of DNA into chromosomes with a specific and regular system serves several functions in life and heredity.

3.1.2 The Chromosomes

W. Waldeyer (1888) coined the term **chromosome** (Gr. chroma = colour, soma = body) for the deeply stained thread-like bodies in nucleus. W.S. Sutton and T. Boveri (1902) suggested that the chromosomes were the carriers of heredity. T.H. Morgan and H. Muller (1922) established that **genes** are linearly located in the chromosome. Dupraw (1965) advocated that the chromosome is made of a *highly folded single molecule of DNA double-helix* associated with proteins. It is called **unineme** model of chromosome. R.D. Kornberg (1974) postulated the *nucleosome model* (See 3.1.3.2) of chromosome. All chromosomes possess the property of self-duplication, which is the key point of the specificity of chromosome number in an organism.

3.1.2.1 Chromosome Number

The number of chromosomes (chromosome set) is constant in each species (Table 3.1). Usually organisms are **diploid**, i.e. they contain two sets (2n) of chromosomes in their somatic cells. 'n' stands for a single or haploid set; the sexually reproducing organisms produce **gametes** (sperm and ova) meiotically. The gametes contain a single set (n) of chromosome and are known as **haploid cells**. The haploid set (n) of chromosome is also known as **genome** of that species.

Sometimes organisms may contain more than two sets of chromosomes in their cells and are known as **polyploid** (Poly = many, ploid = set). Organisms with 3 sets of chromosome are called **triploid** (3n), with 4 sets are called **tetraploid** (4n) and so on.

3.1.2.2 Autosome and Sex-chromosome

In diploid or somatic cells chromosomes usually remain in even number. Each chromosome has a partner which is identical

TABLE 3.1: Chromosome number of some species

Species	Diploid number (2n)
Drosophila melanogaster (fruit fly)	8
Pisum sativum (Garden pea)	14
Zea mays (corn)	20
Canis familiaris (dog)	78
Oryctolagus cuniculus (rabbit)	44
Bos taurus (cattle)	60
Equus calibus (horse)	64
Columba livia (pigeon)	80
Mus musculus (house mouse)	40
Macaca mulatta (Rhesus monkey)	42
Gorilla gorilla (gorilla)	48
Pan troglodytes (chimpanzee)	48
Ascaris megalocephala (Ascaris)	2
Homo sapiens (Man)	46
Felis domesticus (cat)	38
Allium cepa (onion)	16
Oryza sativa (paddy)	24
Triticum aestivum (Wheat)	42
Rattus norvegicus (rat)	42
Musca domestica (house fly)	12
Culex pipiens (mosquito)	6
Lycopersicon esculentum (tomato)	24
Solanum tuberosum (potato)	48
Bombyx mori (silk worm)	56

in all respect; these alike-chromosomes are called **homologous chromosomes** or **homologs** (One of the partners come from father and the other from mother).

But, in all sexually reproducing organisms *one pair of chromosomes* may or may not be alike in shape, structure and function. They are related to the determination of *sex* of the individual. It is universally accepted that the sex is determined by a hereditary mechanism. In majority of organisms there is a pair of **sex-chromosome** which are responsible for sex determination, either male or the female. One of the sexes has a pair of identical **sex-chromosomes** (XX); the other may have a single sex-chromosome which may be **unpaired** (XO) or **paired with** a **Y-chromosome** (XY). These sex-related chromosomes are called **sex-chromosomes, accessory chromosomes, allosomes, heterochromosomes** or **heterosomes**. The rest of the chromosomes have no role in the detemination of sex, and are called **autosomes**.

In humans, there are 23 pairs of chromosomes (established by Tijo and Levan, 1956) of which 22 pairs are autosomes and 1 pair is sex-chromosome. In females, the sex-chromosomes are **XX** and in males, they are **XY**. During gametogenesis (gamete production) the females produce similar types of gametes (ova), i.e., all with **22 autosomes and one X-chromosome** (22AX). Hence they are called **homogametic**. On the otherhand, the males produce dissimilar types of gametes or sperms, i.e., 50% with **22 autosomes and one X chromosome (22AX) and 50% with 22 autosomes and one Y-chromosome** (22AY). Hence, they are called **heterogametic**. Although this is normal in most mammals, the **females** of the birds and reptiles are **heterogametic** and the **males** of them are **homogametic**.

- **X and Y Sex-chromatin in Interphase Nucleus:**

In 1949, **Barr and Bertram** discovered that the **interphase nucleus of female mammals** bears a small **chromatin body** which is absent in males. This body is formed by the **inactivation of one of the two X-chromosomes**. This is called **Sex-chromatin** or the **Barr body**. It remains attached with the nuclear envelope (Fig. 3.1)

On the other hand, in males, a large portion of Y-chromosome is **heterochromatic** (see later) and it appears as a strongly fluorescent body, the so-called **Y-chromatin body**. The number of Y-chromatin bodies is identical to the number of Y-chromosomes. Thus in humans, the interphase nucleus of the normal females (XX) has a single **Barr body** and normal males (XY) have no Barr body, (See Lyon's hypothesis, chapter VIII). The interphase nucleus of males have **Y-chromatin bodies** identical to the number of their Y-chromosomes (as for example, XYY males have 2 Y-chromatin bodies). Usually, the Y-chromatin is not attached to the nuclear membrane.

In the neutrophil leukocytes of human female, the sex-chromatin appears as a small

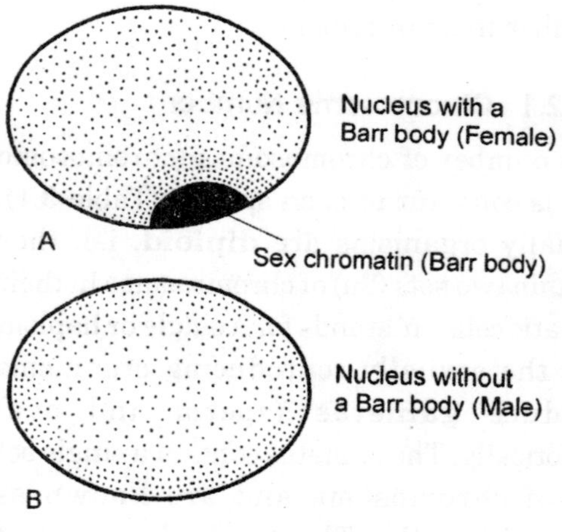

FIGURE 3.1

Sex-Chromatin in normal mammalian cell.

rod-like structure attached to the nucleus called the **drumstick**. (See also 3.1.3.1).

3.1.2.3 Chromosomes: Shape and Types

The chromosomes are best visible at the **metaphase** stage. Each chromosome bears a **centromere** that remains colourless during staining. Its position varies in chromosomes and usually it divides the chromosome in two arms (single arm in telocentric chromosome). The position of the centromere (the unstained, clear region) determines the shape of the chromosomes in 4 types—(Fig. 3.2).

1. **Telocentric** (Telo = end): This type of chromosomes are small rod-shaped having the **centromere at one end.** Consequently these chromosomes have a **single arm.** These are also known as **monobrachial chromosome** or **T-chromosome**.
2. **Acrocentric** (Acro = top): These are also rod-shaped but the **centromere** lies **almost at the end** of the chromosome consequently a very short and a very long arm are produced.
3. **Submetacentric** (Sub - below, meta - after): These are L or j shaped having the **centromere near the middle position but not at the middle**. Consequently two arms are produced of which one is **short** and the other is **long**. The short arm is called 'p'-arm and the long arm is called 'q'-arm.
4. **Metacentric** (Meta = after): These are V-shaped chromosomes having the **centromere** at the middle, resulting two equal arms (p = q).

All of the above chromosome types are called **monocentric** chromosomes, as they have a single centromere. Chromosomes with two centromeres (unusual) are called **dicentric** and without a centromere are called **acentric**. Both are nonviable. Sometimes chromosomes with diffused centromere may occur where no distinct centromere is located. These chromosomes are called **holocentric** or **diffused**.

The size of the chromosomes vary considerably. They may be extremely short as 0.25 µm found in birds or extremely long as 30 µm found in *Trillium*. The average size of human metaphase chromosome is 5 µm.

3.1.2.4 Chromosome Structure

Chromosomes change their structure at different phases of mitosis. Various terminology are used to describe the

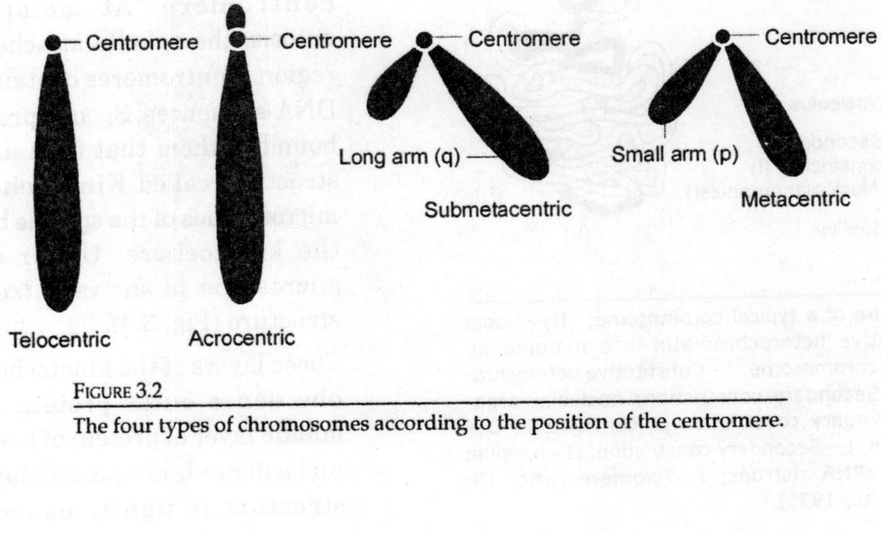

FIGURE 3.2
The four types of chromosomes according to the position of the centromere.

components of the mitotic chromosomes. The mitotic phases are *interphase, prophase, metaphase, anaphase* and *telophase* (see mitosis).

1. **Chromatid:** At **metaphase** each chromosome consists of two symmetrical halves called **chromatids.** Each chromatid is made up of a single DNA molecule and proteins. The two chromatids are attached to each other by the **centromere**, an unstained body (Fig. 3.3).

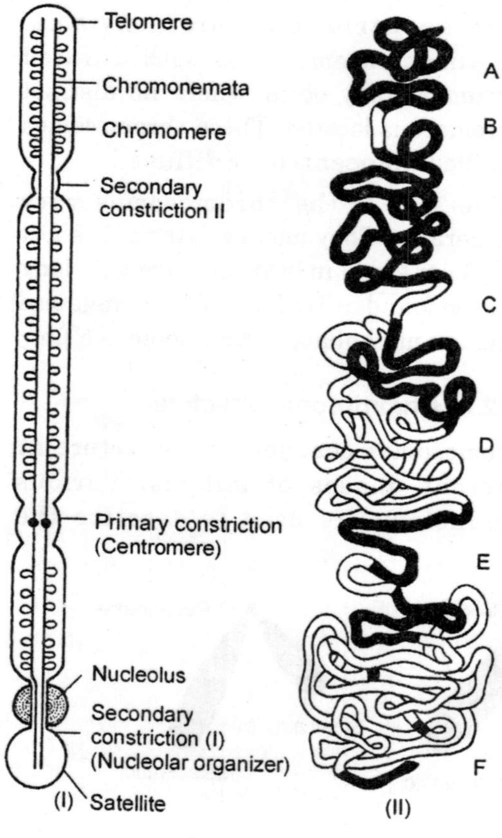

FIGURE 3.3

(I)—Structure of a typical chromosome; (II)—Model of constitutive heterochromatin in a mammalian metaphase chromosome. A—Constitutive heterochromatin; B—Secondary constriction; nucleolar organizer; C—Primary constriction or centromere; D—Euchromatin; E—Secondary constriction. II—possible site of 5S rRNA cistrons; F—Telomere (after De Robertis et al., 1975).

At the start of **anaphase** the chromatids separate from each other and each sister chromatid migrates to opposite poles. These chromatids of anaphase are now called chromosomes. Therefore, the meta-phase chromosomes have two chromatids and two DNA-molecules while anaphase chromosomes have only one chromatid and one DNA-molecule.

2. **Chromonema** (Plural, chromonemata): During **prophase** the chromatin fibres are seen as thin filamentous structures and are called **chromonemata.** They are, in fact, the early stages of chromatids. Each chromonema, like the chromatid, consists of a single DNA-molecule and associated proteins (Fig. 3.3).
3. **Chromomeres:** These are beaded structures (tightly folded regions of DNA) formed by the accumulation of chromatin materials on the **interphase** chromosomes (Fig. 3.3).
4. **Primary constriction, centromere and Kinetochore: Primary constriction** is the thinnest part of the chromosome where lies the **centromere**. At metaphase of mitosis, the spindle attaches to this region. Centromeres contain specific DNA-sequences. Special proteins are bound to them that form a disc-like structure, called **Kinetochore**. The microtubules of the spindle bind with the kinetochore. Under electron microscope is shows a *trilaminar* structure (Fig. 3.4).

Three layers of the kinetochore are—one dense outer protein layer, a middle layer of protein of low density and a dense inner protein layer. This structure is tightly bound to the

centromeric DNA (See 4.5). It provides a centre of assembly for microtubules. The regions of the chromosomal arms that are flanked by the centromere contain highly repetitive DNA.

5. **Telomere**: (Telo = end, mer-part). The tips of the chromosomes are called **telomeres**. They contain the ends of the long linear DNA molecule which are of unusual characteristics (See 4.6). The telomere has a 'polarity' and it does not allow any other chromosomal segment to unite with it.
6. **Secondary constrictions:** Besides primary constriction, there may be **secondary constrictions**, which are constant in position and length in a chromosome. They provide useful tools to identify particular chromosome in a set. They lack a centromere and are distinguished from the primary constriction by the absence of formation of angle at the time of mitotic movement. Certain secondary constrictions may have the role of **nucleolar organizer**.
7. **Nucleolar organizer**: Certain secondary constrictions that contain the genes coding for 18S and 28S ribosomal RNA and that induce the formation of **nucleoli** are called **nucleolar organizers**.
8. **Satellites**: Certain chromosomes have a rounded terminal body beyond the secondary constriction. This is known as **satellite**. This is a morphological entity of chromosome and is completely different from the **satellite-DNA** (See 4.7).

3.1.2.5 Karyotype

Karyotype means the total characteristics that allows the identification of a particular chromosomal set. The characteristics include the number of chromosomes, the relative size, position of the centromere, length of the arms, secondary constrictions and satellites.

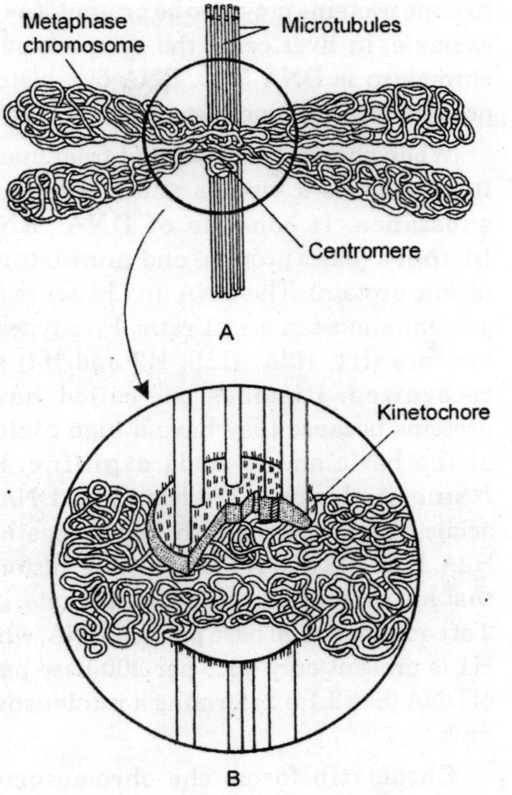

FIGURE 3.4

A. A metaphase chromosome showing the folded fibre structure and the centromere with implanted microtubules; B. A higher magnification, showing the convex electron dense layer.

A diagrammatic expression of a Karyotype is called **idiogram.**

3.1.3 The Chromatin

The term 'chromatin' was coined by Flemming (1878) due to its nature of taking deep colour during staining (chroma = colour). Chemically the **chromatins** are **nucleoprotein** (nucleic acids + proteins) consisting mainly of DNA and histone proteins. A little amount RNA and non-histone proteins may also be present. (As for example, in liver cells, the composition of chromatin is DNA 31%, RNA 5%, histone proteins 36% and non-histone proteins 28%).

When chromatin is isolated from nuclei, it appears as a viscous and a gelatinous substance. It consists of **DNA**, **RNA**, **histones** (basic protein) and **nonhistones** (acidic protein). The DNA and histones are present almost in a 1 : 1 ratio. Five types of histones (H1, H2A, H2B, H3 and H4) are recognized. Histones are called **basic** proteins because they have a high content of the basic amino acids **arginine** and **lysine**. As histones are basic and DNA is acidic, they remain tightly bound together. H2A, H2B, H3 and H4 are the 'core' histones that are present in equimolar amounts, i.e., 2 of each every 200 base pairs of DNA, while H1 is present only once per 200 base pairs of DNA (See 3.1.3.2) forming a **nucleosome** unit.

Chromatin forms the chromosomes. Based on its genic properties and of staining nature, the chromatin is of two types—**euchromatin** and **heterochromatin**.

3.1.3.1 Euchromatin and Heterochromatin

Chromosomes donot stain uniformly. The part of chromosome of an interphase or early prophase stage that stains lightly, contains **partially condensed chromatin**; this is called **euchromatin**. This part contains structural genes—and is **genetically active**. Replication and transcription are seen to occur here.

On the other hand, the darkly stained part of interphase and early prophase chromosome contains **condensed chromatin**; this is called **heterochromatin**. It has a high content of tightly packed repetitive DNA-sequences but very few structural genes. It is **genetically inactive**; very late replication may occur but transcription does not occur here.

At interphase, some condensed heterochromatin occur around the nucleolus, and some inside the nucleolus. The former is called **perinucleolar chromatin** and the latter **intranucleolar chromatin**.

The usually deep-stained heterochromatic regions of the chromosomes may show even more deep and strong stain at places being **positively heteropyknotic** (pyknosis=staining) or may show a comparatively week stain than the euchromatic regions at places being **negatively heteropyknotic**.

Functionally, heterochromatine are of two types—**constitutive** and **facultative**. The constitutive heterochromatin contains *permanently inactive DNA* and it remains condensed throughout the cell-cycle (See later). Usually it is found in the chromosomal parts near the centromere, in the telomere, and in the C-bands. In certain cases, the whole chromosome may remain in heterochromatic state, and is referred to as **B-chromosome** or **Satellite chromosome** (because it contains short repeated sequences of DNA, the **satellite DNA**. See chapter IV). The Y-chromosome of male *Drosophila* is heterochromatic in nature though it contains some genes responsible for male fertility.

On the other hand, the **facultative heterochromatin** contains *periodically active DNA*. It does not stay in a permanently condensed state throughout the cell-cycle. Out of the two X-chromosomes of the mammalian female one remains active and euchromatic while the other is inactive and totally heterochromatic which forms the **Barr body** or **sex-chromatin** on the interphase nuclear membrane. (See X and Y sex-chromatin in Interphase Nucleus, 3.1.2.2).

3.1.3.2 The Nucleosome

The chromatin exists in repeating units. The repeating unit of chromatin is called **nucleosome**. Principally it consists of approximately 200 base pairs (bp) of DNA coiled around an octamer of histone protein.

Roger Kornberg (1974) proposed a model for the nucleosome. He proposed that in each nucleosome four histones (H2A, H2B, H3 and H4) are arranged in *octamers* (2 mols of each, i.e., 2H2A + 2H2B + 2H3 + 2H4 = 8 mols of histones) and are tightly associated with a part of DNA containing approximately 200 bp. The histone octamers are in close contact and the DNA helix is coiled on the outside of the histone-'core'. The DNA-helix combines with the groups of 8-histone molecules to form the structure of **nucleosome** having the appearance of 'beads on a string' under electron microscope. Each 'bead' (nucleosome) appears about 110Å (11 nm) in diameter and 60Å (6 nm) in height, and is joined to the next 'bead' by a thin thread (linker DNA) (Fig. 3.5).

The base-pair number of nucleosome–associated DNA varies according to the degree of digestion of it by **nuclease** enzyme. When the chromatin is partially digested by a nuclease (micrococcal nuclease) an approximately 200-base-pair length of DNA is found to associate with each nucleosome (as only the linker part of DNA is digested). But after a dose of heavy digestion by the nuclease, a 146-bp-long segment of DNA remains present in each nucleosome. This portion is nuclease-resistant and is called **nucleosome core**. In all eukaryotes the *nucleosome core* is formed of a *a 146-bp length of DNA* and *2 molecules of each of H2A, H2B, H3 and H4 histone proteins*. The DNA is wound as $1\frac{3}{4}$ turns of a superhelix around the octamer (Fig. 3.6). The series of 'nucleosome cores' are connected by a linker-DNA which may vary in length from 8-114 base-pairs. In addition to that, one molecule of H1 histone and associated non-histone proteins are also present.

In brief, the complete chromatin subunit consists of 1. **The nucleosome core 2. The linker DNA** 3. **One molecule of H1** and 4. **Non-histone proteins.**

The structure of nucleosomes in euchromatin (genetically active) differs from that of heterochromatin (genetically inactive).

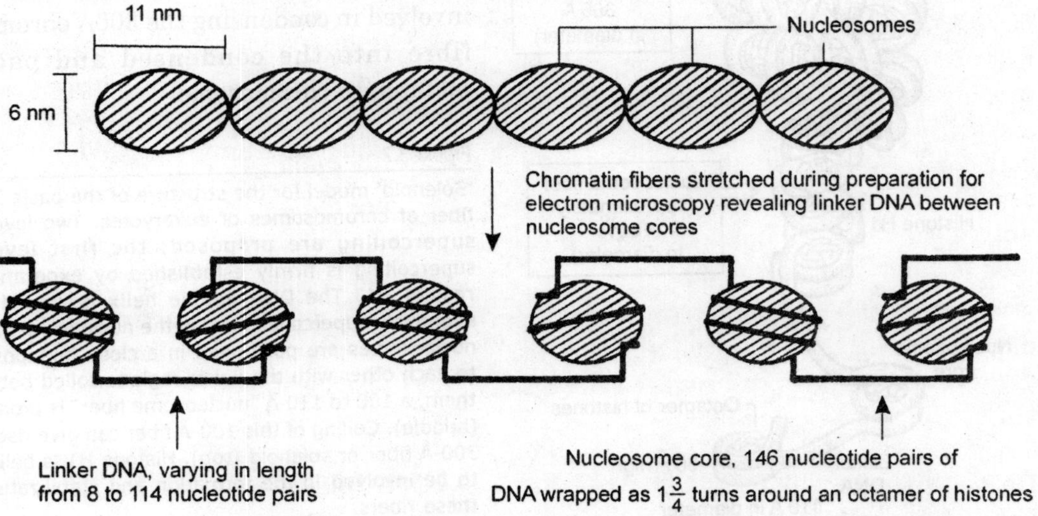

FIGURE 3.5
The nucleosome structure of chromatin.

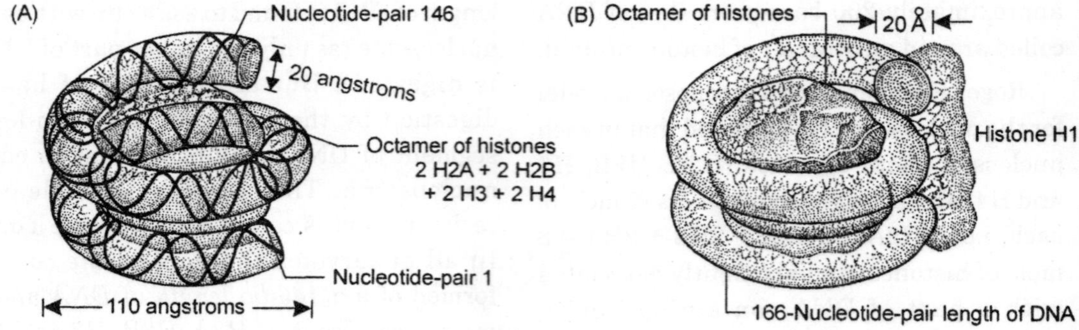

FIGURE 3.6

(a) Structure of the nucleosome core and (b) the proposed role of histone H1 in stabilizing two complete turns of DNA superhelix around the octamer of histones. The nucleosome core consists of a 146-nucleotide-pair-long segment of DNA wound on the surface of a somewhat cylindrical octamer of histones (two each of H2A, H2B, H3, and H4) to yield a roughly ellipsoidal structure.

3.1.3.3 "Solenoid" Model of 300Å Chromatin Fibre

Once more we refer to the chromatin fibre (See 3.1). The tightly coiled or folded **chromatin fibres** of metaphase chromosome have an average diameter of 300 Å.

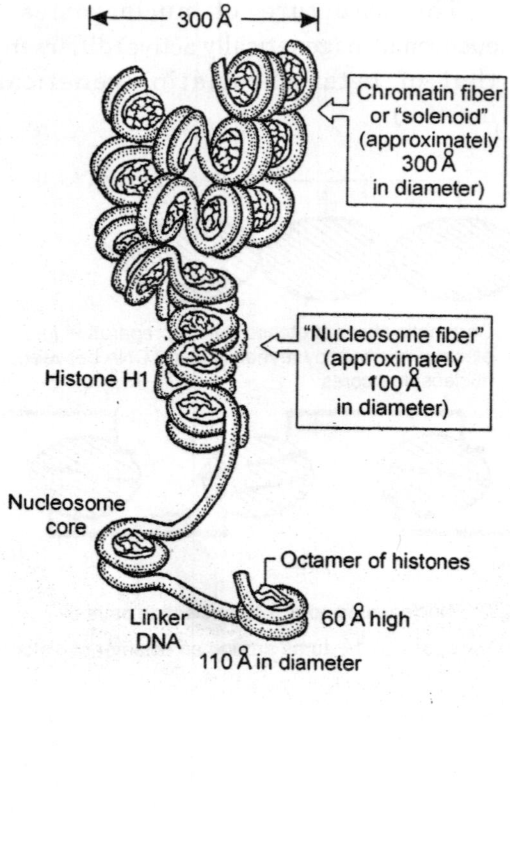

Presumably this is due to that, the nucleosomes lie in direct juxtaposition with each other and approximately a 100 Å-nucleosome fibre is formed. When this 100Å-fibre is wound in a super 'super coil' or a higher order coil or 'solenoid', a 300Å fibre can be resulted (Fig. 3.7).

Histone H1 is involved in the supercoiling of 100Å nucleosome fibre to produce the 300Å chromatin fibre. Finally, the non-histone proteins form a 'scaffold' which is involved in condensing the 300Å chromatin fibre into the condensed and packed metaphase chromosome.

FIGURE 3.7

'Solenoid' model for the structure of the basic 300-Å fiber of chromosomes of eukaryotes. Two levels of supercoiling are proposed; the first level of supercoiling is firmly established by experimental results. (1) The DNA double helix is wound in a negatively supercoiled form in the nucleosome. If the nucleosomes are positioned in a close juxtaposition to each other with the linker regions coiled between them, a 100 to 110-Å "nucleosome fiber" is produced (middle). Coiling of this 100-Å fiber can give rise to a 300-Å fiber or solenoid (top). Histone H1 is believed to be involved in the formation and stabilization of these fibers.

3.2 The Nucleus and the Cell-cycle

Almost all cells are capable of self-reproduction. They are continued to produce series of cells. They divide and multiply. The process of cell-division in the life of a cell is a cyclic phenomenon.

The sequence of events that occurs between one cell-division and the next is called the cell cycle.

Inside a cell, there is a close relationship between the **nuclear mass** and the **cytoplasmic mass**. These two masses remain in a state of optimum equilibrium. This equilibrium is called **Nucleoplasmic Index** (NP). In an optimum condition the NP of a living cell is expressed as $\frac{Vn}{Vc - Vn}$, where Vn is the **nuclear volume** and Vc is the **cytoplasmic volume.**

All cells have two major periods in their life-cycle—the **interphase** (non-division phase) and the **division phase.** These phases occur alternately and repeatedly to form a **cycle** at each cell-generation. The division phase may further be divided into a division of nucleus (mitosis) and a division of cytoplasm (cytokinesis) (Fig. 3.8).

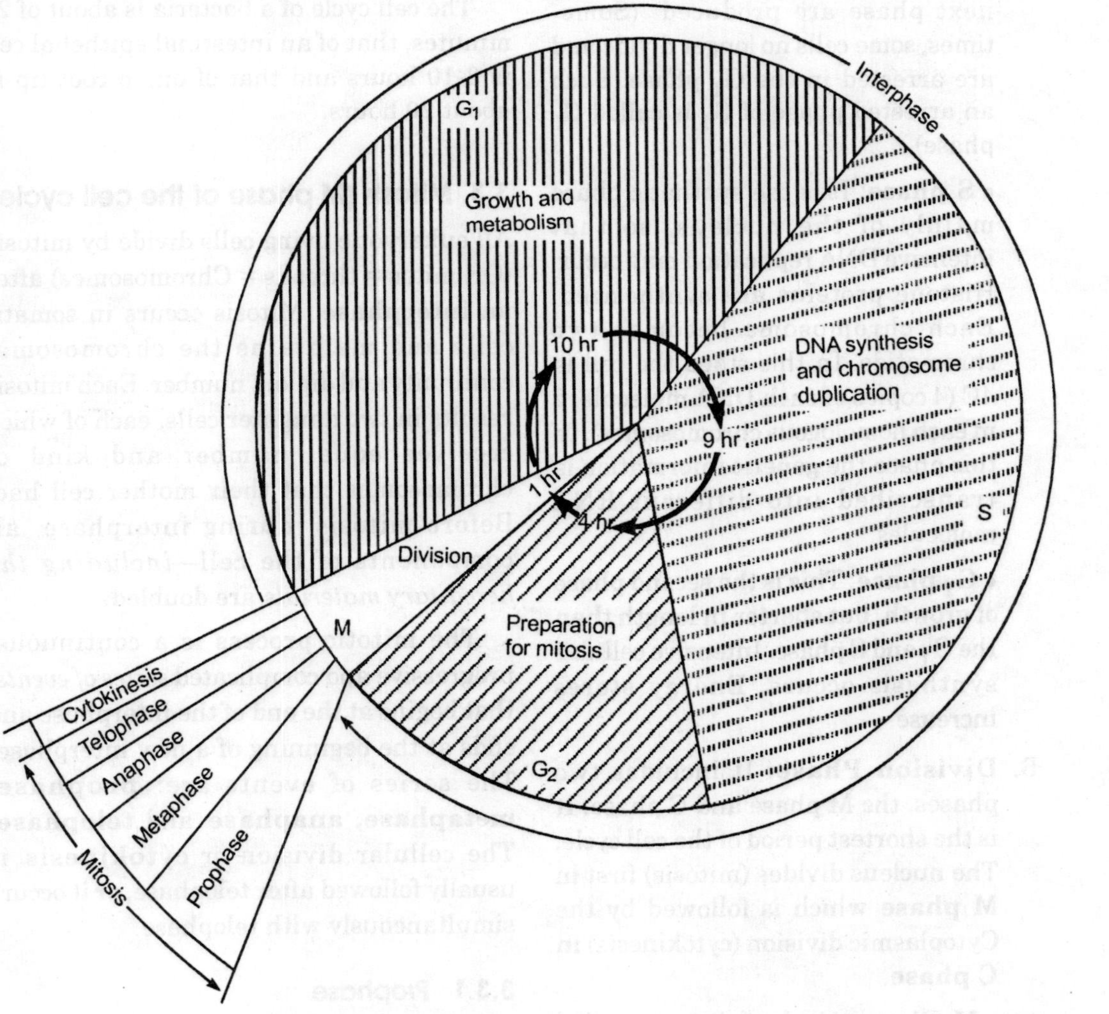

FIGURE 3.8
The eukaryotic cell cycle. This cycle is 24 hours long.

A. **Interphase**: **Synthesis** of cell-materials and **growth** of cell occur in this period. This phase is subdivided into 3 phases—**G_1 phase**, **S-phase** and **G_2 phase**. (The letter 'G' stands for 'gap'. Thus, G_1 is gap-1 and G_2 is gap-2 phase.)

• **G_1 phase**: Intensive cellular synthesis occurs in this phase. It is the longest period when the cytoplasmic organelles are synthesized. Cell-growth and cell-metabolism occur. Substances that can inhibit or stimulate the onset of next phase are produced. (Sometimes, some cells no longer divide and are arrested in the G_1 phase. Such an arrested phase of G_1 is called G_0 phase).

• **S phase**: It is the synthetic phase mainly of the nucleus, so that intensive DNA replication may occur. Histone proteins are synthesized. Each chromosome becomes two chromatids. In this stage the cell is 4C (4 copies of each DNA molecule, 2 in each homologous chromosomes). In this phase the genetic information is transcribed into different RNA molecules.

• **G_2 phase**: This is the second phase of growth, but shorter in length than the G_1 and S phase. Intensive cellular synthesis occurs. Energy stores increase.

B. **Division Phase**: It includes two phases, the M phase and C phase. It is the shortest period of the cell cycle. The nucleus divides (mitosis) first in **M phase** which is followed by the Cytoplasmic division (cytokinesis) in **C phase**.

• **M Phase**: Nuclear division called **mitosis** occurs. There are four stages of mitosis—prophase, metaphase anaphase and telophase (See mitosis).

• **C Phase**: The cell divides and equal distribution of organelles and cytoplasm occurs into each daughter cell.

The span of the cell cycle depends on the type of cell and various external factors like temperature, food, O_2 etc. In a 24-hour-long-cell cycle the G_1 may be approximately of 10 hours, S – 9 hours, G_2 – 4 hours while M and C of 1 hour duration (Fig. 3.8).

The cell cycle of a bacteria is about of 20 minutes, that of an intestinal epithelial cell is 8-10 hours and that of onion root tip is about 20 hours.

3.3 Mitosis (M phase of the cell cycle)

All eukaryotic living cells divide by mitosis (Gr. mitos = threads = Chromosomes) after an **interphase**. Mitosis occurs in somatic cells and maintains the chromosomal continuity and diploid number. Each mitosis results in two daughter cells, each of which receives equal number and kind of chromosomes that their mother cell had. Before mitosis, during interphase, all components of the cell—*including the hereditary materials* are doubled.

The mitotic process is a continuous, progressive and complicated *series of events*, that begins at the end of the interphase and ends at the beginning of a new interphase. The series of events are: **prophase**, **metaphase, anaphase** and **telophase**. The cellular division or **cytokinesis** is usually followed after telophase, or it occurs simultaneously with telophase.

3.3.1 Prophase

The important phenomena of this event are (Fig. 3.9)—

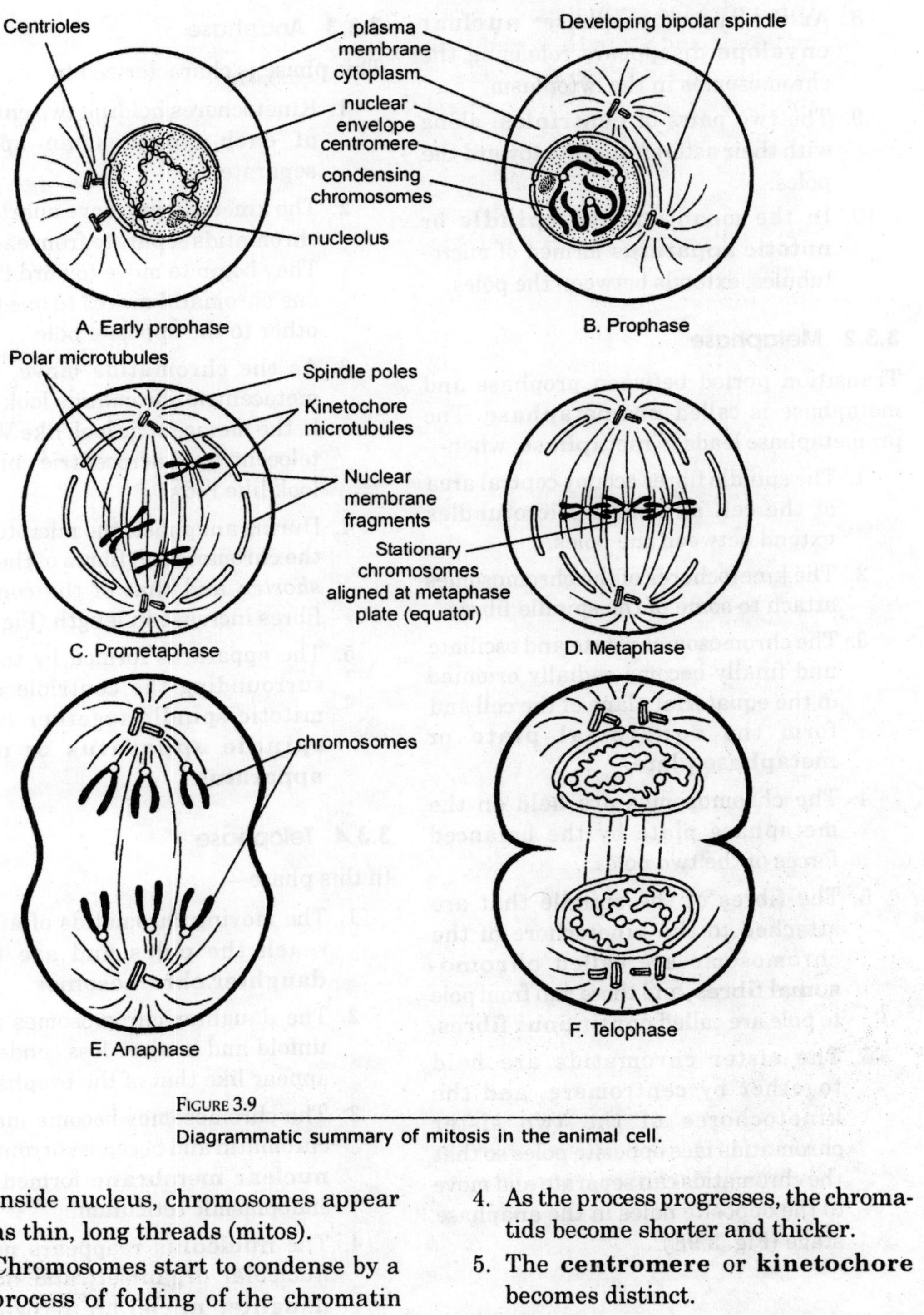

FIGURE 3.9

Diagrammatic summary of mitosis in the animal cell.

1. Inside nucleus, chromosomes appear as thin, long threads (mitos).
2. Chromosomes start to condense by a process of folding of the chromatin fibres.
3. Each chromosome is composed of **two chromatids** (resulted by replication of DNA at S-phase).
4. As the process progresses, the chromatids become shorter and thicker.
5. The **centromere** or **kinetochore** becomes distinct.
6. The chromosomes approach the nuclear envelope.
7. **Nucleolus** reduces in size and gradually disintegrates.

8. At the end of prophase, **nuclear envelope** disappears releasing the chromosomes in the cytoplasm.
9. The two pairs of **centrioles**, along with their asters* migrate toward the poles.
10. In the meantime the **spindle** or **mitotic apparatus** formed of microtubules, extends between the poles.

3.3.2 Metaphase

Transition period between prophase and metaphase is called **prometaphase**. The prometaphase leads to **metaphase**, when—

1. The spindle fibres occupy central area of the cell and their microtubules extend between the poles.
2. The kinetochores of the chromosomes attach to some of the spindle fibres.
3. The chromosomes rotate and oscillate and finally become radially oriented in the equatorial plane of the cell and form the **equatorial plate** or **metaphase plate**.
4. The chromosomes are held on the metaphase plate by the balanced forces of the two poles.
5. The fibres of the spindle that are attached to the kinetochore of the chromosome are called **chromosomal fibres**, and those run from pole to pole are called **continuous fibres**.
6. The sister chromatids are held together by centromere, and the kinetochores of the two sister chromatids face opposite poles so that the chromatids can separate and move to the opposite poles in the anaphase stage (Fig. 3.9E).

3.3.3 Anaphase

This phase is characterized by—

1. Kinetochores holding two chromatids of each chromosome split and separate.
2. The kinetochores move apart, and the chromatids separate from each other. They begin to move toward the poles, one chromatid moves to one pole, the other to the opposite pole.
3. As the chromatids move, the sub metacentric chromatids look like J or L, the metacentric look like V and the telocentric or acrocentric chromatids look like rods.
4. During anaphase, the microtubules of the chromosomal fibres of the spindle *shorten* and that of the continuous fibres increase in length (Fig. 3.9).
5. The apparatus formed by the asters surrounding the centriole and the mitotic spindle together is called **spindle apparatus** or **mitotic apparatus**.

3.3.4 Telophase

In this phase—

1. The moving chromatids of anaphase reach the poles and are termed **daughter chromosomes**.
2. The daughter chromosomes start to unfold and become less condensed to appear like that of the prophase.
3. The chromosomes become masses of chromatin and become surrounded by **nuclear membrane** formed by the endoplasmic reticulum.
4. The **nucleolus** reappears near the nucleolar organizer, and now two **daughter nuclei** are recognized at two poles. They enter into the G_1 phase.
5. Mitotic apparatus, except at the centrioles, disappears (Fig. 3.9).

---

* Mitosis having with asters (around the centriole) forming spindle is called **astral mitosis** (found in animal cells) and that without centriole and asters is called **anastral mitosis** (found in higher plants),.

3.3.5 Cytokinesis

It is the process of partitioning of the cytoplasm dividing the cell into **two daughter cells**. It shows—

1. Both DNA synthesis and mitosis are coupled to **cytokinesis**, which in animal cells, essentially is associated with the formation of a **constriction** of the cell membrane at the equator.
2. The constriction and its inward progress is called **cleavage** and **furrowing** respectively.
3. Cleavage is caused by the contraction of a 'ring' formed by bundle of actin and myosin filaments. This is called **contractile ring**.
4. The contractile ring assembles in early anaphase and develops a *force* which pulls the plasma membrane down into a furrow.
5. During cytokinesis, the cytoplasmic organelles are distributed, and finally, the furrowing separates the two daughter cells.

3.4 Meiosis

Meiosis is a special type of cell division intimately associated with the **sexual reproduction**. Sexual reproduction involves the production of **gametes** (eggs and sperm) which fuse together to form a **zygote** developing the new organism. During zygote formation (fertilization of sperm and ovum), the **hereditary materials** of both the gametes (coming from both parents) are contributed to the zygote (i.e., to the next generation). Thus the next generation-individual carries a **biparental inheritance**. In this mechanism, **meiosis** is the instrument that produces **gametes**, the units of the sexual reproduction. The most important significance of **meiosis** (Gr. meioum = to diminish) is to **reduce** the diploid number (2n) of chromosomes of the gonadal germ cell into a haploid number (n) of chromosomes of the gametes. This is essential, otherwise continuity and maintenance of a diploid number of chromosome of an individual would not be possible. The male gamete (sperm) bears half the number of chromosome of the male parent and the female gamete (ovum) bears half the number of chromosome of the female parent. When the two gametes fuse together through fertilization, the full set of chromosome number (2n) of the species reappear in the zygote. What happens to the chromosome number to a sexually reproducing organism is—

(Diploid) Male	(Diploid) Female
↓	↓
Diploid gonadal germ cell (2n)	Diploid gonadal germ cell (2n)
↓	↓
Meiosis (reduction in chr. no.)	**Meiosis** (reduction in chr. no.)
↓	↓
Haploid gamete sperm (n)	Haploid gamete ovum (n)

↓

Fertilization (n + n)

↓

Zygote (2n)

↓

A diploid male or female organism.

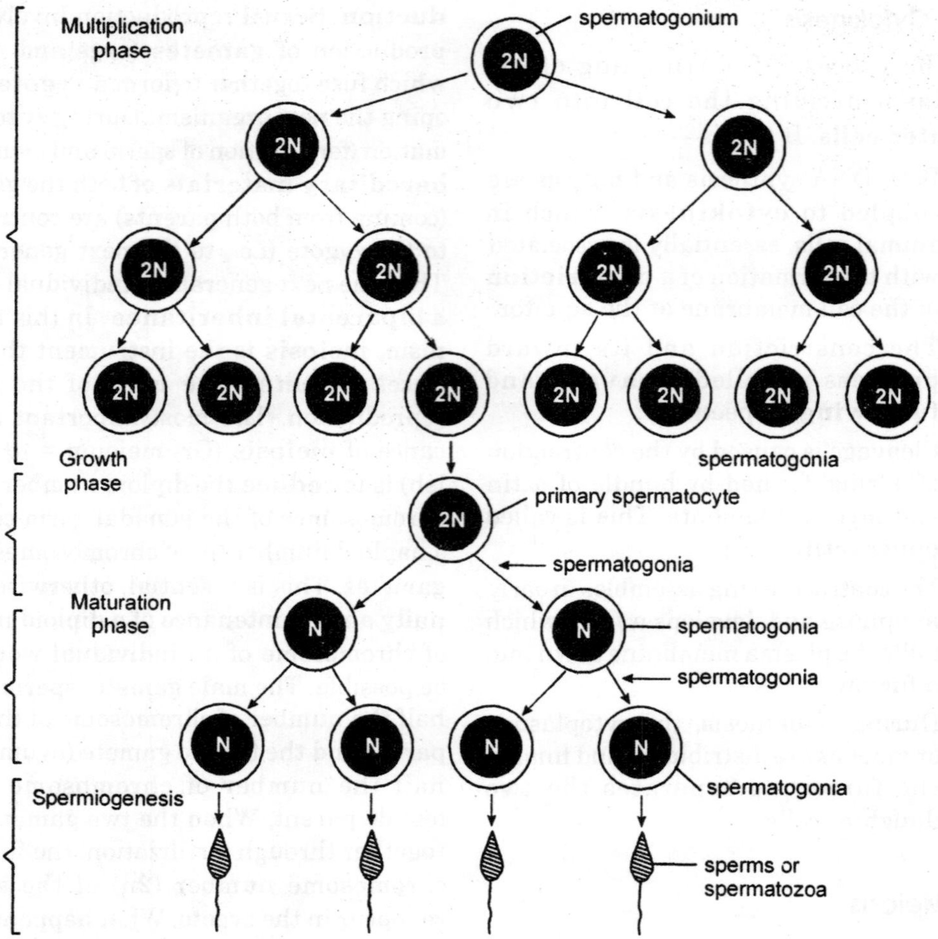

FIGURE 3.10
Spermatogenesis.

The formation of **gametes** from the germ cell is called **gametogenesis**. The process of gametogenesis in males producing **spermatozoa** (sperms) is called **Spermatogenesis**, and that in females producing **ova** (eggs) is called **oogenesis**.

3.4.1 Meiotic Cell-division in Male Gonad (germ cell) or Spermatogenesis

The process of spermatogenesis includes four major phases—**Multiplication phase**, **Growth phase**, **Maturation phase** (including meiosis) and **metamorphosis** of **spermatid to functional sperm** (or spermiogenesis) (Fig. 3.10).

3.4.1.1 Multiplication Phase

The *primordial germ cells* of a mature testis contain chromatin-rich nucleus. They produce innumerable sperm mother cells or **spermatogonia** by repeated **mitotic** cell-divison. The spermatogonia are diploid cells having 2n number of chromosomes.

3.4.1.2 Growth Phase

The spermatogonia grow in size and form mature spermatogonia, called **primary spermatocytes**. These are diploid germ cells that undergo **meiosis**, or **maturation phase**.

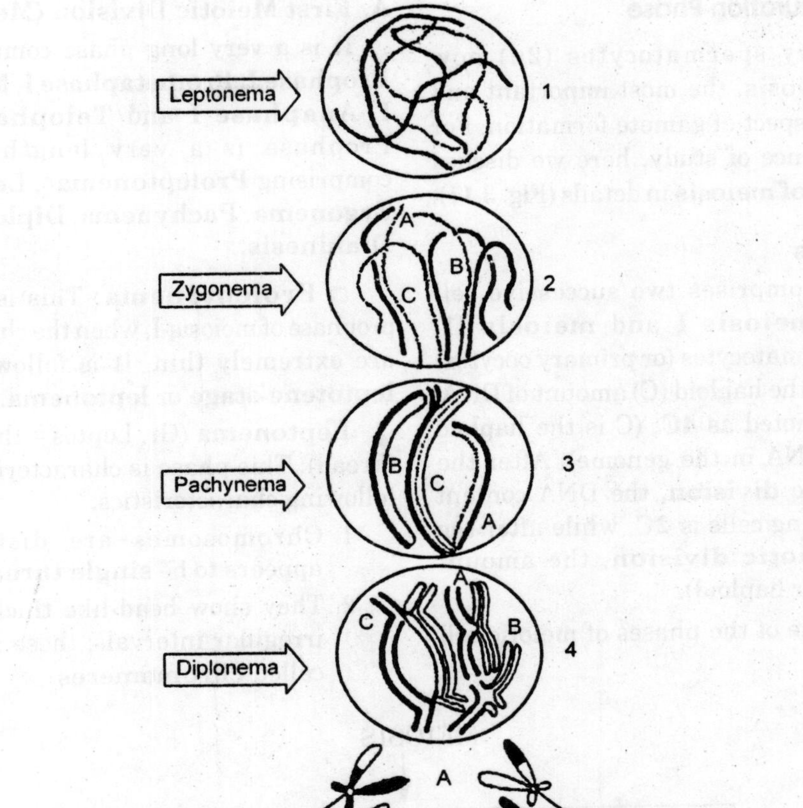

FIGURE 3.11

General diagram of meiosis, illustrating the union, separation, and distribution of the chromosomes.

3.4.1.3 Maturation Phase

The primary spermatocytes (2n) now undergo **meiosis**, the most important and significant aspect of gamete formation. For the convenience of study, here we discuss the **process of meiosis** in details (Fig. 3.11).

- **Meiosis**

Meiosis comprises two successive cell divisions, **meiosis I** and **meiosis II**. Primary spermatocytes (or primary oocytes) have 4 times the haploid (C) amount of DNA. They are denoted as **4C**. (C is the haploid amount of DNA in the genome). After the **first meiotic division**, the DNA content in the resulting cells is **2C**, while after the **second meiotic division**, the amount becomes **C** (or haploid).

The outline of the phases of meiotic cell division is:

A. First Meiotic Division (Meiosis I)

It is a very long phase comprising the **Prophase I, Prometaphase I, Metaphase I, Anaphase I** and **Telophase I**. The Prophase is a very lengthy process comprising **Proleptonema***, **Leptonema, Zygonema, Pachynema, Diplonema** and **Diakinesis**.

□ **Proleptonema**: This is the early prophase of meiosis-I, when the chromosomes are extremely thin. It is followed by the **leptotene-stage** or **leptonema**.

Leptonema (Gr. Leptos = thin, nema = thread): This phase is characterized by the following characteristics:

1. Chromosomes are distinct and appears to be **single thread**.
2. They show bead-like thickenings at irregular intervals; these are the so-called **chromomeres**.

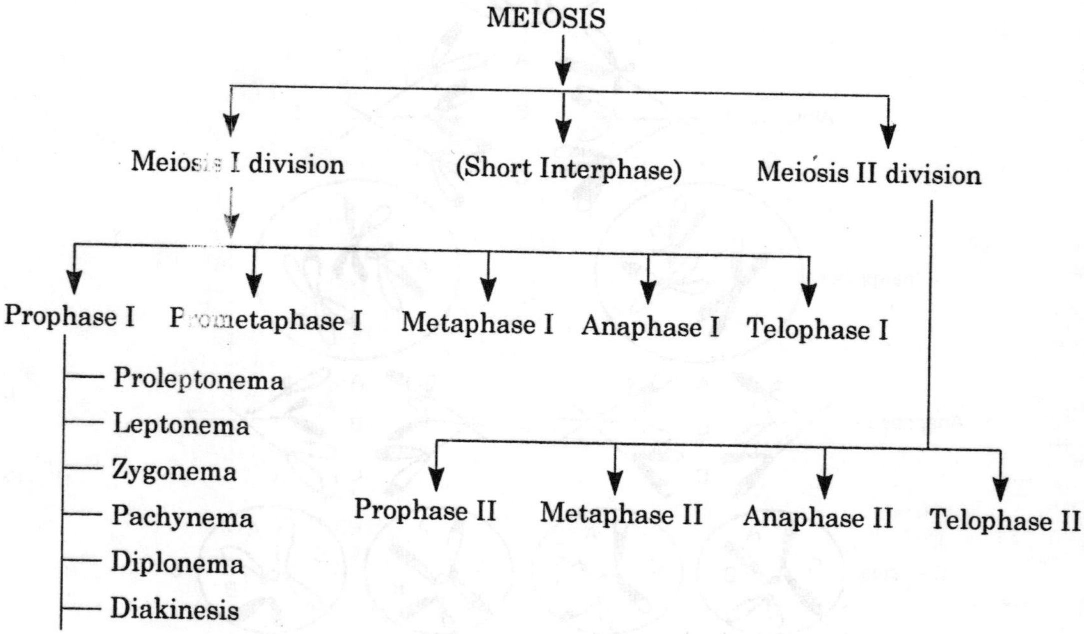

* 'Proleptonema' is the noun-form, while 'Proleptotene' is the adjective form. Thus the stages are Proleptonema, Leptonema, Zygonema, Pachynema, Diplonema or Proleptotene-stage, Leptotene-stage, Zygotene-stage, Pachytene-stage and Diplotene-stage.

3. The chromosomes form loops whose ends are attached to the nuclear membrane. This loop-like chromosomal structures form the characteristic "**bouquet**" stage.

□ **Zygonema** (Gr. zygon = adjoining): The important features are—

1. The homologous chromosomes undergo **pairing** in a zipper-like fashion. The pairing between homolog is called **synapsis**.
2. Pairing of the chromosomes involves the formation of a special structure called **synaptonemal complex**.
3. Pairing is very specific; it occurs point for point and chromomere for chromomere in each homolog.
4. The homologs pair closely but do not fuse together; they remain separated by a gap of 0.15 μm to 0.2 μm which is occupied by the **synaptonemal complex** (S.C.).
5. The S.C. is composed of two lateral and a median components mainly formed of protein.
6. Each lateral component of the S.C. is shared by the two sister chromatids of a homolog.
7. Telomeres of the chromosomes are frequently attached to the nuclear membrane.

□ **Pachynema** (GK. Pachus=thick): Pachytene stage may continue for days, weeks or even years. The events may remain arrested for a very long time. The important features of this stage are—

1. The synapsis is complete.
2. Chromosomes look shorter and thicker.
3. Because of the pairing of homologous chromosomes, now there appears half the number of chromosomes in the nucleus.

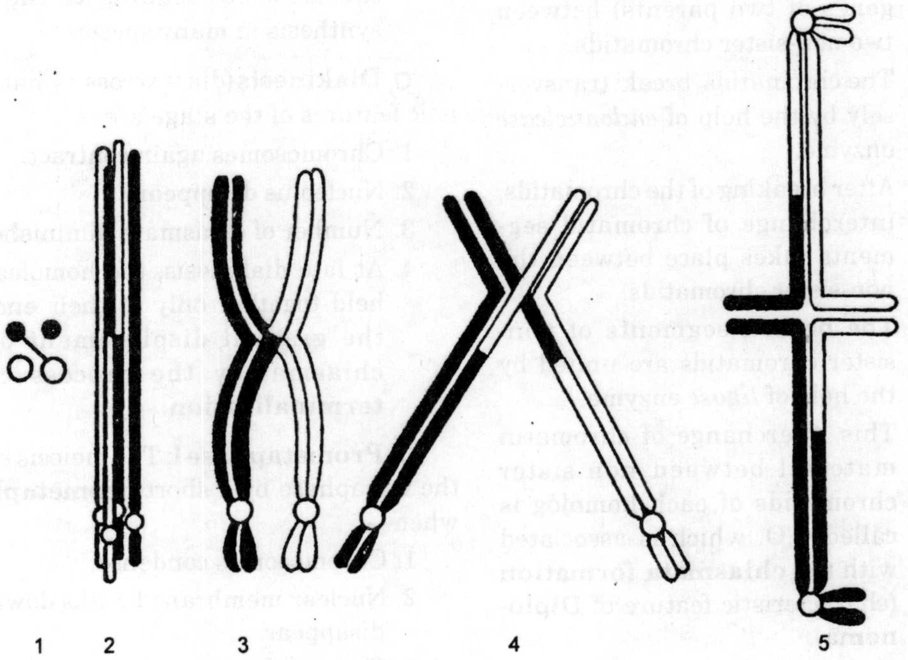

FIGURE 3.12

1 and 2, diagrams showing the process of crossing-over; 3, formation of a chiasma; 4, terminalization; 5, rotation of the chromatids of one bivalent. (After De Robertis).

4. Each pair of chromosomes is called **bivalent** (bi = two) or **tetrad** (as there are 4 chromatids, tetra = 4).
5. There are **4 kinetochores** in a tetrad (each chromatid containing one), *two homologous* and *two sisters*.
6. The two chromatids of each homolog and called **sister chromatids**.
7. **Recombination and Crossing over** take place. **Two of the chromatids of the homologs** (non-sister chromatid) **exchange segments** called **recombination**. It is the most significant event because it allows the genetic exchange and formation of 'new' kind of chromatids that contain certain genes of both the parents. The important features of the **crossing over** (C.O.) are— (Fig. 3.12).
   (a) The C.O. involves reshuffling, redistribution and interchange of segments (containing block of genes of two parents) between two non-sister chromatids.
   (b) The chromatids 'break' transversely by the help of *endonuclease* enzyme.
   (c) After breaking of the chromatids, interchange of chromatid segments takes place between the non-sister chromatids.
   (d) The broken segments of non-sister chromatids are united by the help of *ligase* enzyme.
   (e) This interchange of chromatin material between non-sister chromatids of each homolog is called C.O. which is associated with the **chiasmata formation** (characteristic feature of **Diplonema**).

□ **Diplonema**: Important features of this phase are—

1. The paired chromosomes now repel each other and begin to separate.
2. The homologs remain united at the points of interchange, the **chiasmata** (singular, chaisma = cross piece). These are the places where the C.O. or recombination takes place.
3. At least one chiasma is formed in each bivalent, but the number of chiasma may vary.
4. It is a long-lasting phase and the events may remain arrested in phase for years together (40-45 years in human female. Chromosomes of primary oocyte proceed upto the stage of **Diplonema** and then subsequent stages of meiosis become arrested. This is called **Meiotic I-arrest** or **First Meiotic block** or **Dictyotene stage**).
5. Chromatids are clearly visible but the S.C. disappears.
6. Usually there is an uncoiling of the chromosomes leading to the RNA synthesis in many species.

□ **Diakinesis** (dia = across): Characteristic features of the stage are—

1. Chromosomes again contract.
2. Nucleolus disappears.
3. Number of chiasmata diminishes.
4. At late diakinesis, the homologs are held together only at their ends, by the gradual displacement of the chiasma, by the process called **terminalization**.

□ **Prometaphase I**: The meiosis enters the metaphase by a short **prometaphase**, when—

1. Chromosomes condense.
2. Nuclear membrane breaks down and disappear.
3. The spindle-microtubules are attached to the kinetochores, so that only homologous kinetochores may separate and the two sister chromatids behave as a functional unit.

□ **Metaphase I**: Metaphase I is distinguished by—

1. Chromosomes are arranged at the equator.
2. Homologs are still attached by chiasmata at their ends while the kinetochores are being pulled toward the poles.

□ **Anaphase I**: This stage is characterized by—

1. The sister chromatids of each homolog are united by their own kinetochore.
2. The sister chromatids move toward their respective poles.
3. When the homologous paternal and maternal chromosomes **separate** from each other, two of their chromatids show recombination and the other two remain in parental combination. (It is to be noted here, that the Mendelian principles lie in this phenomenon of reduction division of meiosis. Mendel's first law, the law of segregation is the direct result of separation of homologous chromosomes during Anaphase I. His second law, the law of independent assortment results because each pair of homologous chromosomes resides at the metaphase plate in Metaphase I independently of all other pairs).

□ **Telophase I**: Telophase of Meiosis I can be identified by—

1. Chromatids reach their respective poles.
2. A new nuclear envelope surrounds the chromosomes and a nucleus is formed.
3. Thus, two daughter nuclei are produced.

A short **interphase** follows the Telophase I when **no replication of chromosome occurs**. The chromosomes are **haploid** in number in each nucleus, but each consists of two chromatids.

B. Second Meiotic Division (Meiosis II)

The second meiotic division is of shorter duration with **Prophase II**, **Metaphase II**, **Anaphase II** and **Telophase II** the details of which are similar to that of **mitosis** (See 3.3). The events occur in both of the two daughter nuclei.

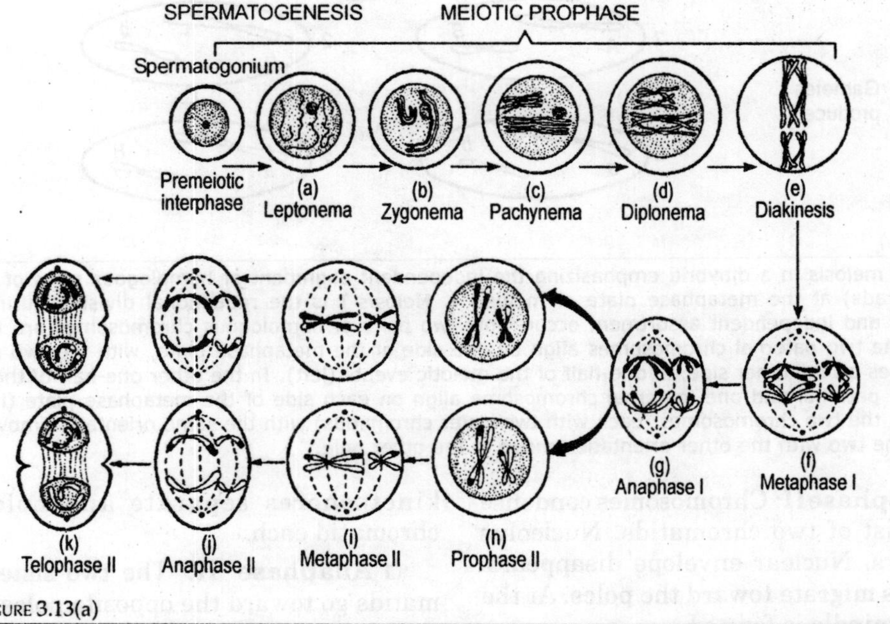

FIGURE 3.13(a)

Diagrams illustrating chromosome changes. Stages in meiosis in an animal with two pairs of chromosomes, one member of each pair having come from the mother and one from the father (a-e) two chromatids, one maternal and one paternal, are shown to undergo crossing-over (breakage and exchange of parts) at two sites.

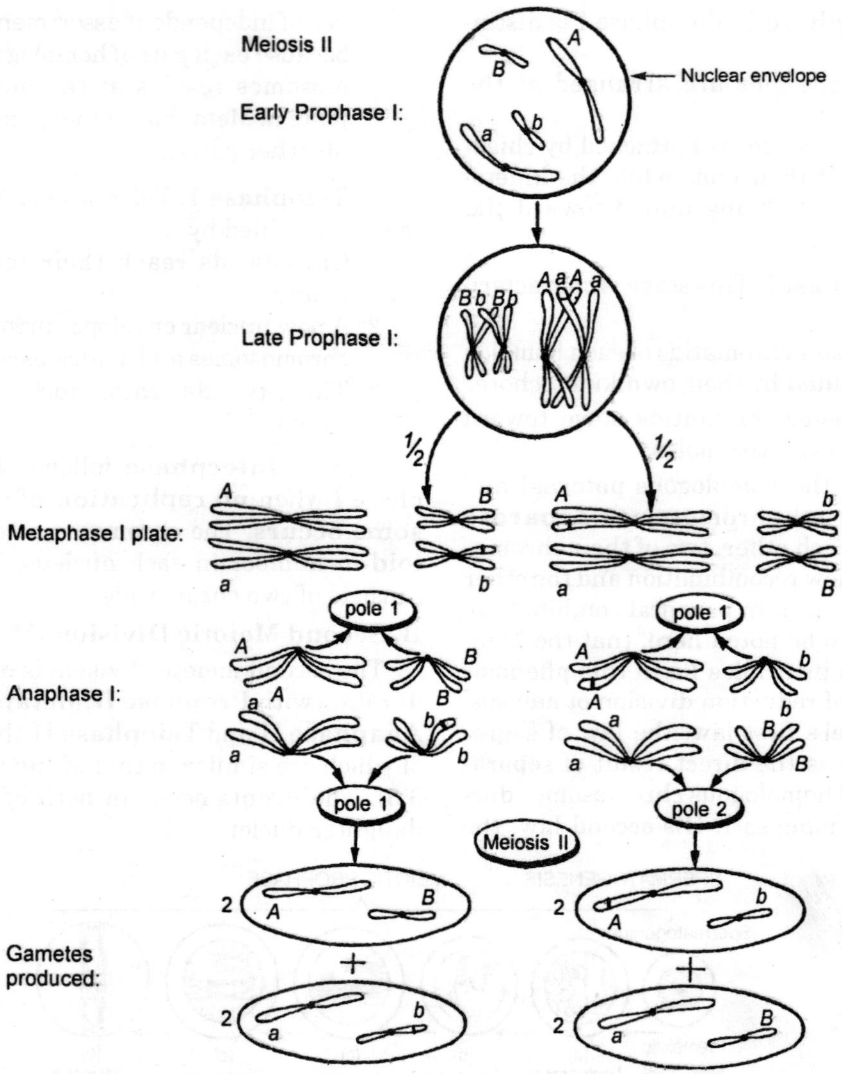

FIGURE 3.13(b)

Diagram of meiosis in a dihybrid emphasizing the independent alignment of homologous pairs of chromosomes (tetrads) at the metaphase plate of meiosis I. Meiosis I is the reductional division during which segregation and independent assortment occur. Only two pairs of homologous chromosomes are included. Note that the two paternal chromosomes align on one side of the metaphase plate, with the two maternal chromosomes on the other side, in one-half of the meiotic events (left). In the other one-half of the meiotic events, one paternal and one maternal chromosome align on each side of the metaphase plate (right). At anaphase I, the two chromosomes (each with two sister chromatids) with the same orientation move to one pole, and the two with the other orientation move to the other pole.

□ **Prophase II**: Chromosomes condense and consist of two chromatids. Nucleolus disappears. Nuclear envelope disappears. Centrioles migrate toward the poles. At the end, the spindle is formed.

□ **Metaphase II**: Chromosomes are arranged on the equatorial plane. The sister kinetochores separate and holds one chromatid each.

□ **Anaphase II**: The two sister chromatids go toward the opposite poles.

□ **Telophase II**: Chromatids are surrounded by nuclear envelope and daughter nuclei are formed. Nucleolus reappears.

Thus, finally, after Telophase II, 4 daughter nuclei (2 from each of the meiotic I product) are formed, *each* of which has a *haploid* number of chromosomes. They produce 4 haploid cells at the end of the Meiosis II and Cytokinesis in gametogenesis. These haploid cells are called *gametes*.

IMPORTANT NOTES:

Meiosis occurs in all sexually reproducing organisms. By meiosis diploid cell or cell nucleus is transformed (through one round of chromosome replication and two round of nuclear divisions) into four haploid cells or nuclei, each with one set of chromosomes. In the first meiotic division synapsis and crossing over of homologous chromosome take place. The crossing over or recombination bring variation in the resultant cells. Meiosis in combination with fertilization conserves the specificity in the number of chromosomes in a species generation to generation. Both paternal and maternal chromosomes are involved in meiotic process and contribute their genes to the gametes through crossing over.

Once again we return to the process involved in the *maturation phase* of *spermatogenesis* (3.4.1.3) [Fig. 3.13(a) and 3.13(b)].

In spermatogenesis, the *primary spermatocytes* undergo *First Meiotic Division*, followed by cytokinesis. As a result two *haploid secondary spermatocytes* are produced. Both the secondary spermatocytes now enter into the *Meiosis II* (which is simply a mitotic cell division) and produce *four haploid spermatids*.

Maturation phase is significant because during this phase each diploid primary spermatocyte (2n) produces four haploid (n) spermatids. And, *as there occurs C.O. in the Meiosis I, the spermatids possess genetic recombination in their chromosomes contributed by both parents.*

The spermatids are ovoid immobile structure and are not capable of functioning as male gametes. These cells have to undergo a special type of *differentiation*, i.e., the *metamorphosis of spermatid to functional sperm* called *spermiogenesis* or *spermioteliosis*.

3.4.1.4 Spermiogenesis (Metamorphosis of Spermatid)

The non-motile spermatids become motile and functional spermatozoa (achieving a *head, middle piece* and *tail*) through a series of changes.

Head:

1. The chromatin materials become closely packed inside a shrunk nucleus that loses its water content, nucleolus, RNA and many proteins, and become light.
2. The haploid amount of DNA persists. A new protein, protamine may associate with it. The shape of nucleus changes from spherical to elongate.
3. An anterior prolongation of the spermatid called *acrosome* develops. It releases enzyme that help the sperm to penetrate the egg-membrane. Acrosome is formed by the golgi-apparatus of the spermatid. During its formation, *pro-acrosomal granule* condenses inside the enlarging vacuole of the golgi-complex. The granule increases and transforms into *acrosomal granule*. This is the core of the acrosome. The wall of the vacuole spreads over the acrosomal granule and in front of the nucleus, covering them with a double-membrane forming an *acrosomal cap*.

Middle piece:

4. The two cylindrical centrioles move apart and lie behind the nucleus. One of them lies in a pocket at the posterior of nucleus and is called *proximal centriole*, while the other lying behind it is called *distal centriole*. The *axial filament* of the flagellum of the sperm develops from the distal centriole.

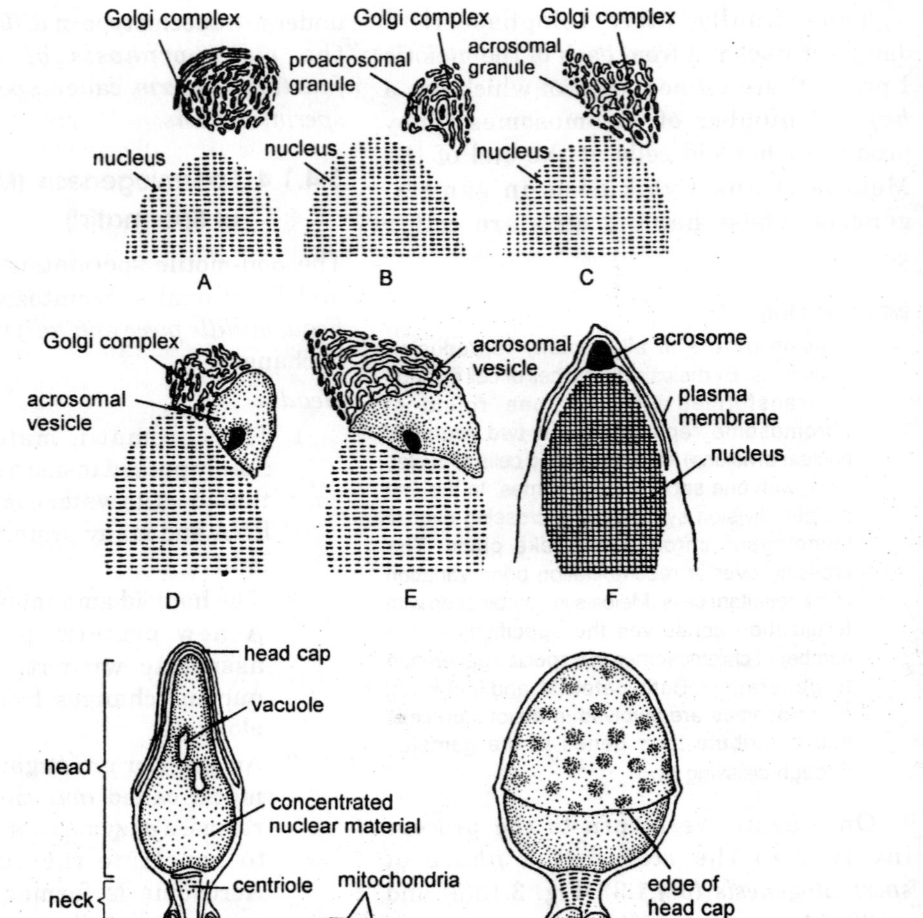

FIGURE 3.14

Top: Formation of acrosome.

Below: Structure of human spermatozoon. (E.M. structure) D, E, F, G and H are cross sections of the relevant parts.

5. The dispersed *mitochondria* of the spermatid concentrate around the proximal part of axial filament. Gradually all the mitochondria fuse together. Outside the mitochondrial sheath, a condensed cytoplasmic sheath, rich in microtubule is present. This sheath is called *manchette.*
6. Most of the cytoplasm of the spermatid is lost. Only a condensed layer, rich in microtubules is formed around the periphery of the sperm. This is also called *manchette,* which surrounds the posterior part of head and the whole middle part.
7. *Tail:* The *tail* or *flagellum* helps spermatozoon to swim with head forward. It is developed by the axial filament with 9 + 2 microtubular arrangement that forms the *central core.* The rest of the tail has two parts—a *mid piece* and an *end piece.* The mid piece is formed by *9 longitudinal* and the end piece only by the *axial filament.* A very little cytoplasm and a plasma membrane cover the tail (Fig. 3.14).

Spermatogenesis is a continuous process occurring constantly or periodically in the testis of sexually mature male animals. It produces many millions of sperms. In mammals the cycle of spermatogenesis may take weeks or months to complete. The developmental time from spermatogonia to mature sperm (gamete) is about 74 days in human. Each sperm contains 22 autosomes and a Y-chromosome or 22 autosomes and an X-chromosome. Hence the human males are *heterogametic.*

3.4.2 Oogenesis or Formation of Female Gamete, Ovum

The process of gamete (ovum or egg) formation in female is called *oogenesis* (Fig. 3.15).

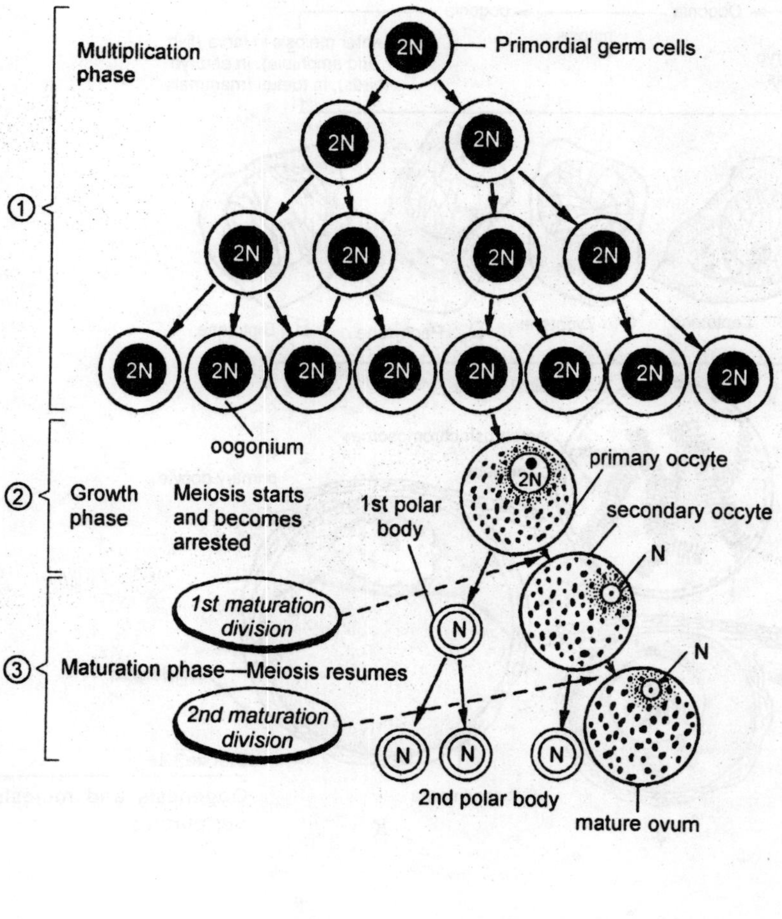

FIGURE 3.15
Process of oogenesis.

The process of oogenesis is identical to the spermatogenesis in nuclear division, but it is quite different in cytoplasmic events. Huge amount of RNA, proteins and yolk matters are incorporated into the oocyte during *growth phase* and *meiosis I.* These matters are required for the developing embryo.

The process of oogenesis consists of 3 major phases—*Multiplication phase, Growth phase* and *Maturation phase.*

3.4.2.1 Multiplication Phase

The primordial germ cells (2n) of the germinal epithelial of the ovary undergo series of *mitosis* and produce large number of diploid *Oogonia.*

This is also known as *proliferative phase.* In humans, other mammals and birds, the oogonia are produced once in the life of that organism. It occurs long before their sexual maturity. After production of a huge stock of oogonia in this phase, the primordial germ cells never produce new oogonia in adult ovary. But in fish and amphibians the multiplication phase and mitotic division of primordial germ cells occur throughout the adult life producing oogonia in each mating season.

3.4.2.2 Growth Phase

The egg requires huge reserve food and other cytoplasmic organelles for the development of future embryo. As the oogonia *grows* the reserve food (yolks) and other materials accumulate in the egg cytoplasm; hence the growth phase becomes a very lengthy and elaborate phase in most animals, specially in those with *mesolecithal* and *macrolecithal* (with moderate and large amount of yolk respectively). As for example, in frog, growth phase lasts for 3 years (Fig. 3.16).

During the elaborate phase of *growth* of oogonium the following changes are noticed in it:

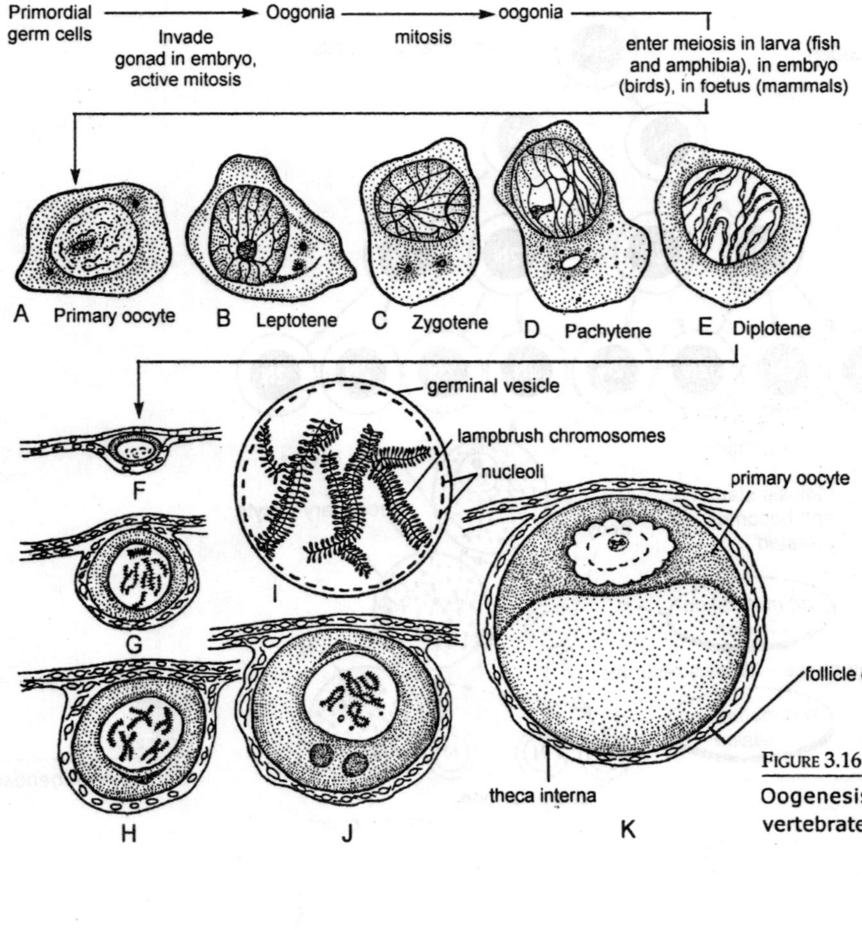

FIGURE 3.16
Oogenesis and meiosis in vertebrates.

1. Oogonium enters *meiosis I* and is called *primary oocyte.* Meiosis progresses upto *diplotene* and then the process is arrested for a considerable time (for many months to years). This event is called *meiosis I arrest* or *first meiotic block.*
2. The nucleus of the primary oocyte increases in size and volume and is called *germinal vesicle.* Inside the germinal vesicle, rapid *transcription* (RNA synthesis) goes on. In piscian, amphibian, reptilian and avian germinal vesicle *lampbrush chromosomes* are formed by great extensions of the *side loops* of the chromatids. The loops are the active site for various *RNA synthesis* (in the nucleus) that leads to the various *protein synthesis* in the cytoplasm of the primary oocyte.
3. The *nucleolus* inside the germinal vesicle enlarges by volume (and increases in number also in some species) and synthesizes ribosomes. This is associated with the *selective gene amplification* that enhances the number of genes that code for ribosomal RNA.
4. Cytoplasm of the primary oocyte increases by volume by accumulating large number of *mitochondria* (which aggregate together to form "*mitochondrial cloud*", numerous *ribosomes, smooth* and *rough endoplasmic reticulum* (SER and RER), large number of *golgi bodies,* many *proteins, amino acids* and other cytoplasmic inclusions.
5. The primary oocyte enters into a process called *vitellogenesis,* which is significant for synthesis and accumulation of *yolk* in the cyto-plasm. Yolk is the reserve food for future embryo. Accumulation of huge amount of yolk renders rapid growth of the primary oocyte. It contains proteins, phospholipids and neutral fats. *Granular yolk, yolk platelets* and *yolk sphere* develop in the cytoplasm.
6. At the time of growth, the primary oocyte attains differentiation and definite pattern of organisation. It attains *polarity,* i.e., an *animal* and a *vegetal* pole are differentiated. An animal-vegetal axis is established. Dorsal and ventral surfaces are differentiated. A *hyaline cortical layer* is differentiated from the ooplasm and lie below the plasma membrane as a *cortex.*

3.4.2.3 Maturation Phase

The meiosis resumes and continues through this phase. As a result, one large haploid *ovum* and three small haploid *polar bodies* (polocytes) are produced. The ovum is the functional gamete, while the polar bodies are nonfunctional. After a *long stay in diplotene* (dictyotene) in the growth phase, the chromosomes of germinal vesicle reaches the diakinesis stage of meiosis. The nuclear membrane of the germinal vesicle breaks up, microtubules are formed; the contracted chromosomes assemble to the periphery of the animal pole of the oocyte where a spindle is formed by the microtubules. The spindle is arranged in perpendicular position to the cell membrane. The microtubules of the spindle are attached to a narrow region around the animal pole. After formation of the polar bodies, they are expelled through the animal pole. Before that, during anaphase of meiotic I division, a *bulge* appears at the surface of the animal pole, where aggregate the haploid set of chromosomes (Fig. 3.17).

During telophase I, the bulged portion is pinched off from the rest of the oocyte. The very small pinched off nonfunctional cell is called *First Polar body* or *Polocyte,* and the remaining large oocyte with another haploid set of chromosomes is called *secondary oocyte*

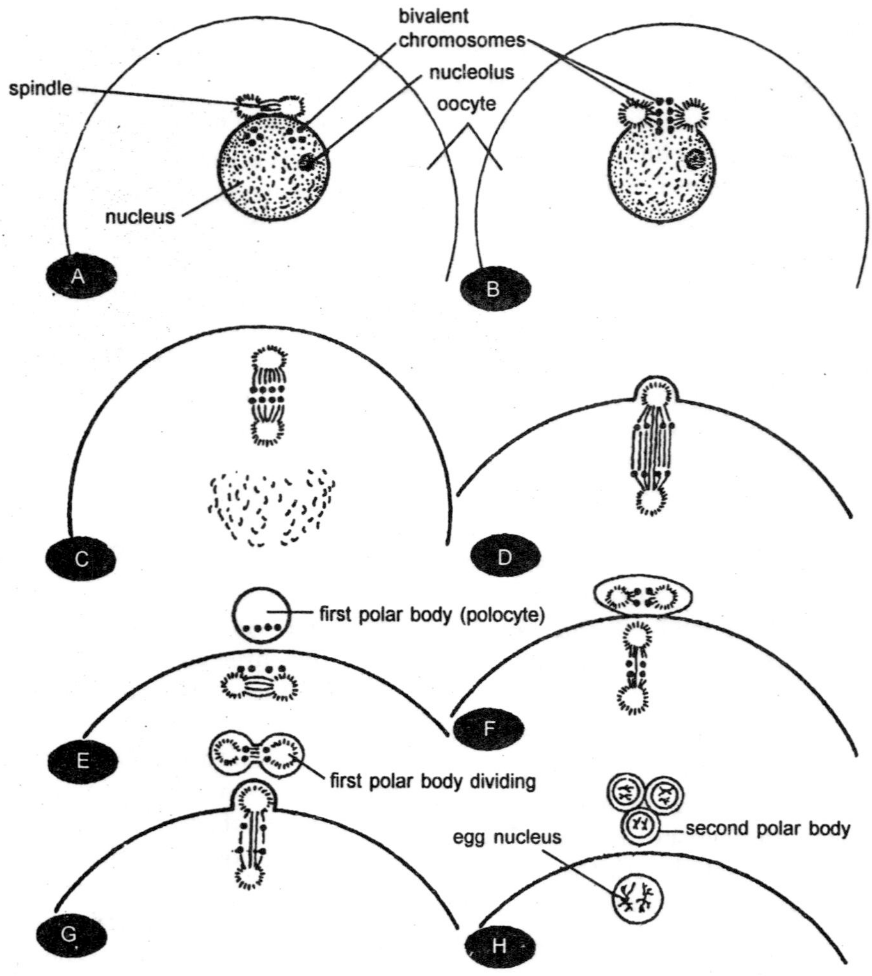

FIGURE 3.17

Reduction divisions or meiosis in an oocyte. A—Oocyte immediately before the meiotic division: B, C; D—First meiotic division; E—First polar body separated, preparation for second meiotic division; F, G—Second meiotic division and simultaneously, division of first polar body; H—Meiosis completed.

or *ootid*. The size remaining almost same, the secondary oocyte undergoes Meiosis II. It is also a phase of equal nuclear and unequal cytoplasmic division so that a very small haploid nonfunctional *secondary polar body* or *polocyte* and a large haploid functional ovum are produced. The previously produced first polar body also undergoes meiosis II and two more secondary polar bodies are produced.

Thus, at the end of maturation division, there are four meiotic products, of which only one is large and mature functional *female gamete* or *ovum* and three are small polocytes without any function. Gradually the polocytes disintegrate.

The developmental span from oogonia to ovum may be from the sixth week of embryonic age to the menopausal age (about 45-50 yrs) in human female. A mature ovum always contains 22 autosomes and 1 X-chromosome. As a female produces all gametes of similar chromosomal constituent, they are called *homogametic*.

3.4.3 Comparison of Spermatogenesis and Oogenesis

Spermatogenesis	*Oogenesis*
1. Occurs in testis.	1. Occurs in ovary.
2. Spermatocytes have no yolk and reserve food materials.	2. Oocytes have large amount of yolk as reserve food.
3. Growth phase is not so significant.	3. Growth phase is very significant.
4. At the end of spermatogenesis 4 equal sized spermatids are produced from each primary spermatocyte.	4. Four unequal cells are produced at the end. Out of the 4 meiotic products, one is large and functional ovum and three others are small non-functional polocytes.
5. All spermatids undergo spermiogenesis and each develops into a mature spermatozoa.	5. No such metamorphosis occurs and only one female gametes the ovum is produced from an ootid.
6. Polar bodies do not develop.	6. Three polar bodies are produced.
7. The whole process takes some weeks or months to complete in most organisms.	7. The process is too long, and may take many years to complete.
8. The spermatozoa are capable to fertilize egg sometimes after their differentiation.	8. The ovum is capable of being fertilized sometime before they have completed their meiotic division. They can be fertilized even before the polar bodies are released.

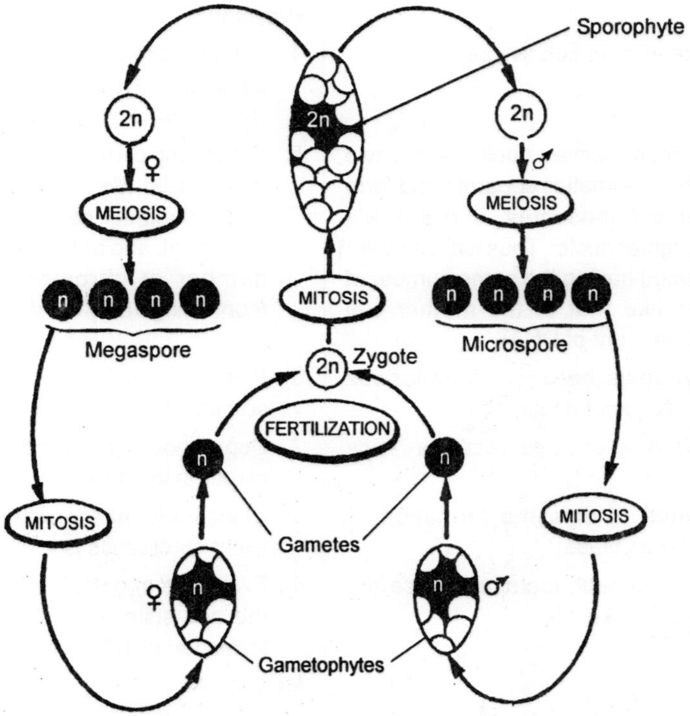

FIGURE 3.18

Alternation of generations between haploid gametophytes and diploid sporophytes in plants. In the lower plants, the gametophyte is dominant, whereas in the higher plants, the sporophyte is dominant.

3.5 Spore Formation in Plants

In plants, gametes are not the direct result of meiosis, but intervened by a process of *sporogenesis*. Sporogenesis produces *spores* which are **not** sex-cells. The spores produce *gametophytes*.

The *gametes* are produced from gametocytes, in which the chromosome number is already reduced to haploid numbers. In plants, an *alternation of generation* is present between a haploid and a diploid phase.

The diploid plants are called *sporophytes* which undergo meiosis to produce *spores*; the spores contain reduced (haploid) chromosome number. Spores develop into haploid *gametophytes*. The gametophytes produce haploid *gametes* which are capable of fertilization. Fertilization results into *zygote* (2n) which develop into the sporophyte plants again (Fig. 3.18).

3.6 Comparison between Mitosis and Meiosis

Mitosis	*Meiosis*
1. Mitosis occurs continuously in the somatic cells having diploid amount of DNA (2C).	1. Meiosis occurs at the time of gametogenesis in the germ cells having diploid amount of DNA (2C).
2. Only one sequence of division occurs.	2. Two successive divisions occur, one reduction division and the other homotypic division.
3. Comparatively shorter in duration.	3. Longer in duration, comprising meiosis I and meiosis II.
Prophase	**Prophase**
4. Short prophase with no subdivision.	4. Long prophase with at least 5 substages, viz., leptonema, zygonema, pachynema, diplonema, diakinesis.
5. Homologous chromosomes duplicate into two chromatids. The chromatids separate and form new daughter chromosomes, in the newly formed two daughter nuclei. Thus two daughter cells are produced having the same number of chromosomes like that of the mother cell (containing 2C amount of DNA).	5. Separating chromosomes and their chromatids may not be identical due to crossing over. A single daughter cell receives only one kind of chromatid, and after the end of the process, the number of chromosomes becomes half (containing 1C amount of DNA) the mother cell.
6. Pairing or synapsis between homologous chromosomes does not occur.	6. Pairing or synapsis of homologous chromosomes takes place (zygotene stage).
7. Duplication of chromosomes occurs in early prophase.	7. Duplication and splitting of chromosomes takes places in the pachytene stage of late prophase.
8. No phenomenon of chiasma formation or crossing over takes place.	8. Chiasma formation with crossing over of non-sister chromatids is a significant phenomenon.
9. No exchange of genetic material; hence the daughter cells are alike.	9. Exchange of genetic material occurs between the non-sister chromatids, resulting into the formation of different kinds of daughter cells.
Metaphase	**Metaphase**
10. The chromatids are in the form of a *dyad*. Pairs of chromatids line up on the equator of the spindle.	10. The chromatids remain as a *tetrad*. They line up on the equator.

Contd.

Mitosis	*Meiosis*
11. The kinetochores of the chromosomes are directed toward the equator and the arms are directed toward the poles.	11. The kinetochores are directed toward the poles and the arms toward the equator.
Anaphase	**Anaphase**
12. The dyad splits and form monad having a single chromatid.	12. The tetrad splits and forms dyads having two chromatids attached to a single centromere.
Telophase	**Telophase**
13. After reaching the poles, chromatids start to unfold and appear like the chromosomes of the prophase.	13. The chromatids form short and thick chromosomes.
Significance	**Significance**
14. Responsible for cell multiplication and growth of body.	14. Responsible for producing gametes in sexually reproducing organisms.
15. Two daughter cells result after each mitotic division.	15. (a) 4 daughter cells called spermatids are produced after meiotic division in males. Each one develops into spermatozoa and acts as a functional gamete. Or (b) 4 daughter cells, one very large and 3 very small called ovum and polar bodies respectively are produced after meiotic division. Only the ovum acts as a gamete.
16. Daughter cells are diploid (2n) with 2C amount of DNA.	16. Daughter cells are haploid (n) with 1C amount of DNA.
17. No genetic variability results.	17. Genetic variability is the most significant event (Fig. 3.19).

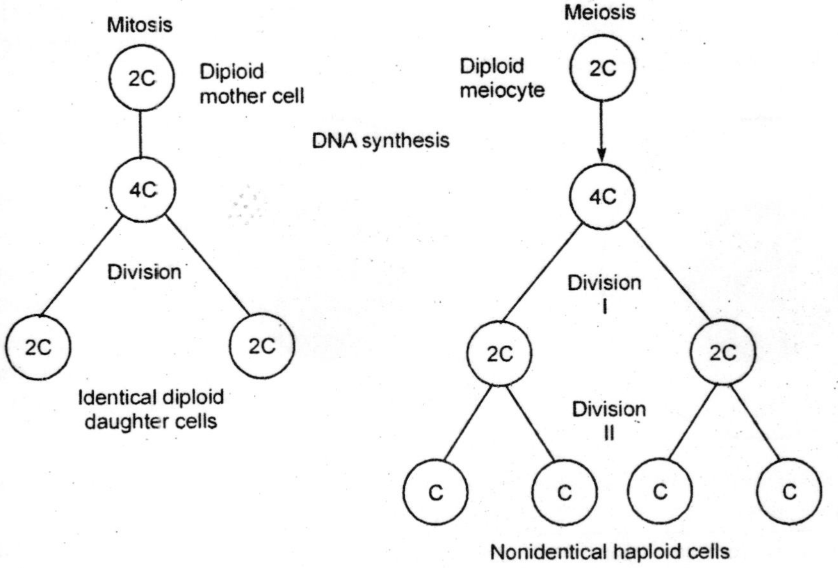

FIGURE 3.19

Comparison between Mitosis and Meiosis. C is the haploid amount of DNA in the genome.

3.7 Meiosis and Mendel's Principles of Heredity

The Mendelian principle of *segregation* and principle of *independent assortment* are the direct result *meiotic events*. Let us consider the contrasting phenotypes of plant-height, seed-colour and seed-shape in a trihybrid of F_1-peas (Fig. 3.20), the genes of which are located on *separate* homologous pair of chromosomes (chromosome-4, 1 and 7 respectively; in pea, there are 7 pairs of chromosomes, out of which only 3 have been mentioned here to explain the segregation and independent assortment). Thus, one of the maternal 4th homologous chromosome contains dominant **D** and its paternal counterpart contains the recessive allele **d**. Similarly, one of the 1st homologous pair contains **G** and the other **g**, while one of the 7th pair contains **W** and the other **w**.

During zygotene stage of meiosis, the six chromosomes *synapse* to form three homologous pairs. During synapsis, each chromosome finds its counterpart and alleles come together in corresponding positions. During gametogenesis, at metaphase I, *random assortment* of chromosome occurs. At second meiotic division *segregation of sister chromatids* occurs. At that time, the 4th, 1st and the 7th chromatid (in this example) *separate* or *assort* independently along with their own alleles. As a result, 8 types of gametes may be produced, and they all will possess haploid number of chromosomes having the genes DGW, DGw, DgW, Dgw, dGW, dGw, dgW and dgw. (As genes remained in separate chromosomes, no linkage occurred and Mendel could postulate his principle of independent assortment). The above example clearly demonstrates the parallelism between meiosis and Mendelism. See (Fig. 2.20).

IMPORTANT NOTES:

1. Chromosome carries genetic information. Large amount of DNA remains tightly packed and coiled within the nucleus forming chromatin structures and eventually the chromosomes.
2. Number of chromosome is constant in each species. A set of chromosome is called ploidy. Thus cells containing a single set of chromosome is called haploid, with two sets is called diploid, with 3 sets triploid and so on. The diploid set is the usual chromosome number of a species. The non-sex chromosomes, are called autosomes; they show no sexual differences and are alike in both males and females. Sex-chromosomes are responsible for sex-determination. One of the sexes has a pair of identical sex-chromosomes (XX) and the other may have a single sex-chromosome which may be unpaired (XO) or paired with a Y-chromosome (XY). XX-constitution are homogametic, XO and XY-constitution are heterogametic.

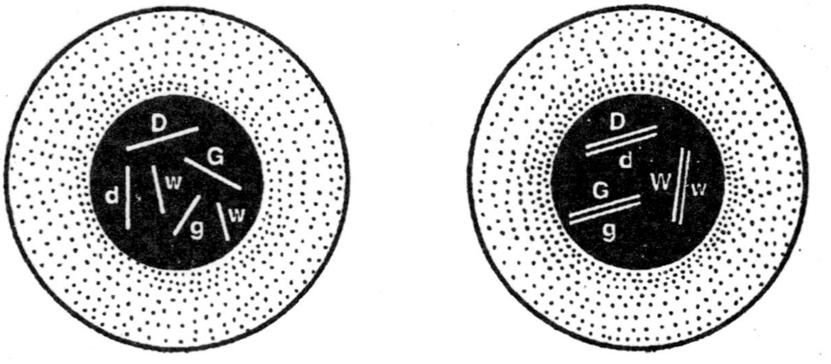

A B

FIGURE 3.20

Cells showing 3 chromosome pairs: (A) before synapsis and (B) after synapsis.

3. The chromatins are the nucleoproteins consisting mainly of DNA and histone proteins. RNA and other proteins may also be associated in little amount. Histones and DNA are present almost in 1 : 1 ratio. Histones are of five types—H1, H2A, H2B, H3 and H4. Two of each of H2A, H2B, H3 and H4 are present per 200 bp of DNA. This octamer and H-1 form a nucleosome unit.
4. Partially condensed chromatin is called euchromatin which is genetically active. Tightly condensed chromatin is called heterochromatin which is genetically inactive. Certain chromosomes are totally heterochromatic, as for example the Y chromosome of male *Drosophila*. One of the two X-chromosomes of female mammal forms the sex-chromatin or Barr body which is heterochromatic, while the other remains euchromatic.
5. The sequence of events that occurs between one cell division and the next is called the cell cycle. It contains G_1-phase, S-phase, G_2-phase and M-phase.
6. Mitosis and meiosis are two types of cell division. Mitosis occurs in somatic cells and multiplies the number of cells. Each cell divides into two daughter cells which are identical to their mother-cell. Meiosis occurs in germ cells and produces gametes in sexually reproducing organisms. Four haploid daughter cells are produced from a single mother cell at the end of meiosis. Resultant cells are with 1C amount of DNA and show genetic variability.
7. Formation of gametes in sexually reproducing organisms is called gametogenesis. Gametogenesis in males is called spermatogenesis, as it produces sperms, and that in females is called oogenesis, as it produces ova. At the end of spermatogenesis, 4 equal sized spermatids are produced from each primary spermatocyte; each spermatid undergoes spermiogenesis to form mature functional spermatozoon. At the end of oogenesis, 4 unequal (1 large, 3 small) cells are produced of which the large one is the functional ovum and the other small three cells are polocytes. The process of oogenesis takes a long time to be completed.

3.8 Giant Chromosomes (Polytene and Lampbrush)

Very large chromosomes found in certain cells at certain stages of the cell-cycle are called *giant chromosomes*. The chromosomes in the salivary gland cells of *Drosophila* are about 100 times larger than chromosomes of other body cells. T. Painter, E. Heitz and H. Bauer discovered these chromosomes in the salivary gland cells of *Drosophila* and called them *salivary gland chromosomes*. These chromosomes contain many (poly) threads of chromonemata (tene) or DNA side by side, and are now named as *polytene chromosomes*.

The dipteran insects like *Drosophila*, *Chironomus* etc. contain polytene chromosomes in their salivary gland cells. These cells contain high amount of DNA; they replicate their chromosomal DNA several times without separation of them (called *endomitosis*). The resultant daughter chromatids (DNA strands) remain lined up parallely side by side attaining *polyteny* or size of a thick chromosome. The polytene chromosome of *Drosophila* bears about 1000 DNA strands lying side by side. It can develop from 10 rounds of DNA replication without separation. This parallel replication magnify differences in density along their length (e.g. chromomeres). These density differences produce 'banding' (Fig. 3.21). The numbers of bands are constant in a particular species. In *Drosophila melanogaster* about 5000 bonds are present. Possibly each band is a functional genetic region. Bands are alternated with interbands. In polyteny the two homologous chromosomes of each diploid pair are also often lined up side by side. Certain bands may change in appearance in different tissues at different times. When they appear in expanding forms, they are called '*puffs*' (previously called *Balbiani Rings*). The polytene chromosomes are visible during interphase and prophase. The highly folded DNA in the *bands* are alternated with

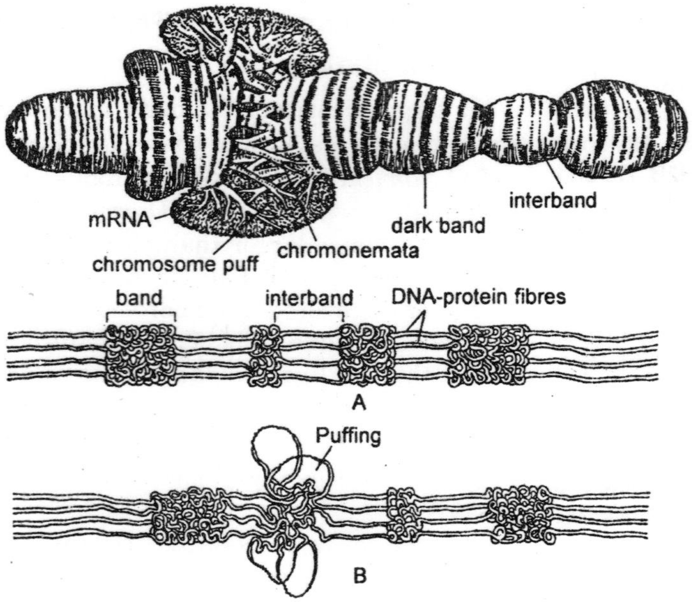

FIGURE 3.21

Polytene chromosome of an insect, showing bands and interbands and a puff or Balbiani ring.

loosely folded DNA in the *interbands*. The 5000 bands of *D. melanogaster* are alternated with 5000 interbands. The bands and puffs are related to the RNA synthesis. Due to the association of the homologous pairs together (somatic pairing), it appears that the chromosome number is half the normal number in the salivary gland cells. Thus in *D. melanogaster* only 4 polytene chromosomes (instead of 8) are found. All the 4 chromosomes are joined together by their centromeric region, forming a common centromere called *chromocentre*. Most part of the 4th autosome and the whole Y-chromosome diffuses into the chromocentre, (Fig. 3.22).

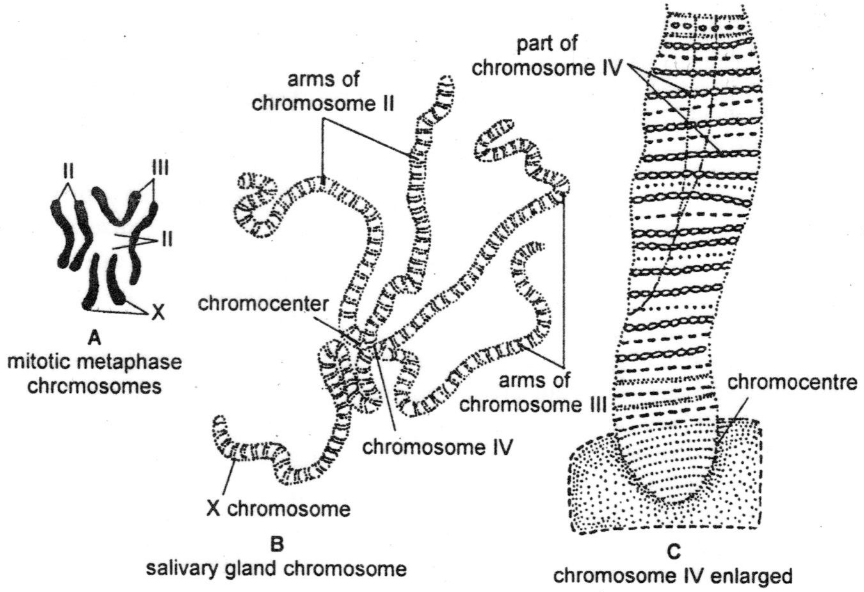

FIGURE 3.22

Normal and polytene chromosomes of *Drosophila melanogaster*. A—Normal mitotic chromosomes. B—Polytene chromosomes of female, C—An enlarged IV chromosome.

• Lampbrush Chromosome

In some amphibia (salamander), the meiotic prophase chromsomes of the *primary oocyte* may reach about 1000 μm in length; these large chromosomes have long lateral loops giving them a hairy "lamp-brush" look; these are called *lampbrush chromosomes*. Lampbrush chromosomes are larger than the polytene chromosome.

In the prolonged diplonema of meiotic prophase I, the lampbrush chromosomes are found to occur in *bivalents,* in which the homologous chromosomes are held together by chiasmata. Each bivalent has 4 chromatids, two in each homolog. The characteristic *lateral loops* extend from the *chromomeres* of the axis of each homolog. Each chromosome has two loops, one arising from each chromatid. Each loop of the two sides are symmetrical but they may differ in size and thickness according to phase of action. The loops have a constant position in the chromosome (See chromosome mapping). In *Triturus,* there are about 10,000 loops per chromosome in the oocyte. The *axis* of the loop is made of single DNA molecule. Rapid and intense RNA synthesis occurs in the unfolded DNA of these loops. **RNA polymerase** are present in the DNA-axis of the loop for **transcription** (See chapter 10). The lateral loop has a *matrix* consisting of RNA and proteins. The matrices are thicker at one end of the loop (Fig. 3.24) and thinner at the other end. RNA synthesis (transcription) begins in the thinner end and progresses toward the thicker end. The lampbrush chromosomes are probably responsible for 'gene amplification' (See oogenesis) during the growth phase of the oocytes. Many of the loops of lampbrush chromosome may act like a single gene while others may contain several genes.

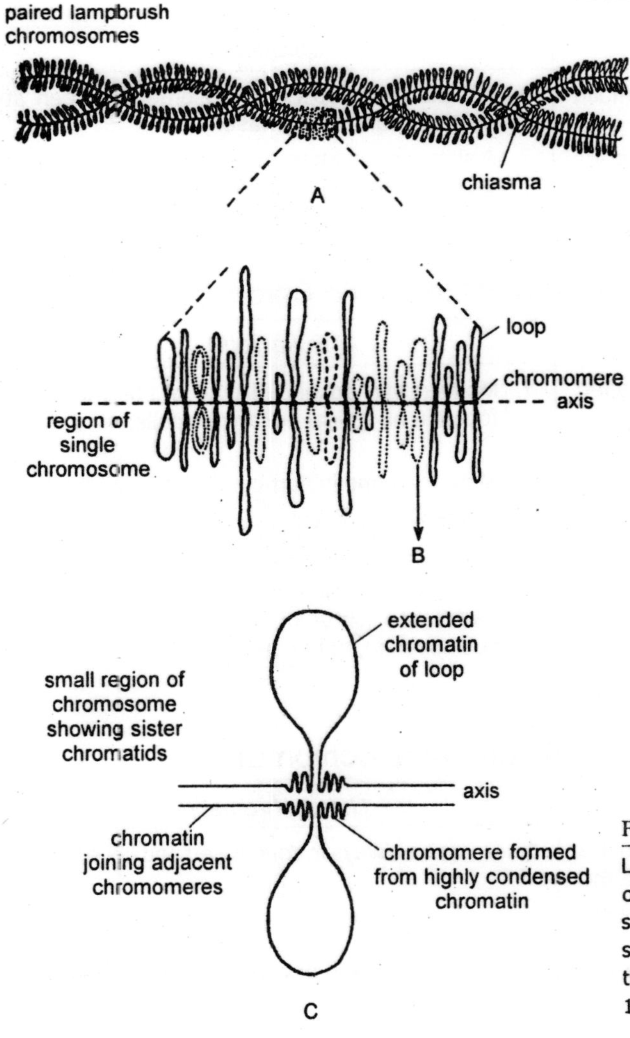

FIGURE 3.23

Lampbrush chromosome structure. A—Bivalent or paired homologous chromosomes in pairing showing chiasmata, B—A part of one homolog showing paired loops given out by two chromatids; C—Single pair of loop (after Alberts *et al.* 1989).

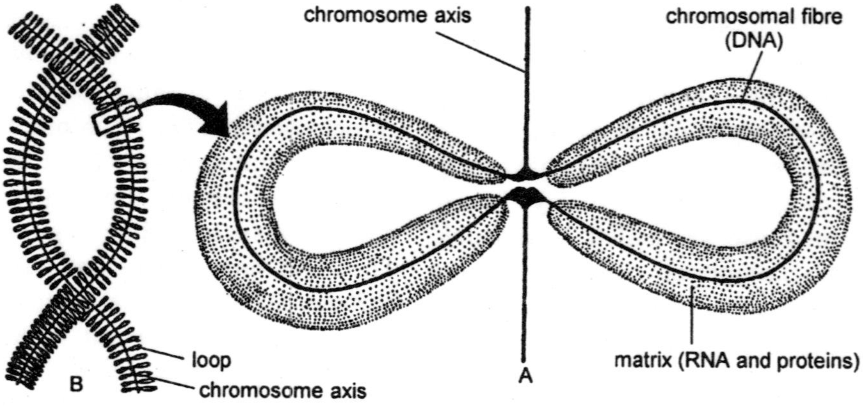

FIGURE 3.24

Lampbrush chromosome. A—At low magnification, B—Loop magnified.

3.9 Comparison between Chromatin and Chromatid

Chromatin	Chromatid
1. Component of chromosome seen as chromatin thread in interphase and prophase.	1. Longitudinally split chromosome-part seen at prophase and metaphase.
2. Constitutes chromosome. Made of nucleic acids and protein.	2. A sister chromosome, having all the features of that from which it has been duplicated.

3.10 Comparison between Prokaryotic and Eukaryotic Chromosome

Prokaryotic	Eukaryotic
1. Primitive type of chromosome; also called prochromosome. Found in bacteria, blue-green algae, and mycoplasma.	1. Modern and advanced type of chromosome, found in all typical eukaryotic cells.
2. Single in number.	2. More than one in number. Usually paired.
3. Circular DNA present.	3. Linear DNA present.
4. DNA is not attached with histone.	4. DNA is always in association with histone.
5. Not covered by a nuclear membrane; thus chromosomes have direct link with the cytoplasm.	5. Firmly and closely encircled by the nuclear membrane; chromosomes are suspended in nucleoplasm, not in the cytoplasm.

3.11 Comparison between Euchromatin and Heterochromatin

Euchromatin	Heterochromatin
1. Found uncoiled or in stretched condition in the interphase nucleus; but in division phase the euchromatin remains in coiled condition in the chromosome.	1. Remains always in condensed and coiled state.
2. Euchromatin stains lightly in interphase but deeply in division phase.	2. Heterochromatin always stains very deeply.

Contd.

Euchromatin	*Heterochromatin*
3. Genetically active.	3. Genetically inactive.
4. Transcription takes place.	4. Transcription does not occur.
5. Crossing over occurs.	5. Crossing over does not occur.
6. Normal DNA replication occurs.	6. Late DNA replication occurs.
7. Most part of the chromosome carry euchromatin.	7. Sex-chromatin (Barr body) and most of the Y-chromosome contain heterochromatin.

CHAPTER 4

GENETIC MATERIAL

Nucleic acids, Nuclein, DNA is the genetic material, Transforming Principle, Transforming material is the DNA, Hershey-Chase experiment, RNA is the genetic material in some viruses, Structure of DNA, Watson-Crick model, Alternative forms of DNA, DNA-replication, Centromeric DNA, Telomeric DNA, Satellite DNA, RNA, Structure of RNA, mRNA, tRNA, rRNA, Comparison between DNA and RNA.

In chapter 3, we have learned that chromosomes carry hereditary materials or genetic information. We have also learned that chromosomes are composed of nucleoproteins which consists of nucleic acids (deoxyribonucleic acids or DNA and ribonucleic acids or RNA) and proteins (histones and nonhistones). On the basis of this knowledge, it is obvious to assume that any one of the DNA, RNA or protein in the chromosome must be the genetic material, (i.e., must carry the genetic information). After many years of researches and experiments (1928-44), finally it was clearly established that in all organisms, *the genetic information resides in the nucleic acids, not in the proteins;* and *in most cases the genetic information is stored in the DNA*. Only in some retroviruses (RNA-viruses), RNA plays *the genetic role.* In 1944 and onwards, it was confirmed that the DNA (deoxyribonucleic acid) is the genetic material.

4.1 Nucleic Acids: Properties and Chemistry

All living organisms contain *nucleic acids* in the form of *deoxyribonucleic acid* (DNA) and *ribonucleic acid* (RNA). DNA, in most organisms, stores the genetic information which is *transcribed* into RNA molecules. The 'information' remains 'encoded' in the specific nucleotide-sequences of RNA called *codons.*The codons 'code' for specific amino acid sequences and are *translated* into proteins.

These two phenomena, *transcription* and *translation* are important in genetic transmission. The phenomenon of *transcription of DNA into RNA,* and *translation of RNA into protein* is called *Central Dogma* of molecular biology.

The genetic information flows through central dogma.

$$\text{DNA} \xrightarrow{\text{transcription}} \text{RNA} \xrightarrow{\text{translation}} \text{Protein}$$

The exception of the central dogma is observed in RNA-viruses, where the RNA acts as genetic material and is transcribed into DNA by *reverse transcriptase enzyme.* This DNA then follows the usual path of central dogma. This phenomenon is called *reverse central dogma.*

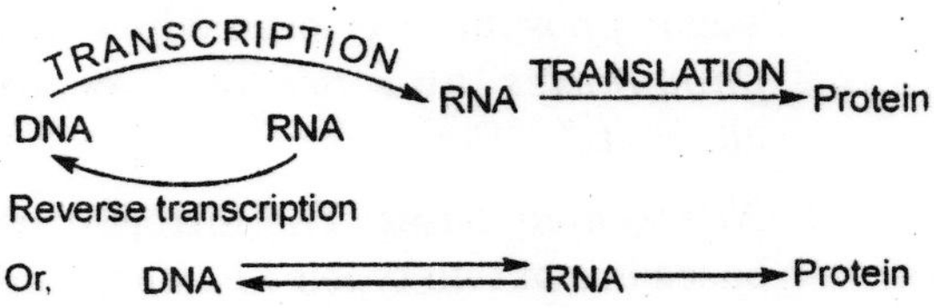

To understand the function of nucleic acids, the *chemical structure* of them must be known.

• Chemical Structure of Nucleic Acid

All nucleic acids are composed of *pentose, phosphoric acid* and *nitrogenous bases.* The unit of nucleic acid is called *mononucleotide.* A mononucleotide has two parts—*nucleoside* and *phosphoric acid.* The nucleoside is composed of *nitrogen base* and *pentose* (5-carbon sugar). Nitrogenous bases are of two types: *purine* and *pyrimidine.* Purine is of two kinds—*Adenine* and *Guanine* while Pyrimidine is of three kinds—*Cytosine, Thymine* and *Uracil. Pentose* is of two types: *ribose* and *deoxyribose* (Table 4.1).

Let us consider the structures one by one. *Nucleotide,* the unit of nucleic acid has two parts—the *nucleoside* and the *phosphoric acid.* The nucleoside has two parts—*Pentose* and *Nitrogenous bases.*

A. *Pentose*: Pentoses are 5-carbon sugars. They are of two types: *ribose* and *deoxyribose.* These two are all alike but for only difference that the deoxyribose contains *one less oxygen atom* (de-oxy) at its second carbon position (Fig. 4.1). In DNA, the sugar is *deoxyribose* and in RNA, the sugar is *ribose.*

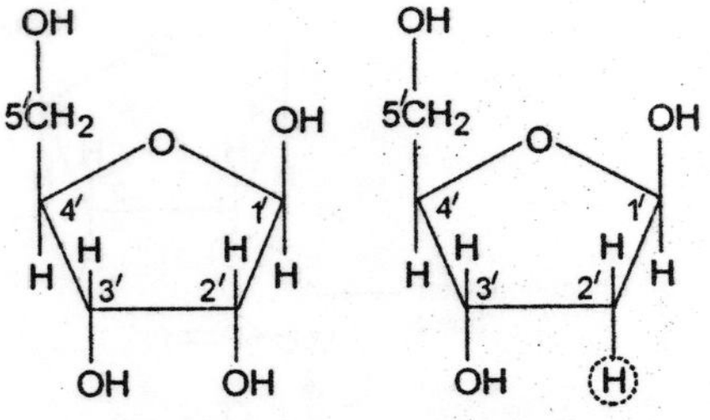

FIGURE 4.1

Pentoses found in nucleic acids.

TABLE 4.1

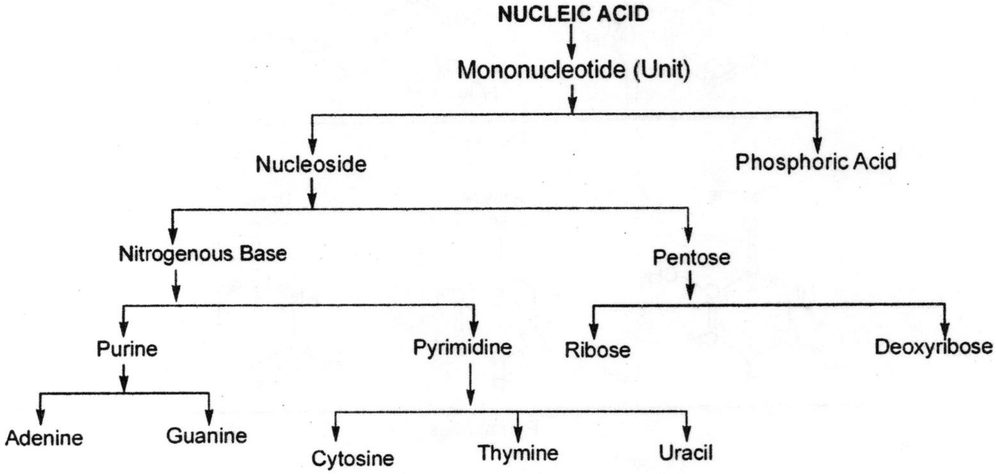

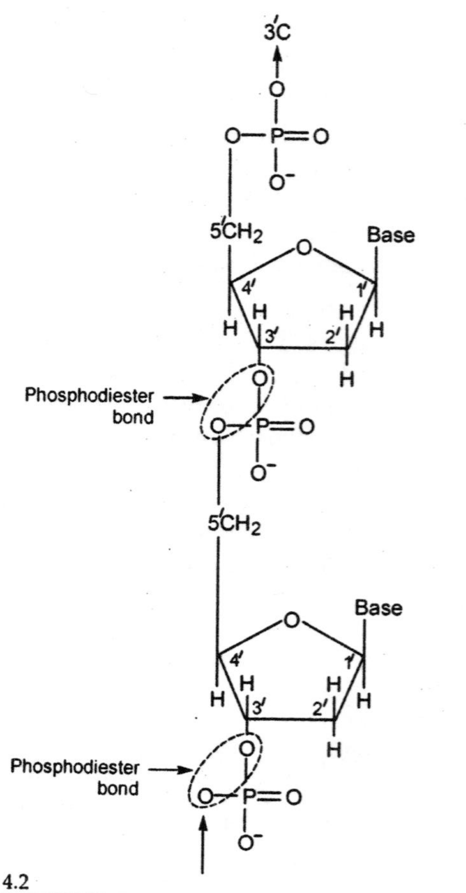

FIGURE 4.2

Backbone of sugar-phosphate with phosphodiester bond.

B. *Phosphoric Acid* (Phosphate): The sugars of the successive nucleotides are linked together by means of *phosphodiester* bonds. These bonds link the 3′ carbon in the pentose of one nucleotide to the 5′ carbon in the pentose of the next nucleotide (See 4.3.1). Thus, a ribbon or backbone of 'sugar-phosphate' is formed in which *phosphates* and *pentoses* occur alternately (Fig. 4.2).

C. *Nitrogenous bases*: The nitrogenous bases are of two types—*purines* and *pyrimidines.* The purine contains two fused rings while the pyrimidine contains a single heterocyclic ring (Fig. 4.3).

In both DNA and RNA, the purines are same, i.e., both contain *adenine* and *guanine,* but the pyrimidines are different. The pyrimidines of DNA are *cytosine* and *thymine* while those of RNA are *cytosine* and *uracil.* (See 4.3, structure of DNA and RNA).

4.1.1 Nucleoside and Nucleotide

In nucleic acids, the monomer unit is called *nucleotide* which is formed by the

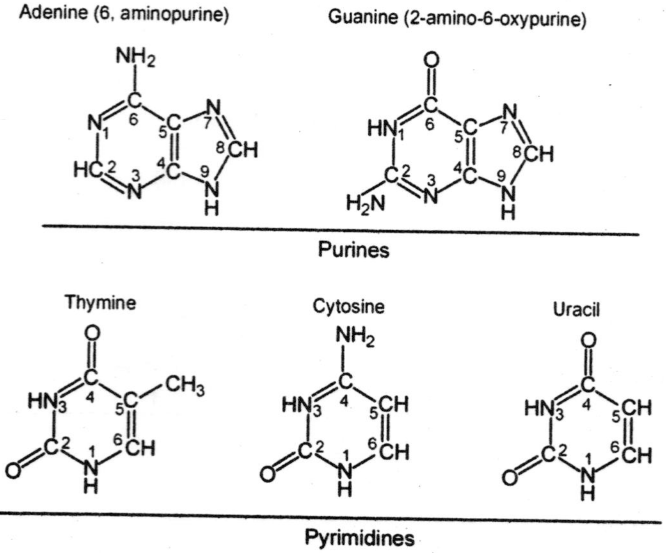

FIGURE 4.3

The bases of DNA and RNA.

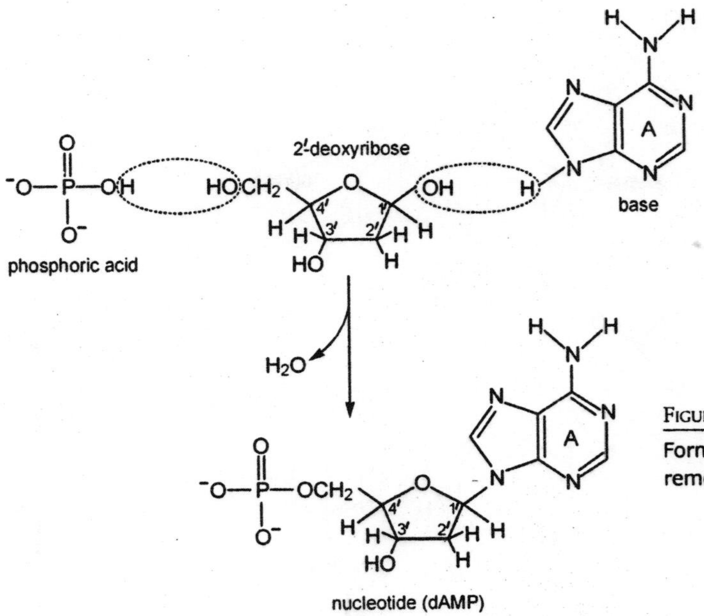

FIGURE 4.4(a)

Formation of nucleotide by removal of water.

combination of *nitrogenous base, pentose* and *phosphate*. When the phosphate is excluded from this combination, it is called *nucleoside* (i.e., the nucleoside is formed by base and pentose only).

Base + Pentose = Nucleoside

Base + Pentose + Phosphate = Nucleotide.

The table 4.2 shows some examples of nucleosides and nucleotides.

There is a difference between dAMP and AMP. When the purine base *adenine* links with *ribose* sugar (instead of deoxyribose), it forms the nucleoside called *adenosine*. When adenosine links with a single phosphate, it forms *adenosine monophosphate* or *AMP* which is a nucleotide. Similarly adenosine diphosphate (ADP) and adenosine triphosphate (ATP) are also nucleotides.

Figure 4.5 shows a structue of ATP and its components. Table 4.3 shows the nucleotides (base + sugar + phosphoric acid) of the ribose series.

TABLE 4.2

Nitrogenous base	base + sugar (deoxyribose) = Deoxyribonucleoside	Deoxyribonucleoside + Phosphate = Deoxyribonucleotide
Adenine (A)	Deoxyadenosine	Deoxyadenosine monophosphate (dAMP)
Guanine (G)	Deoxyguanosine	Deoxyguanosine monophosphate (dGMP)
Cytosine (C)	Deoxycytidine	Deoxycytidine monophosphate (dCMP)
Thymine (T)	Deoxythymidine	Deoxythymidine monophosphate (dTMP)

TABLE 4.3

Base	Nucleotide	Abbreviation
Adenine (A)	Adenylic acid (adenylate)	AMP
Uracil (U)	Uridylic acid (uridylate)	UMP
Cytosine (C)	Cytidylic acid (cytidylate)	CMP
Guanine (G)	Guanylic acid (guanylate)	GMP

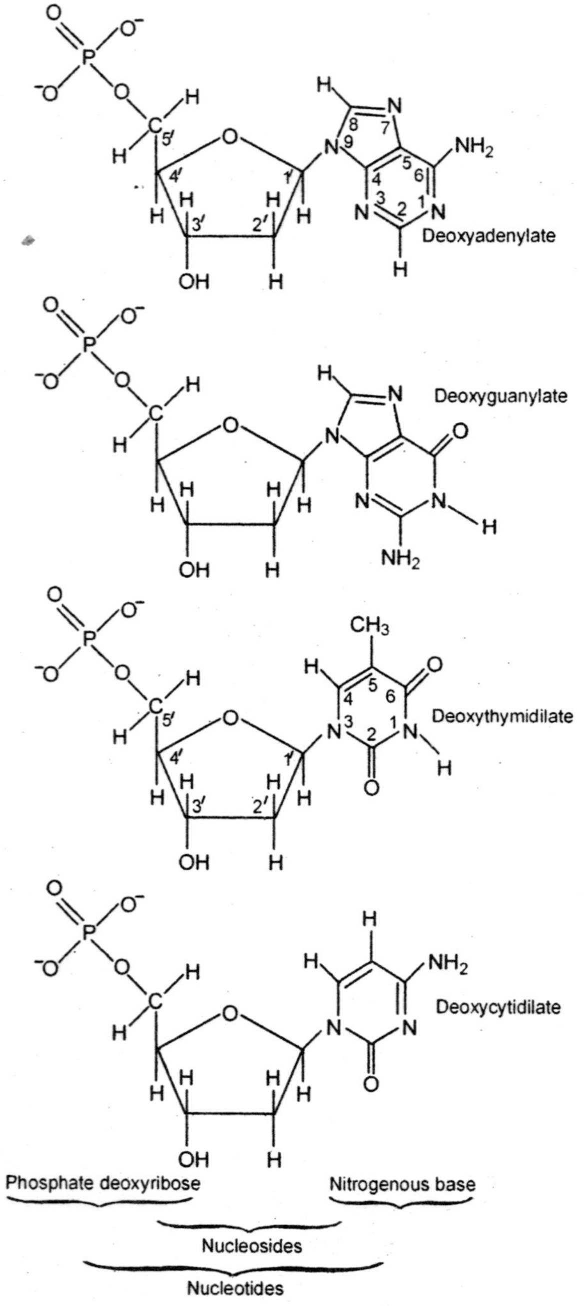

FIGURE 4.4(b)

The four common deoxyribonucleotides. Each of the nitrogenous bases is attached to deoxyribose by a bond that connects a ring nitrogen atom of the base to the number 1 atom of the deoxyribose. The suffix "ate" equals "acid".

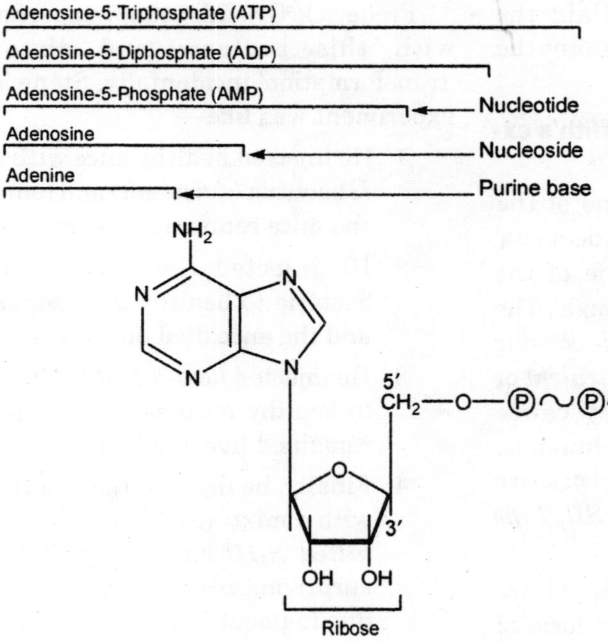

FIGURE 4.5

A, structure of ATP and its components.

4.1.2 Nuclein

Nuclein was discovered by J.F. Miescher in 1868 from the human pus-cell, but its importance was not realized at that time. It is now recognized that this acidic material contained polynucleotide chain (DNA) or nucleic acid.

4.2 Evidences to Suggest DNA as Genetic Material

So far we have learned that there are evidences enough to suggest the nucleic acid as the genetic material. We have also learned the basic nature of nucleic acids. The indirect evidences that suggest DNA as genetic material are—

1. Most of the DNA is located in the chromosome, and chromosomes are the carriers of heredity. On the other hand, RNA and proteins are abundantly distributed in cytoplasm also (but cytoplasm plays little or no role in heredity).
2. A strong correlation exists between the amount of DNA per cell and the number of sets of chromosomes per cell. (Diploid cells have twice the amount of DNA as the haploid gametes of the same species).
3. Almost universally, the molecular composition of the DNA is the same in all the cells of an organism (composition of RNA and protein is highly variable in various cell types).
4. Regarding stability, the DNA proves to be more stable than RNA or protein.

4.2.1 Proof that DNA is the Genetic Material

It is now well established that the genetic information remains stored in DNA, i.e., DNA is the genetic material in most organisms. (In retroviruses only, the RNA is the genetic material).

The role of DNA in mediation of transformation was indicated in an experiment of F. Griffith in 1928. It was the first important experiment with pneumococcus bacteria that led him to the discovery of 'transforming principle'. It laid the foundation of the proof that DNA stores the genetic information.

- **Transforming Principle (Griffith's experiment)**

There are some different forms of the pneumonia-bacteria, 'pneumococcus' (*Streptococcus pneumoniae*). One of the forms is 'smooth' and the other, 'rough'. The smooth types have a *polysaccharide capsule* around the bacterial cell and are *virulent* or pathogenic in nature, i.e., they cause pneumonia in affected mice and humans. They are called *Type S*. These types are further classified as *Type SI, Type SII, Type SIII* etc..

The other forms of pneumococcus are 'rough'-type which are the mutant form of the S-type. They lack a polysaccharide capsule and are *avirulent* or nonpathogenic, i.e., they cannot cause pneumonia. They are called R Type, and are further classified as *Type R-I, R-II, R-III* etc. The type S are larger than the Type R in size.

Frederick Griffith (1928) experimented with this bacteria and discovered 'transformation' incidentally. Stepwise, his experiment was like—

1. He injected healthy mice with *live R-II* bacteria (avirulent) and found that the mice remained live and healthy.
2. He injected *live S-III* (virulent) bacteria to healthy mice separately, and the mice died due to pneumonia.
3. He injected *heat-killed S-III* bacteria to healthy mice and found the mice remained live and healthy.
4. Finally, he injected the healthy mice with a mixture of *live R-II* and *heat-killed S-III* bacteria, and found a surprizing result that the mice *died* due to pneumonia ! (Fig. 4.6).

The postmortem of these dead mice (result of the 4th step) showed that their blood had both R-II and S-III living pneumococcus. Griffith's conclusion was that *some component* of the heat-killed S-III *must have converted living R-II* to virulent S-III.

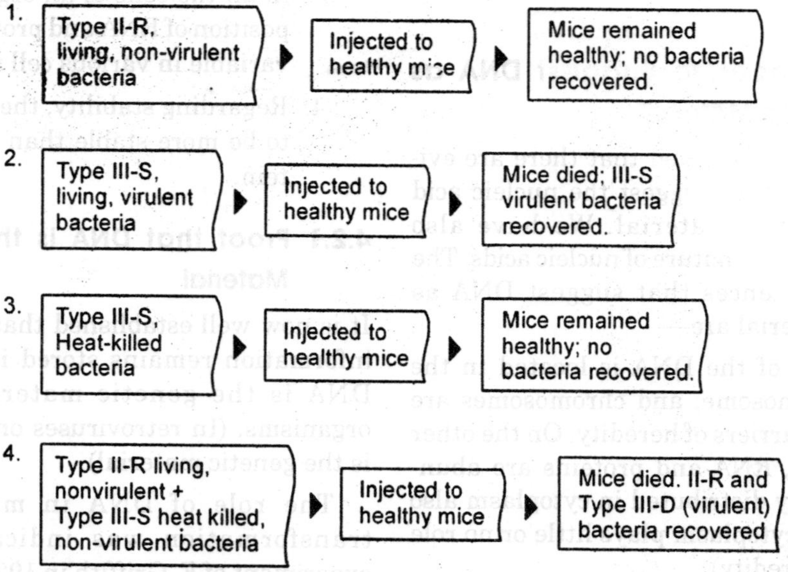

FIGURE 4.6

Griffith's transformation experiment.

This component, what he called the "transforming principle", caused transformation in *streptococcus pneumonia.* (The transformation of avirulent R-II to virulent S-III is not due to mutation, since mutation can result from type II to type II only, not type III).

Griffith's experiment and conclusion was supported by an *in vitro* experiment by Sia and Dawson in 1931 which showed that the transformation was not mediated in any way by the host, the living mice.

4.2.1.1 Transformation not due to living host: Sia and Dawson's Experiment

R. Sia and M. Dawson (1931) performed an experiment *in vitro,* i.e., there was no host specimen like mice; they applied the principle of Griffith's experiment in test tube and found that—

when, 1. R-II bacteria without capsule were grown in culture medium they produced only R-II bacteria colonies.

2. Heat-killed S-III bacteria with capsule were grown in culture medium; they could produce no bacteria colony.
3. Living R-II plus heat-killed S-III were grown in culture medium and then treated with serum for precipitation; the result showed both R-II bacteria (in precipitation) and live S-III bacteria colony.

The experiment proved that "something" from the heat-killed S-III bacteria—"the transforming principle"—had converted some living R-II cells to S-III types.

4.2.1.2 *"Transforming material" is the DNA: Avery, Macleod and McCarty's Experiment* (or, Proof that DNA is the genetic material)

After 16 years of Griffith's experiment, *the first proof that genetic information is stored in DNA* came from the experiment of Oswald Avery, Colin Macleod and Maclyn McCarty in 1944. In this experiment they proved beyond doubt that DNA is the only component of the S-III *S. pneumoniae* bacterial cells that could transform R-II cells to S-III types. Stepwise, their experiment was like—

1. R-II colonies were developed from R-II bacteria.
2. Highly purified DNA was extracted from heat-killed Type S-III. (The DNA could produce no colony).
3. Type R-II cells were mixed with DNA from heat-killed S-III cells. This mixture was treated with serum that precipitates R-II cells from mixture. S-III colony developed.
4. R-II cells were mixed with DNA from heat-killed S-III cells in a separate test tube and was treated with serum that precipitates R-II cells. ***Protease*** enzyme was mixed to this mixture to digest only *protein.* S-III colony developed.
5. R-II cells were mixed with DNA from heat-killed S-III bacteria in another test tube, and then it was treated with serum that precipitates R-II cells. ***RNase*** enzyme was mixed to it to digest *RNA only.* S-III colony developed.
6. Finally, the R-II cells were mixed with DNA from heat-killed S-III cells in another test tube. This mixture was treated with serum that precipitates R-II cells. ***DNase*** enzyme was mixed to it to digest *DNA only. No colony developed.*

The experiment clearly proved that as soon as the mixture was treated with *DNase,* it destroyed the 'transforming activity' of the DNA of the S-III bacterial cells. Thus, it could not transform the R-II to S-III, and colony did not develop. This experiment

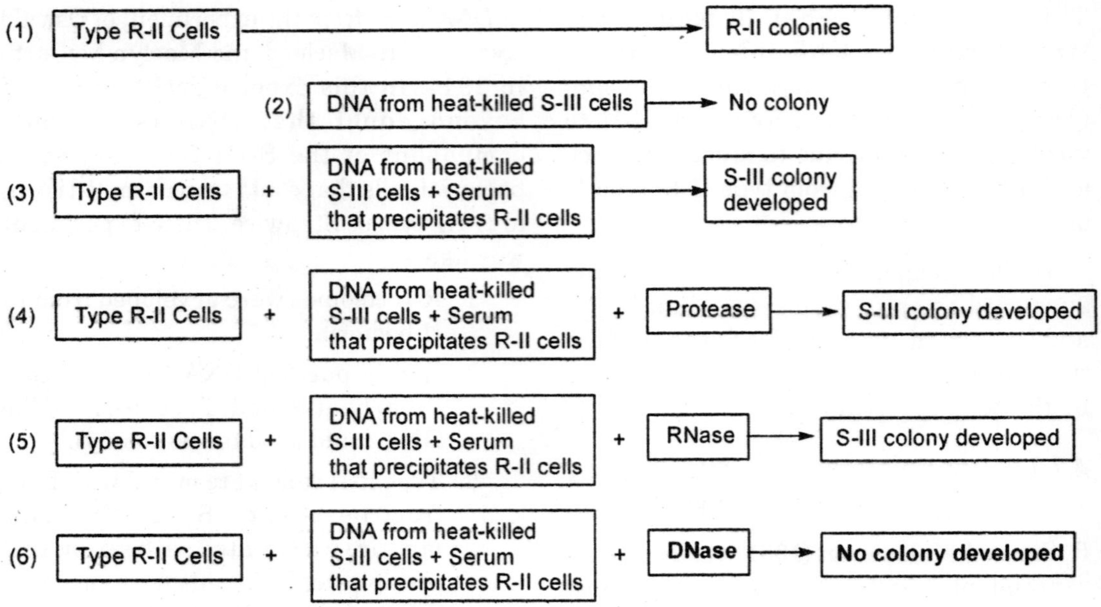

FIGURE 4.7

Avery, Macleod and McCarty's experiment to proof that "transforming principle" is the DNA.

established that *DNA is the transforming material.* In other words, it established that genetic information resides in the DNA, i.e., *DNA is the genetic material.*

[In the experiment of Avery, MacLeod and McCarty, the DNA transferred the gene coding for the enzyme, *UDP-glucose dehydrogenase,* required for the synthesis of the *polysaccharide capsule,* the characteristic feature of the virulent S-III type. This gene becomes incorporated into the chromosome of the nonvirulent R-II type and *transforms* it to virulent S-III type].

4.2.1.3 'Hershey-Chase' *Experiment to prove DNA as Genetic Material* (in virus, bacteriophage T_2)

In 1952, Alfred Hershey and Martha Chase made a very simple but elegant experiment on the virus, *Bacteriophage* T_2 to demonstrate that DNA is the genetic material in the virus too. The virus *Bacteriophage* T_2 (T_2-virus) infects the bacteria *Escherichia coli* and reproduces within the host cell. The virus contains approximately 50% DNA in the core and 50% protein in the coat. Hershey and Chase showed that, when T_2-virus infects an *E. coli*, only the DNA of the virus enters the cell while the protein remains outside. It implied that the genetic information for reproduction of the virus must be in the DNA (since only DNA has entered the cell !).

Their experiment was based on the fact that the DNA contains phosphorus but no sulfur, and the protein contains sulfur but no phosphorus. What did they do is that—

1. They labelled the T_2-viral-DNA by the radioactive isotope of phosphorus ^{32}P (by growing them in a medium containing radioactive isotope of phosphorus, ^{32}P, in place of the normal isotope ^{31}P).
2. They labelled the viral-protein-coats by radioactive isotope of sulfur, ^{35}S. (by growing them in a medium containing radioactive sulfur, ^{35}S, in place of normal isotope, ^{32}S).

Next, the experiment was conducted in two phases: (A) with T_2-phage labelled with ^{32}P-DNA, and (B) with T_2-phage labelled with ^{35}S-protein coat.

(A) 1. At first *E.coli* bacteria and ^{32}P-DNA labelled T_2-phage were mixed together and allowed sometime so that the bacteria are infected.

2. The phage-infected bacteria were then agitated in a Waring blender so that the phage coats are sheared off.

3. The sheared off phage coats (proteins) without radioactivity were separated from the bacteria by centrifugation. As a result, bacterial cells were sedimented as pellets and the phage-particles were suspended. Radioactivity was noticed in the sedimented pellets of cells, as the radioactive DNA of the phage entered into the cells.

(B) 1. In next phase, *E.coli* cells and ^{35}S-protein-coat-labelled T_2-phage were mixed together and allowed sometime so that the bacteria are infected.

2. The phage-infected bacteria were then subjected to shearing forces in a Waring blender. It removed most of the protein coats (and the radioactivity) from the cells (but phage-DNA without radioactivity have entered).

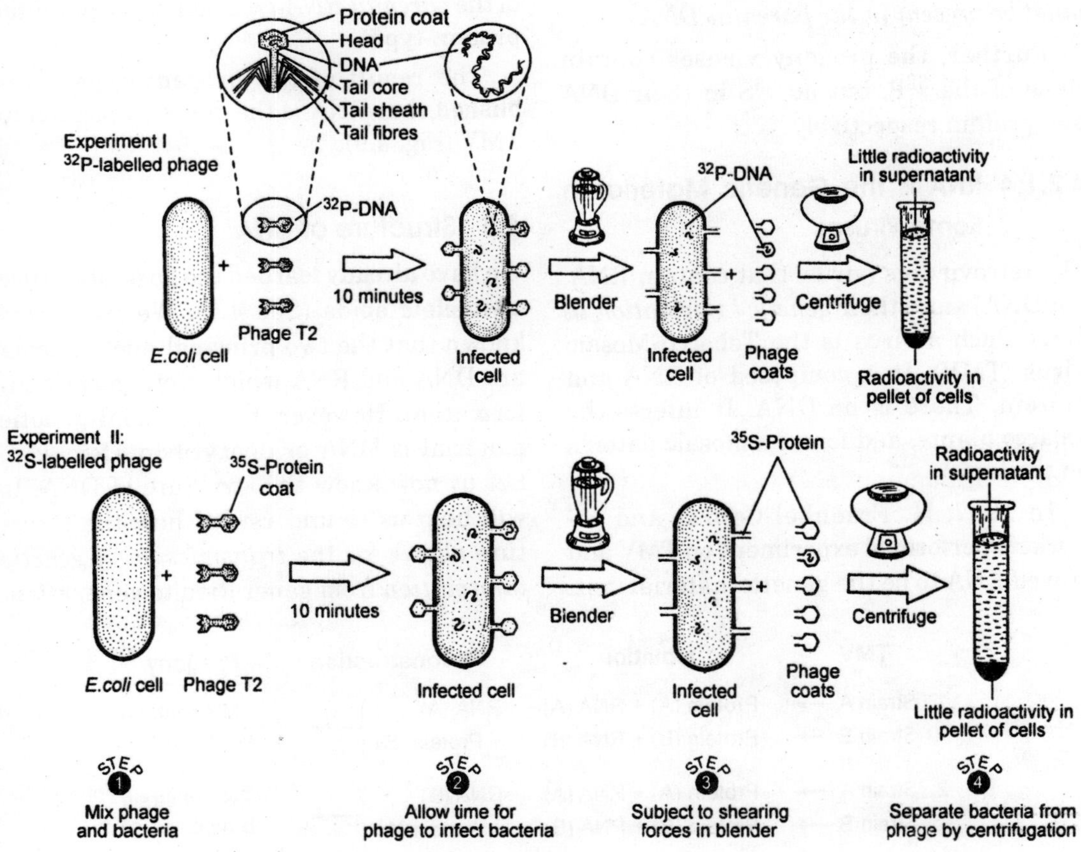

FIGURE 4.8

Demonstration by Hershey and Chase that the genetic information of bacteriophage T_2 resides in its DNA.

3. The sheared off phage-coats (proteins) with radioactivity were separated from the infected cells by centrifugation. As a result cells were sedimented as pellets and the phage particles were suspended. Radioactivity was noticed in the supernatant (as the radioactive proteins of the phage did not enter into the cell) (Fig. 4.8).

The result of Hershey-Chase experiment indicated that the *DNA of the T_2-phage enters the bacterial cells,* the protein-coat remains outside, but it did not affect phage-reproduction. The progeny viruses are produced *inside the cell,* proving that *the genetic information* for production of both DNA and protein coat of the progeny viruses *must be present in the parental DNA.*

Further, the progeny viruses contain some of the ^{32}P, but no ^{35}S in their DNA and protein respectively.

4.2.1.4 RNA is the Genetic Material in Some Viruses

The retroviruses (virus, that contain RNA, not DNA) store their *genetic information in RNA.* Such a virus is the Tobacco Mosaic Virus (TMV). It is composed of RNA and protein. There is no DNA. It infects the tobacco plants, and forms a mosaic pattern on the leaf.

In 1957, H. Fraenkel-Conrat and co-workers performed experiment on TMV and proved RNA to be the genetic material in it.

Fraenkel-Conrat and B. Singer first isolated the RNA and the proteins from the TMV. Then they used the TMV-protein alone to infect the tobacco plants. Plants remained healthy. Next, they used the TMV-RNA only to infect the plants and became successful; the plants became infected. It proved that *the RNA alone was responsible* for causing infection.

Their next experiment was called "reconstitution experiment". There are different strains of TMV that differ in the chemical nature of their protein coats. They "reconstituted" viruses which contain protein from one strain and RNA from another. Such 'hybrid' viruses were applied to tobacco plants for infection. Plants were infected and developed progeny viruses. They found that all the progeny viruses were of the *parental RNA-type,* not of the parental protein-type.

The result of the progeny-type established that RNA is the genetic material in TMV (Fig. 4.9).

4.3 Structure of DNA

We have already learned the basic structure of nucleic acids (See 4.1). We have also known that the two principal nucleic acids are DNA and RNA which store genetic information. However, the principal genetic material is DNA or deoxyribonucleic acid. Let us now know the structure of DNA. It will help us to understand how the structure allows for the *transmission of genetic information* from generation to generation.

TMV	Isolation	Reconstruction	Progeny
1. Strain A →	Protein (A) + RNA (A)	RNA (A)	
Strain B →	Protein (B) + RNA (B)	+ Protein (B) →	TMV of strain A type.
2. Strain A →	Protein (A) + RNA (A)	RNA (B)	
Strain B →	Protein (B) + RNA (B)	+ Protein (A) →	TMV of strain B type.

FIGURE 4.9

Proof that RNA is the genetic material in TMV-Reconstitution experiment.

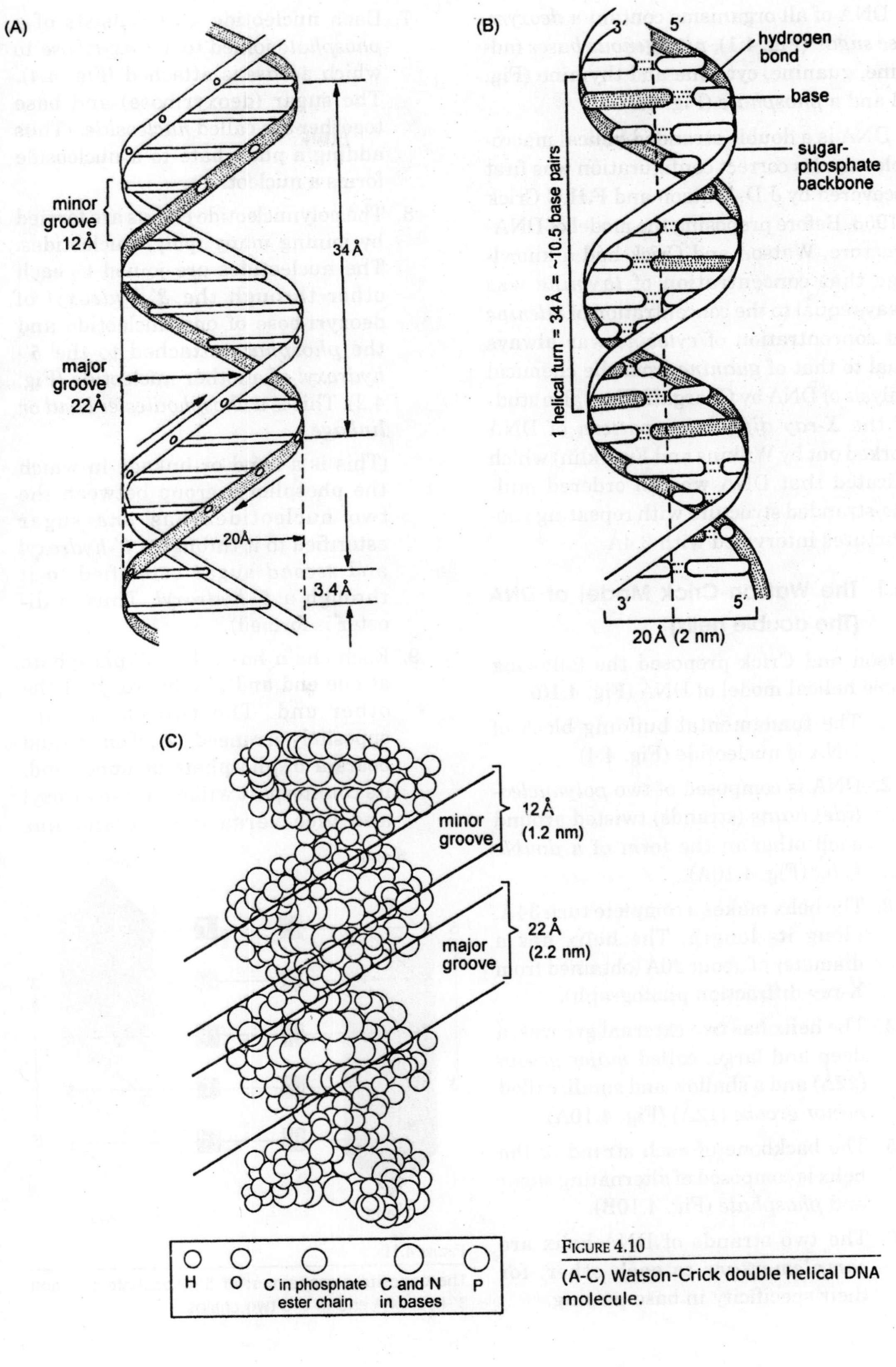

FIGURE 4.10

(A-C) Watson-Crick double helical DNA molecule.

DNA of all organisms contain a *deoxyribose sugar* (Fig. 4.1), *nitrogenous bases* (adenine, guanine, cytosine and thymine (Fig. 4.3 and a *phosphate* (Fig. 4.2).

DNA is a double stranded helical macromolecule. Its correct configuration was first discovered by J.D. Watson and F.H.C Crick in 1953. Before proposing the model of DNA-structure, Watson and Crick had a knowledge that concentration of *thymine* was always equal to the concentration of *adenine* and concentration of *cytosine* was always equal to that of *guanine* from the chemical analysis of DNA by Chargaff. They also studied the *X-ray diffraction pattern* of DNA (worked out by Wilkins and Franklin) which indicated that DNA was an ordered multiple-stranded structure with repeating substructures intervaled with 3.4Å.

4.3.1 The Watson-Crick Model of DNA (The double helix)

Watson and Crick proposed the following double helical model of DNA (Fig. 4.10).

1. The fundamental building block of DNA is nucleotide (Fig. 4.4).
2. DNA is composed of two *polynucleotide chains* (strands) twisted around each other in the form of a *double helix* (Fig. 4.10A).
3. The helix makes a complete turn 34Å. along its length. The helix has a diameter of about 20Å (obtained from X-ray diffraction photograph).
4. The helix has two external grooves; a deep and large, called *major groove* (22Å) and a shallow and small, called *minor groove* (12Å) (Fig. 4.10A).
5. The backbone of each strand of the helix is composed of alternating *sugar* and *phosphate* (Fig. 4.10B).
6. The two strands of DNA-helix are *complementary* to each other, for their specificity in base-pairing.
7. Each nucleotide unit consists of a *phosphate* joined to a *deoxyribose* to which a *base* is attached (Fig. 4.4). The sugar (deoxyribose) and base together are called *nucleoside*. (Thus adding a phosphate to a nucleoside forms a nucleotide).
8. The polynucleotide chains are formed by joining many (poly) nucleotides. The nucleotides are joined to each other through the *3′-hydroxyl* of deoxyri-bose of one nucleotide and the *phosphate* attached to the *5′-hydroxyl* of another nucleotide (Fig. 4.2). This is a *phosphodiester bond or linkage.*

   (This is a bond or linkage in which the phosphoryl group between the two nucleotides has *one* sugar esterified to it through a *3′-hydroxyl* and *second* sugar esterified to it through a *5′-hydroxyl*. Thus, a diester is formed).
9. Each chain has a free 5′-phosphate at one end and a 3′-hydroxyl at the other end. The two chains are oppositely arranged, i.e., if one strand bears a 5′ phosphate at upper end, its counterpart will bear 3′-hydroxyl and vice versa (one strand has

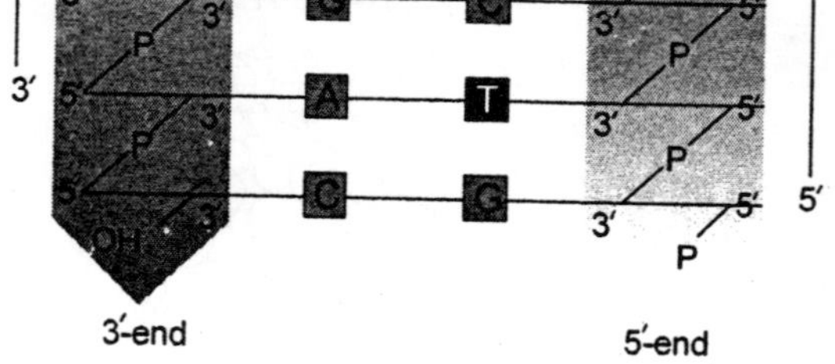

FIGURE 4.11

The opposite arrangement of 5′-phosphate end and 3′-hydroxyl end in the two chains.

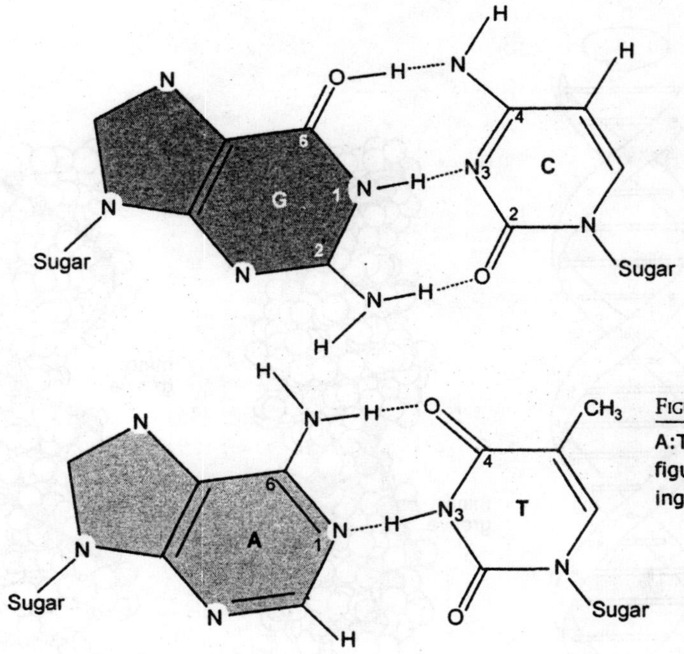

FIGURE 4.12

A:T and G:C base pairs. The figure shows hydrogen bonding between the bases.

phosphodiester linkage in 5′→3′ direction while the other has the linkage in 3′→5′ direction) (Fig. 4.11).

10. The two strands of the helix are held together by weak, non-covalent hydrogen bonds between each pair of bases. The resulting base pairs are stacked between the two strands perpendicular to the axis like the steps of a spiral stair case (Fig. 4.10A, B).
11. *Adenine* of one strand is always paired with *thymine* of the other strand. Similarly, *Guanine* is always paired with *cytosine.* This is called Watson-Crick base pairing. (The base at the 3′end of the other strand) (Fig. 4.11).
12. The specificity of base-pairing results in a complementary relationship between the sequence of bases on the chains (Fig. 4.12).
13. The hydrogen bonds between the complementary bases contribute to the stability to the helix and the *specificity* of the base pairing. For this reason adenine always pairs with thymine by two hydrogen bonds while cytosine pairs with guanine by three hydrogen bonds.
14. The complementarity between the bases gives DNA the unique feature of storing and transmitting of the genetic information. It also ensures the self-encoding character of DNA.
15. The base pairs in DNA are stacked in 3.4Å apart with 10 base pairs per 360° (one full turn) of the double helix (The Mica experiment shows that DNA has 10.5 base pairs per turn of the helix in solution).
16. The sugar-phosphate backbones of the two complementary strands are *antiparallel*. It means the strands have *opposite chemical polarity*.
17. The opposite polarity of the strands is useful in the DNA replication process (see 4.4).
18. DNA is a highly stable molecule. This stability is due to— (a) large number of H-bonds and (b) hydrophobic bonding (or 'stacking forces') between the stacked base pairs.

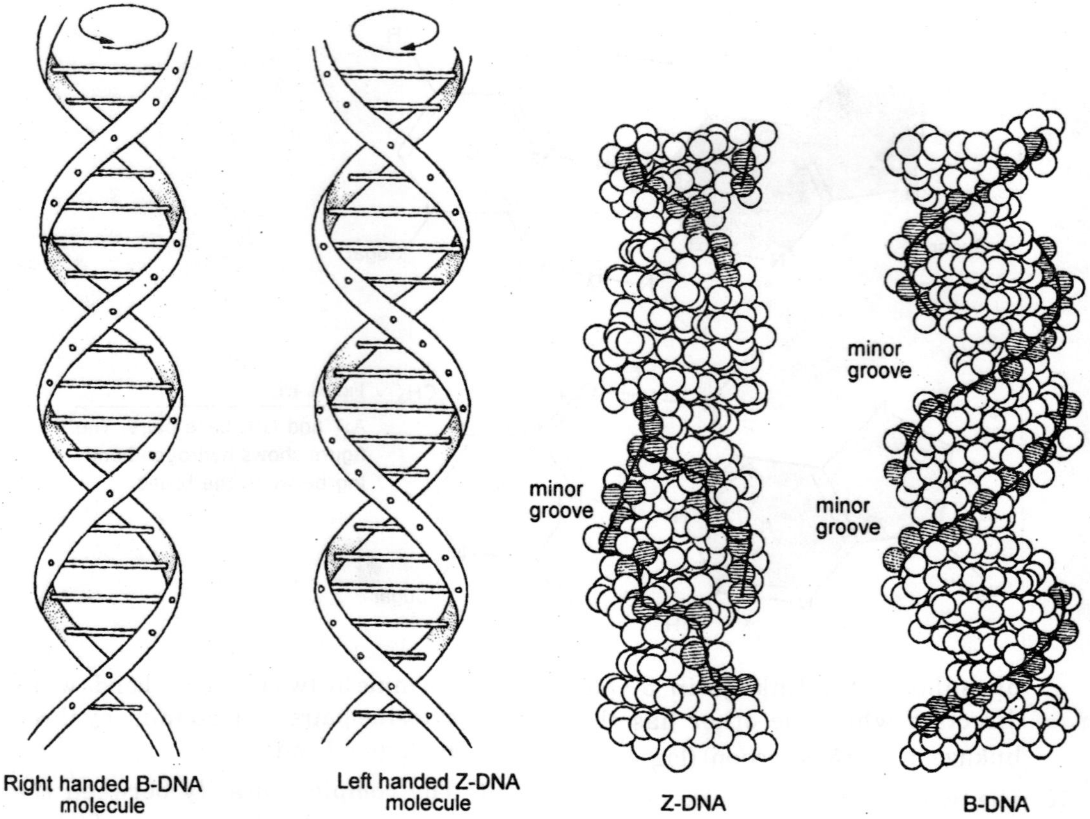

FIGURE 4.13

Two molecular forms of DNA double helix, right-handed B-DNA and left-handed Z-DNA. The heavy black line in each molecule goes from phosphate group to phosphate group, indicating a smooth right-handed helix in B-DNA and irregular or zigzag left-handed helix in Z-DNA (after De Robertis and De Robertis Jr., 1987).

4.3.2 Alternative Forms of DNA

DNA may be flexible in their structure depending on their environmental interactions. There are three such alternative forms of DNA, viz., *B-form*, *A-form* and *Z-form* DNA.

TABLE 4.4: Comparison of 3 types of DNA

Property	A-DNA	B-DNA	Z-DNA
Helix direction	Right handed	Right handed	Left handed
Base pairs per 360°	10.9	10.4	12
Helix diameter	2.2 nm	2.0 nm	1.8 nm
Major groove	Extremely narrow and very deep	Wide, not very deep	Flattended
Minor groove	Wide and shallow	Narrow, moderately deep	Extremely narrow and very deep
Helix axis location	Major groove	Through base pairs	Minor groove
Overall morphology	short and wide	Long and thin	Elongated and thin

The vast majority of DNA molecules found in the aqueous protoplasm are of *B-form* which follow the 'Watson-Crick model' structure. The *B-form DNA* is a right-handed helix with an average of 10.4 nucleotide pairs per turn (rather than 10). The *A-form DNA* is found in a medium of high concentration of salts or in a dehydrated environment. Usually it is not found in a living system. It is also a right-handed helix with 11 nucleotide pairs per turn. This conformation is found in the DNA-RNA heteroduplexes. *The Z-form* or *Z-DNA* is a left handed double helical structure. The sugar-phosphate backbone shows a *zigzagged path* (Z for zigzagged). They occur in very light salt concentration and they have 12 nucleotide pairs per turn (Fig. 4.13). Table 4.4 shows a comparison of the three types of DNA.

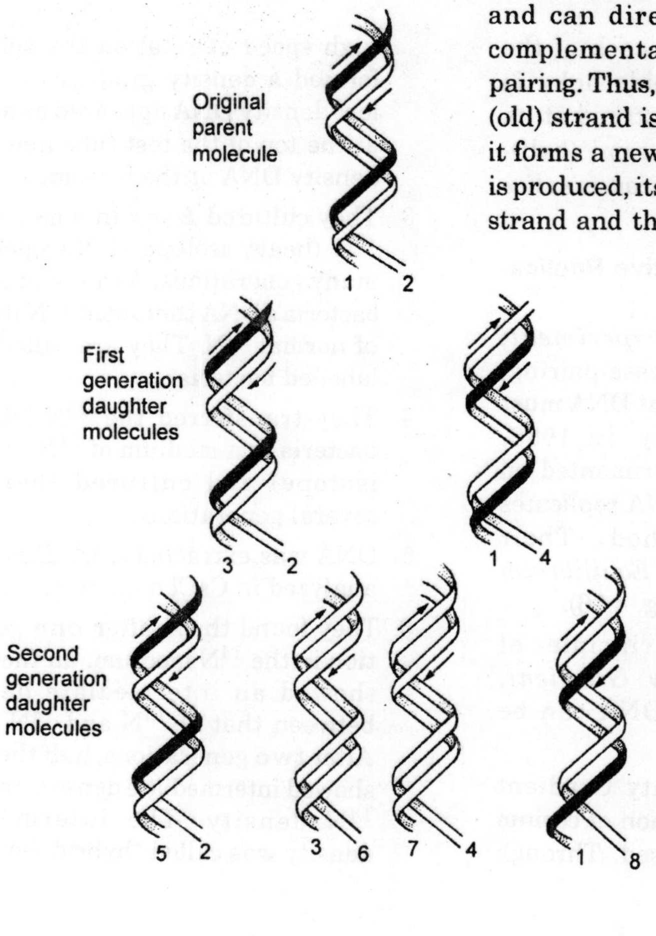

FIGURE 4.14

Semiconservative DNA replication.

4.4 DNA Replication

DNA replication is a template-directed nucleic acid synthesis reaction where the template and the nascent strand are the same type of nucleic acid. It means, DNA synthesizes itself from an existing one to an exactly another one based on its template-strands. DNA replicates very fast and with great fidelity. It is a complex process and involves many cellular functions and several verification procedures to ensure the fidelity in replication.

4.4.1 Principles of DNA Replication

Replication of DNA is a *semiconservative* process. DNA double helix has two complementary strands. Their base pairing is very specific. Each of strands, when separated, can function as a *template* (on which the new strand will be synthesized) and can direct the synthesis of a new complementary strand by specific base pairing. Thus, during synthesis one parental (old) strand is conserved (as template) and it forms a new strand; when a double helix is produced, its half (semi) is the old parental strand and the other half is a new one. In

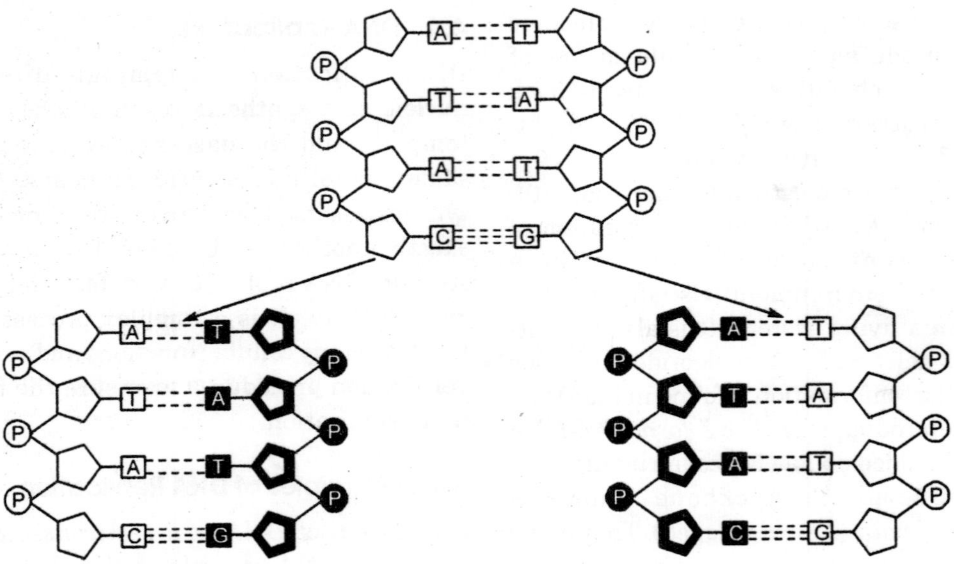

FIGURE 4.15

Guided synthesis of new strands of DNA upon old strands. An adenine in the original strand is replaced by a thymine in the new strand, while a guanine is replaced by a cytosine, and so forth.

this way during replication, each of the strands of the parental double helix is *conserved* synthesizing a new strand at its side, i.e., the double helix is semi (half)-conserved. This is called *semiconservative replication* (Fig. 4.14 and 4.15).

- **Proof of Semiconservative Replication**

Meselson and Stahl's experiment: Observing the specificity of base-pairing, Watson and Crick proposed that DNA must replicate semiconservatively. In 1958, *M.Meselson* and *F.Stahl* experimented on *E.coli* and established that DNA replicates in *semiconservative* method. Their experiment was based on the *Equilibrium Density Gradient* principle (Fig. 4.16).

1. By adopting the principle of *Equilibrium Density Gradient*, different density of DNA can be separated.
2. In Equilibrium Density Gradient Centrifugation, a solution of cesium chloride (CsCl) was used. Through high speed of rotation the solution formed a density gradient, i.e., the low-density DNA appeared as a band at the top of the test tube and high-density DNA at the bottom.
3. They cultured *E.coli* in a medium of ^{15}N (heavy isotope of Nitrogen) for many generations. As a result all the bacteria's DNA contained ^{15}N instead of normal ^{14}N. They are called ^{15}N-labelled bacteria.
4. They transferred the ^{15}N-labelled bacteria to a medium of ^{14}N (normal isotope) and cultured there for several generations.
5. DNA was extracted from *E.coli* and analyzed in CsCl density gradient.
6. They found that—after **one** generation in the ^{14}N medium, all the DNA showed an intermediate density between that of ^{15}N and ^{14}N DNA. After **two** generations, half the DNA showed intermediate density and half ^{14}N density. The intermediate density was called "hybrid density".

7. From this result, they inferred that there was a "DNA-hybridization", i.e., half of the DNA strands remained 'old' and the half was synthesized 'a new'. This is conformative to the prediction of Watson and Crick—that the DNA-replication must be *semi-conservative.*

   It was soon established that not only in the *E.coli,* but in all other prokaryotes and eukaryotes the replication of DNA occurs by a semiconservative method.

4.4.2 Factors for DNA Replication (DNA replication process needs specific enzymes and other components)

A. *DNA Polymerases*: Kornberg and his co-workers were able to synthesize DNA **invitro** for the first time in 1957, and found the important role of an enzyme, called *DNA polymerase* (or, Kornberg's enzyme) at that time. It is now known as *DNA Polymerase-I* or *DNA pol I.* They isolated the enzyme from *E.coli.*

   Soon it was found that there are four other *DNA polymerases*; *DNA polymerase II, DNA polymerase III, DNA polymerase IV* and *DNA polymerase V.*

   DNA polymerase I performs *three* important functions in *E.coli,* playing a key role in chromosome replication and in repairing damaged DNA. It has a *5′→3′ polymerase activity* along with two *exonuclease activities.*

   Its *exonuclease* activity is unique—that is, *DNA polymerase I* contains both *5′→3′ exonuclease activity* (which cuts back DNA strands starting at 5′-end), and *3′→5′ exonuclease activity* (which cuts off mononucleotides from the 3′-end) (Fig. 4.17). *DNA polymerase II* is a DNA repair enzyme. It has also a 5′→3′ polymerase and only a 3′→5′ exonuclease activity. DNA polymerase III has 5′→3′ polymerase and 3′→5′ exonuclease activities. Both DNA polymerase I and DNA polymerase III replicate DNA in 5′→3′ direction. Both enzymes also have 3′→5′ exonuclease activity. They can remove nucleotides from 3′ end of a chain. Exonuclease activity of them is used in the 'proof-reading' and DNA-error correction mechanism. *DNA polymerases IV and V* function in DNA repair.

B. *Four d NTPs*: The 4 deoxyribonucleoside 5′ -triphosphates, include d ATP, d GTP, d TTP and d CTP, and are compulsory in DNA replication. These 4 nucleotides are collectively called d NTP (deoxyribose nucleoside triphosphate). In absence of any of them DNA replication would not occur.

C. A pre-existing *DNA* is necessary requisite for DNA replication. Its strands act as *DNA template* on which the new strand would be synthesized.

D. Mg^{++} are needed for the optimal activity of the DNA polymerase.

E. An *initiator protein* is necessary which binds to the replicator and is responsible for initiation of replication. It is called *dna A* protein in *E.coli.*

F. Several other proteins, called *IHF, FIS, dna B, SSB, TBP* etc. are also necessary.

G. *Helicase* enzyme is responsible for untwisting of DNA (in both direction from the origin of the replication).

H. *DNA primase* enzyme is essential to form an *RNA primer.*

I. *DNA ligase* enzyme is responsible for sealing the gaps between *Okazaki fragments* (See later).

J. *Gyrase* or *Topoisomerase II* enzyme is needed for clearing both DNA strands and avoid tangling of DNA.

K. *Topoisomerase I* enzyme cleaves one strand of DNA and relax supercoiled structure.

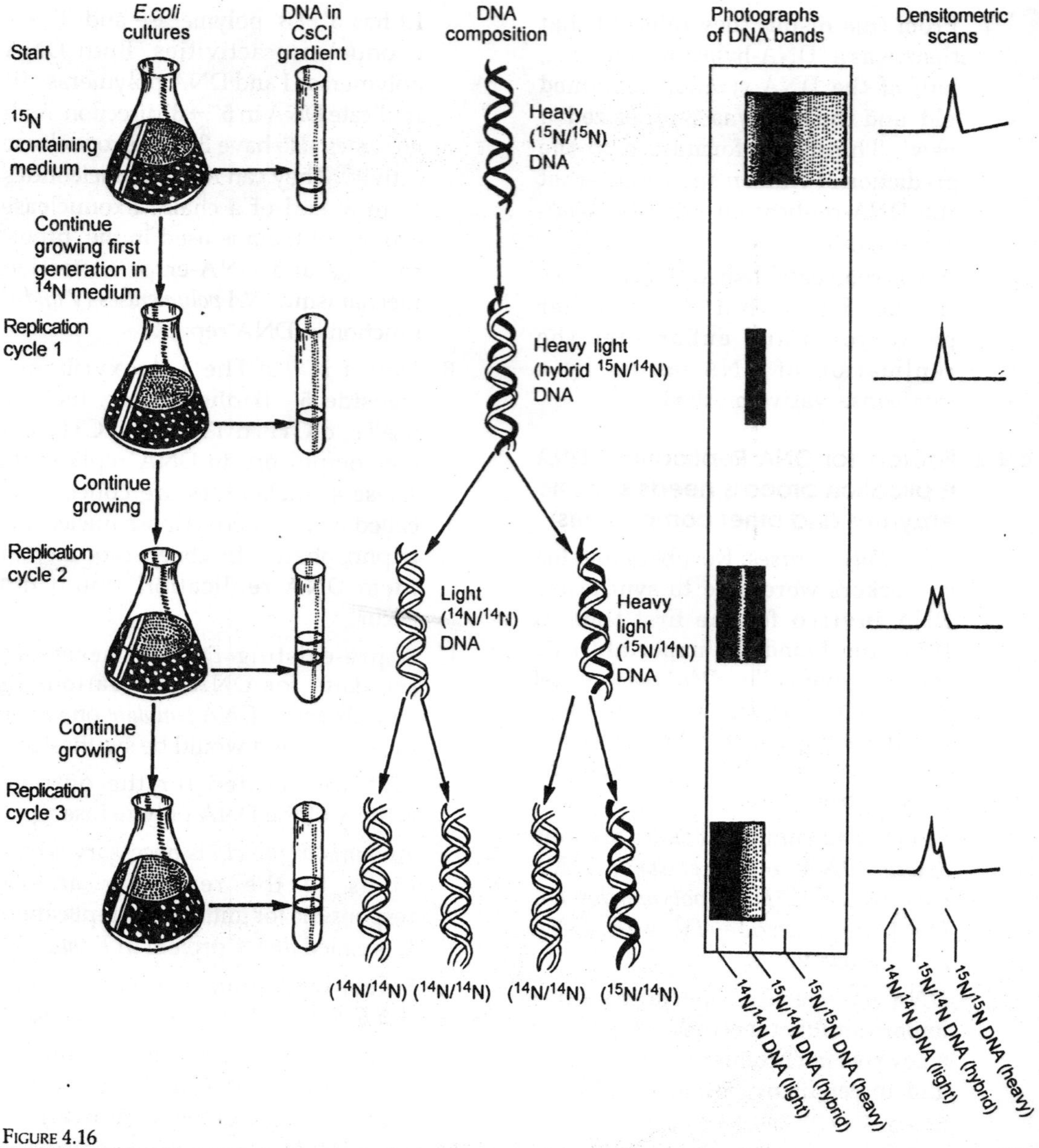

FIGURE 4.16

The Meselson-Stahl experiment: The demonstration of semiconservative replication in *E.coli*. Cells were grown in a ^{15}-N-containing medium for several replication cycles and then were transferred to a ^{14}N-containing medium. At various times over several replication cycles, samples were taken: the DNA was extracted and analyzed by CsCl equilibrium density gradient centrifugation. Shown in the figure are a schematic interpretation of the DNA composition after various replication cycles photographs of the DNA bands and densitometric scans of the bands (After Russell).

4.4.3 Process of DNA Replication in Prokaryotes (in *E.coli*)

The synthesis of DNA in *E.coli* (a prokaryote) involves 3 steps: *Initiation, Elongation of chain* and *Termination*.

Initiation:

1. The initiation of DNA replication involves the *denaturation of DNA*. An initiator protein, *Dna A* binds to a particular DNA-sequence called

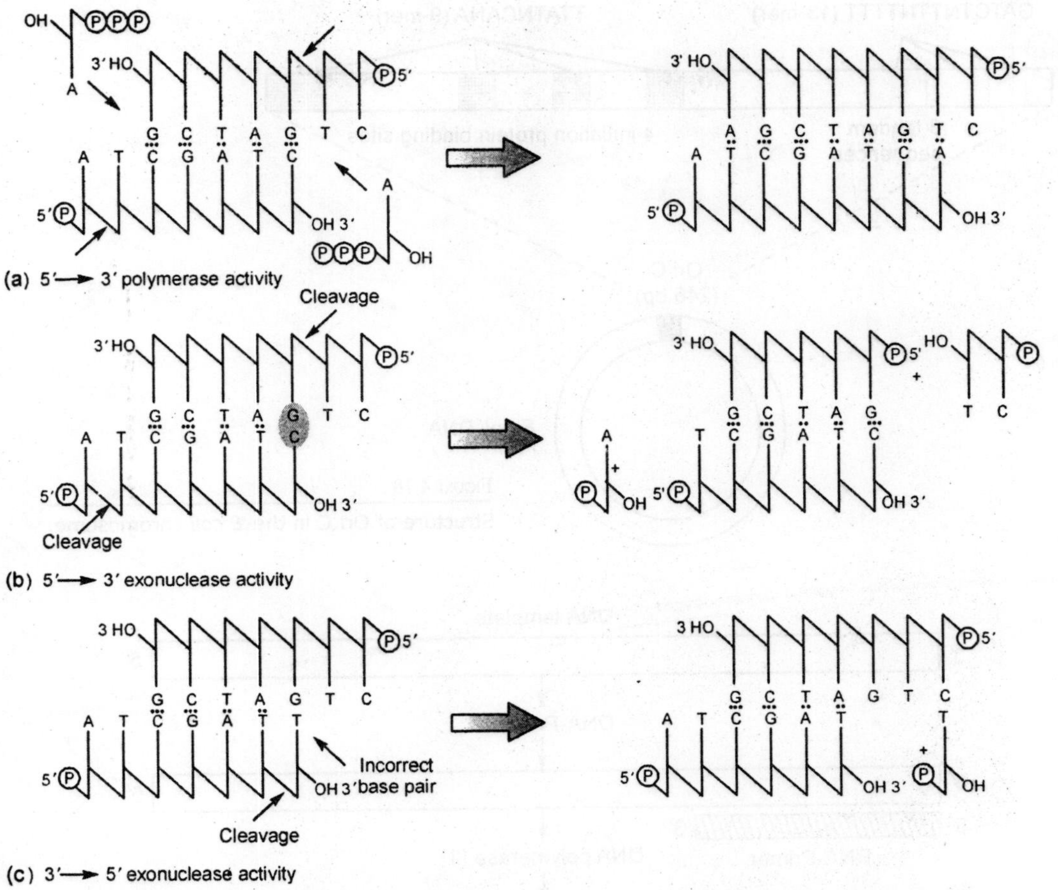

FIGURE 4.17

The three activities of DNA polymerase I of *E.coli.* All three activities play important roles in the replication of the *E.coli* chromosome.

replicator (*Ori C* in *E.coli*), and stimulates denaturation of DNA.

2. Initiation commences within the 2 single strands of denaturated DNA-double helix. This locally denatured area is called *replication bubble*, and the single strands on which the new DNA-strands are synthesized are called *DNA templates*.
3. Due to uncoiling of the two strands (denaturation) from each other, a Y-shaped structure called *replication fork* is formed, the two bifurcated arms of which act as templates.
4. The fork moves toward the uncoiling of the DNA. When denaturation occurs in the middle of the DNA, two replication forks are formed and replication proceeds *bidirectionally*. When there is a single fork, however, the replication may be *unidirectional*. The new DNA-strand is always synthesized in *5′ → 3′ direction.*
5. The *Ori C* or *replicator* of *E.coli* is a 245 base pair (bp) region that contains 3 copies of 13-bp AT-rich sequence and 4 copies of 9-bp sequence. At the time of *prepriming of DNA replication*, the *Dna A* protein binds to the four 9-bp repeats in Ori C. Then additional molecules of *Dna A* protein bind and form a complex with ori C. This

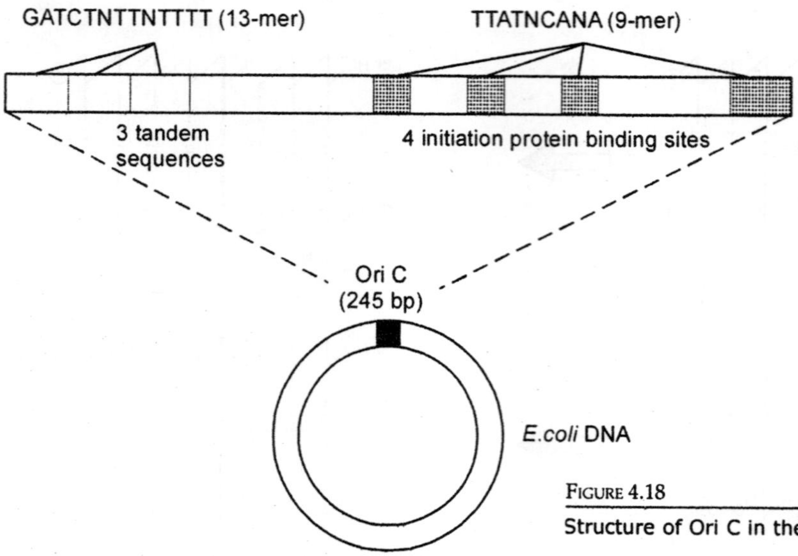

FIGURE 4.18

Structure of Ori C in the *E.coli* chromosome.

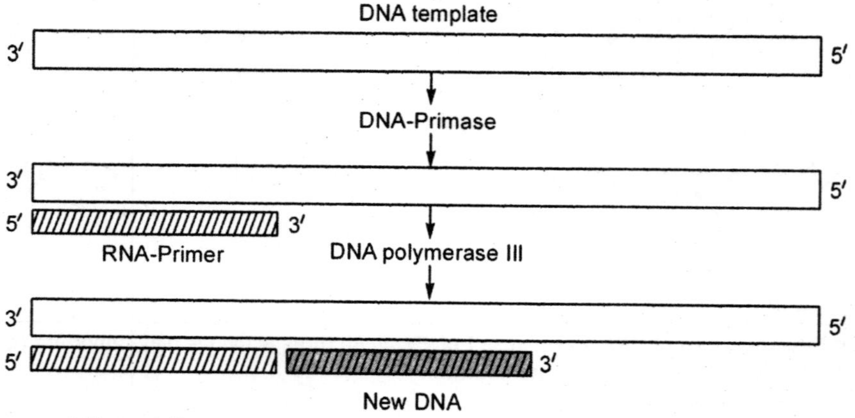

FIGURE 4.19

Formation of RNA-primer by the action of DNA-primase and new DNA nucleotide by the action of DNA polymerase III.

complex is wrapped on the surface and separates the strands at 13-bp repeats. This is followed by joining of the *Dna B* protein (helicase) and *Dna C* protein at the initiation complex, producing a *replication bubble.* The helicase moves along a single strand by the help of ATP (Fig. 4.18).

6. Next enzyme involved in the initiation is the *DNA primase* which binds to the *helicase* and to the single DNA-strand to form a body called *primosome.* It synthesizes a very short *RNA primer* (of 5-10 nucleotides) (Fig. 4.19).

Chain Elongation:

7. New nucleotides are added to the *RNA-primer* in 5′→3′ direction by the help of *DNA polymerase.* The primer acts as a substrate for the DNA polymerase.
8. *Single-strand DNA binding (SSB) proteins* bind to each single strand

(template) of the replication fork to keep the two strands separated.

9. The *DNA primase* at the fork synthesizes another RNA primer for the other template-strand.

10. The two DNA templates are in opposite polarity (i.e., one runs in 5′→3′ direction and the other in 3′→5′ direction).

The new DNA replica always grows in 5′→3′ direction. Thus, to maintain the 5′→3′ polarity of DNA synthesis the new DNA nucleotides are polymerized in opposite directions on their respective templates.

As the new DNA is made in the 5′→3′ direction, its strands grow *continuously* on one template strand and *discontinuously* on the other template strand.

11. The new DNA strand which grows *continuously* (along the direction of the movement of the replication-fork) is called *leading strand.* It requires a *single RNA primer* for its synthesis. The other strand that grows *discontinuously* (in the opposite direction of the movement of the replication fork) is called *lagging strand.* It needs *series of RNA primers* because the lagging strand is formed in several *fragments* (Fig. 4.20).

12. Along with the movement of the replication fork, the *helicase* uncoils more DNA. The enzyme *gyrase* (or topoisomerase II) helps the process.

13. The leading strand of DNA grows on the template continuously in 5′→3′ direction without pause or break but the *lagging strand* behaves differently.

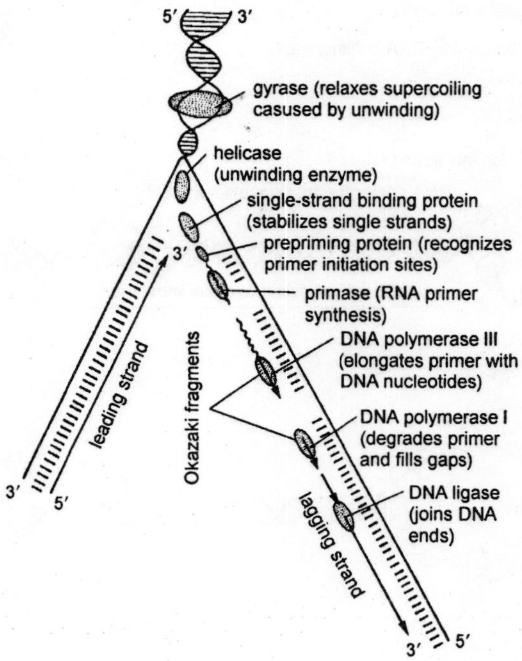

FIGURE 4.20

Some of the major protein and enzyme activities occurring on the lagging strand at or near the replication fork of a replicating DNA double helix.

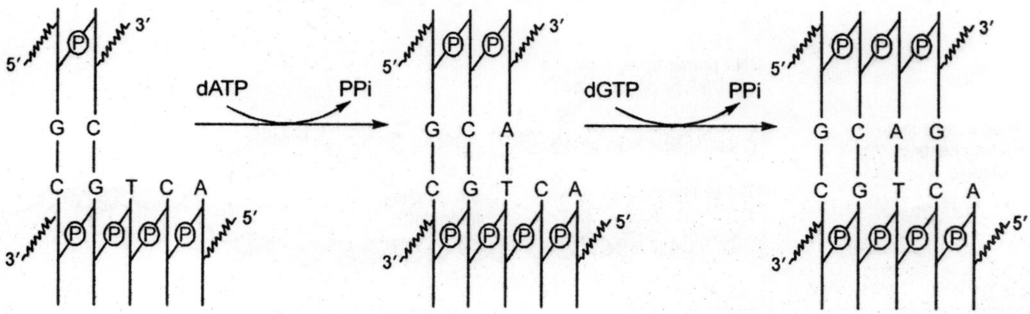

FIGURE 4.21

Polymerization by DNA polymerases. Process of adding of nucleotides step by step. Chain elongation process.

It grows a little at first, and then stops. For its further elongation, a new initiation with the formation of a new RNA-primer is compulsory. Afterwards, the process of adding of nucleotides to the RNA primer with the help of DNA polymerase III restarts (Fig. 4.21).

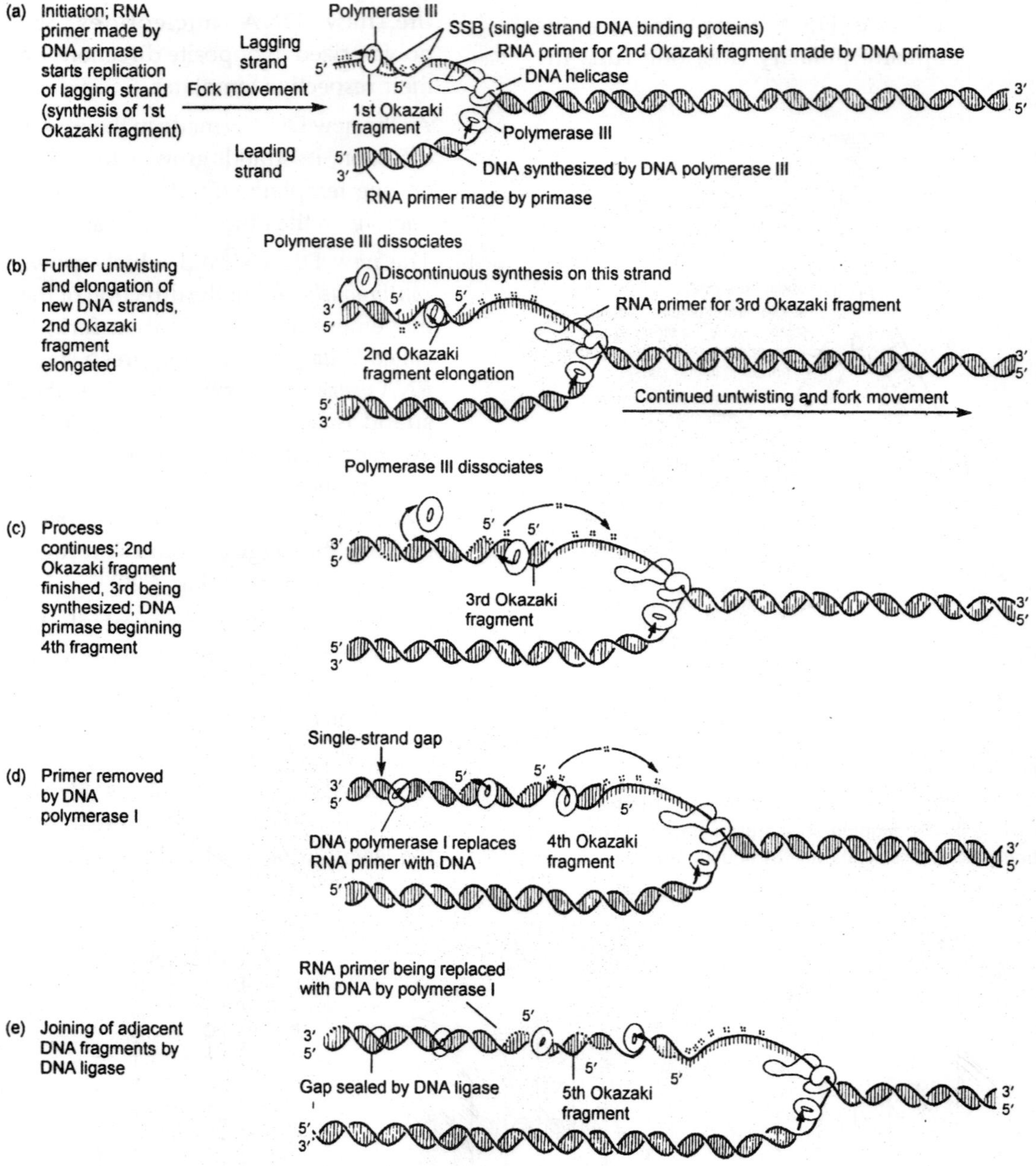

FIGURE 4.22

Model for the events occurring around a single replication fork of the *E.coli* chromosome. (a) Initiation, (b) Further untwisting and elongation of the new DNA strands, (c) Further untwisting and continued DNA synthesis, (d) Removal of the primer by DNA polymerase, (e) Joining of adjacent DNA fragments by the action of DNA ligase (After Russell).

A small fragment of lagging strand is again formed and the process stops. Then again an RNA primer is formed and the process is repeated. Thus, the lagging strand is formed piece by piece. The little fragments of the lagging strand were discovered by Okazaki and his colleagues, and so these are called *Okazaki fragments* (Fig. 4.22).

14. These lagging strand fragments (Okazaki fragments) are subsequently **joined together** to form a continuous lagging strand by the help of enzymes, *ligase* and *DNA polymerase I*.
15. When two or more Okazaki fragments are formed, the 3′-end of the newer fragment remains adjacent to the primer at the 5′-end of the previously made fragment. The 5′→3′ exonuclease activity of the DNA pol I removes the RNA primer of the older fragment. It again continues the 5′→3′ synthesis of DNA so that the RNA primers are replaced with DNA. When all the RNA primers are replaced by the DNA nucleotides, the fragments are joined by *DNA ligase* by sealing the single-strand nick by catalyzing the formation of a phosphodiester bond between the 3′ OH and 5′ phosphate of the older and newer fragments respectively (Fig. 4.23).
16. **Replisome**: In the very much complex process of DNA replication, the essential proteins (enzymes) are closely associated together to form a multienzyme complex called *replisome* or *replication apparatus*. In this apparatus, the two DNA polymerases III of the lagging strand and the leading strand, and the other associated polypeptides form a complex called the *DNA pol III holoenzyme*.

Proofreading or *correction* or *editing of the chain:*

17. DNA-polymerization machine acts with great fidelity. It is possible by a 'proofreading' or 'editing' mechanism. Here, 'proofreading' or 'editing' means the removal of wrongly paired nucleotide, if any.
18. Both the polymerization (adding of nucleotide in the elongating chain) and the 'editing' are catalyzed by the *DNA pol I enzyme.* A mispaired nucleotide is removed by the exonuclease activity of *DNA-pol-I* (Fig. 4.24).

Chain termination:

19. In *E.coli,* termination of replication occurs at *ter A* and *ter B* sites. They block the movement of the replication fork.
20. Finally, *DNA topoisomerase* involves in separation of the nascent DNA molecules.

IMPORTANT NOTE

DNA replication occurs in semi-conservative method. DNA replication is catalyzed by various types of DNA polymerases. All DNA polymerases synthesize DNA in 5'→3' direction. DNA polymerases may also have other activities, such as 'proofreading' (editing) and removing unwanted

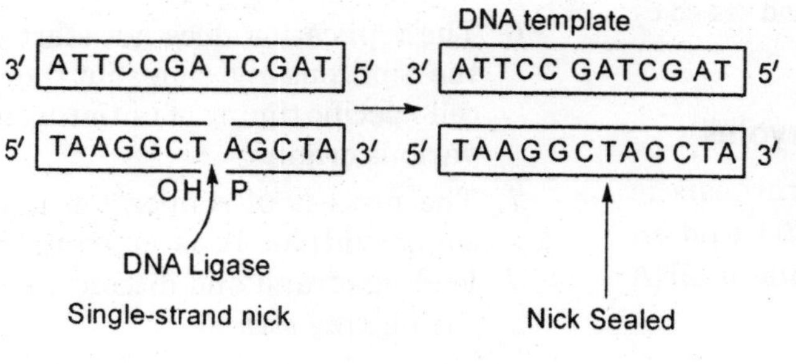

FIGURE 4.23

Action of DNA ligase and nick-sealing.

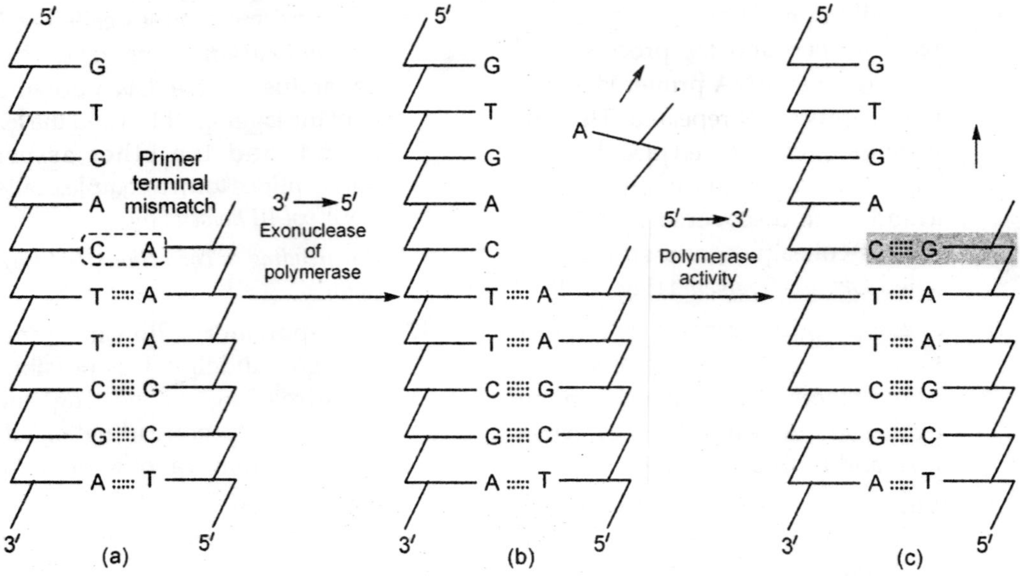

FIGURE 4.24

"Proof-reading" by the $3' \rightarrow 5'$ exonuclease activity of DNA polymerases during DNA replication. If DNA polymerase is presented with a template and primer containing a 3′ primer terminal mismatch (a). It will not catalyze covalent extension ("polymerization"). Instead, the $3' \rightarrow 5'$ exonuclease activity, an integral part of many DNA polymerases, will cleave off the terminal mismatched nucleotide (b). Then, presented with a correctly base-paired primer terminus, DNA polymerase will catalyze $5' \rightarrow 3'$ covalent extension of the primer strand (c). (After Gardner, Simmons and Snustad).

nucleotides (removal of RNA primer). DNA replication starts by denaturation of DNA-double strand with the help of helicase. Other enzyme, DNA primase, binds to the helicase and synthesizes a very short RNA primer. New nucleotides of DNA are synthesized in $5' \rightarrow 3'$ direction and the chain elongates, after the formation of RNA-primer. Later, the RNA primer is removed by the exonuclease activity of DNA polymerase I. As new DNA strand is always synthesized in $5' \rightarrow 3'$ direction, chain growth in one strand (leading strand) is continuous and in the other (lagging strand) is discontinuous. The fragments of small DNA of the lagging strand are called Okazaki fragments which are finally joined and sealed by DNA-ligase.

4.4.4 DNA Replication in Eukaryotes

The overall process of DNA-synthesis is similar in prokaryotes (*E.coli*) and in *eukaryotes*. The important events of DNA replication in eukaryotes are—

1. Each chromosome has a linear DNA double helix that acts as the templates.
2. The process of replication in eukaryotes is much faster.
3. *Initiation* occurs at *many* origins (instead of one), where the DNA denatures (uncoiling, separation and formation of replication fork).
4. Replication proceeds bidirectionally.
5. The extension of DNA from the origin to the two terminal points (where the two forks fuse) on each side of the origin is called a *replicon.*
6. The replication does not start in all the replicons at a time, but there is a cell-specific timing of initiating at the various origins.
7. The process of replication is semi-conservative. It is continuous in leading strand and discontinuous in the lagging strand.

8. 'Okazaki fragments' are formed in the lagging strand; they are very small having only 100-200 nucleotides.
9. The newly replicated DNA in eukaryotes is packaged into nucleosomes very rapidly.
10. The replicon or 'replication apparatus' is highly complex.
11. Four distinct *DNA polymerases*, α, β, δ and γ take part in the process. The polymerase α, β and δ are present in nuclei while the polymerase γ is present in mitochondria and chloroplast.
12. Polymerase δ is involved in the polymerization of the leading strand.
13. The other proteins or "replication factors" are much complex and are identified as RF-A, RF-B, RF-C etc.
14. The protein PCNA (Proliferating Celi Nuclear Antigen) is a cofactor of polymerase δ. More proteins involved in the process are yet to be identified.
15. No 'proofreading' protein has been identified.
16. The entire process of DNA replication in eukaryotes have not yet been clearly understood.

4.4.5 "Rolling Circle" Replication in Virus

The circular DNA of bacteriophage shows "rolling circle" model of replication (Fig. 4.25).

In this type of DNA synthesis, what happen are—

1. A *nick* appears in one of the two strands of the circular DNA double-helix. It allows the 5′ end to remain attached to a membrane of the host bacteria while frees the 3′ end to act as a *primer*.
2. Thus DNA synthesis forms a complement to the unbroken parental strand.
3. This time the intact parental strand "rolls" and unwinds to extend the template.
4. Simultaneously, the 5′ end-strand (which is attached to the bacterial membrane) is also lengthened and

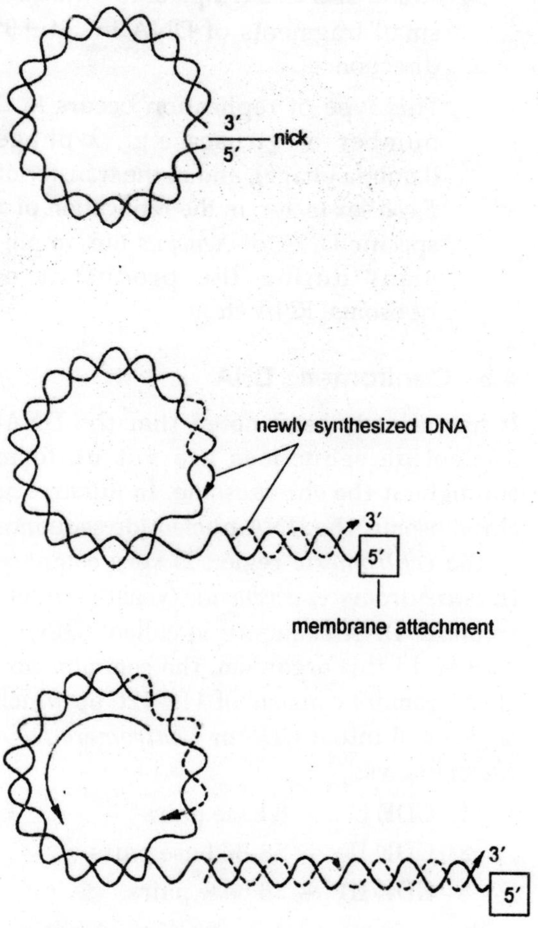

FIGURE 4.25

Rolling circle replication model of circular DNA double helix. A nick in one of the strands allows the 5′ end to become attached to a membrane while the other end is free to act as a primer so that DNA synthesis forms a complement to the unbroken parental strand. As synthesis proceeds the unbroken parental DNA rolls and unwinds, thus extending the template upon which new DNA is synthesized. Meanwhile the "tail" strand attached to the membrane is also lengthened during this process and provides a template upon which complementary DNA can be synthesized. Note that since the complement to the tail strand has its 3′ end toward the membrane. Synthesis must proceed in discontinuous fragments (After Strickberger).

serves as a *template* upon which new DNA is synthesized. It is important to note that synthesis on this strand is *discontinuous* by forming *DNA fragments*. This template-strand is called *tail-strand*. The uncoiling of this strand acts as a template to synthsize small fragments of DNA in 5′→3′ direction.

This type of replication occurs in a number of viruses e.g., λ-phage (lambda phage), and in the transfer of *E.coli* sex factor, in the replication of a specific DNA of *Xenopus* (an amphibian) during the producton of ribosomal RNA etc.

4.5 Centromeric DNA

It has already been noted that the DNA-nucleotide sequences are not uniform throughout the chromosome. In eukaryotic chromosome, the DNA-nucleotide sequence in the *centromeric* region is very complex. In *Saccharomyces cerevisiae* (yeast), the centromeric DNA sequence is called *CEN sequence*. In this organism, the common core of centromere consists of 112-120 bp which is grouped into 3 *CDE* or *Centromere DNA Elements*; viz.,

1. CDE I – 8 base pairs
2. CDE II – 78-86 base pairs
3. CDE III – 26 base pairs.

The CDE II and CDE III have mainly A-T base pairs while the base pairs of CDEI consists of specifically *RTCACRTG*, where *R* is a purine base.

The centromere is responsible for accurate segregation of the replicated chromosomes to the daughter cells during mitosis and meiosis.

4.6 Telomeric DNA

The *telomeres* of eukaryotic chromosomes are always *heterochromatic*. The telomere helps replication and gives stability to the chromosome. Telomeric DNA sequences are of two types: *simple* and *telomere-associated*.

The *simple* base-sequences are found at the extreme end of the chromosomal DNA. They are *species-specific* and give stability to the chromosome. They contain *simple repeated DNA sequences one after the other* (i.e., tandemly repeated). In human, the telomeric repeated DNA sequences is 5′-TTAGGG -3′. The telomeric DNA is not double stranded at the end, but one strand forms a loop on itself called *t-loop*, and the other forms a *D-loop*.

The *telomere-associated* sequences of the DNA are very complex, formed by repetition of thousands of base pairs, present inward from the chromosome end.

4.7 Satellite DNA (Repetitive DNA)

Usually the prokaryotic chromosomes bear non-repeated base-pairs. But in eukaryotic chromosomes, certain base-sequences are *repeated* for many times. These are called *repetitive DNA*. When eukaryotic chromosomes are centrifuged at high speed for a long time in cesium chloride (CsCl) solution, it reveals that the DNA forms a *large main band* and one to many *small bands* in the centrifuge tube. These small bands of DNA are called *satellite bands* (satelles = subordinate) and the DNAs present in these bands are called *satellite DNAs*. They contain *repetitive sequence* of base pairs.

The repetitive DNA may be *moderately repetitive* (a few to about 10^5 repetitions) or *highly repetitive* (10^5-10^7 repetition). Both of them appear many times in a genome. The repetition of the base pairs in the DNA may be arranged *dispersely* or *tandemly*. Tandemly repeated (one after the other) sequences are more common in eukaryotic genomes. The simple telomeric sequences (See 4.6) contain mostly tandemly repeated

DNA. Centromeres (See 4.5) also contain huge amount of tandemly repeated DNA.

The *dispersed repeated DNA* (also called interspersed repeated DNA) are distributed at irregular intervals. The interspersion of sequences may be *long* (LINE = *L*ong *int*erspersed *el*ement) or short (SINE = *s*hort *in*terspersed *el*ement). The dispersed or interspersed repetitive DNA contribute to a significant amount of moderately repetitive DNA in the genome.

4.8 RNA

We have already learnt that the nucleic acids are of two types: DNA (deoxyribonucleic acid) and RNA (ribonucleic acid).

Now we shall see the importance of RNA in biological systems. The role of RNA has two main aspects—*genetic* and *nongenetic*. The *genetic RNA* performs the role of hereditary material, as we have already discussed in the case of Tobacco Mosaic Virus (TMV, see 4.2.1.4). The *nongenetic* RNA, however, does not take direct part in the transmission of hereditary material but it helps *Central Dogma* (See 4.1) of molecular biology. That is, all DNA is first *transcribed into RNA,* and the RNA is *translated into protein.* This RNA is *nongenetic-RNA*. The nongenetic RNAs are of 3 types: *mRNA* (or, messenger RNA), *tRNA* (or, transfer RNA or soluble RNA) and *rRNA* (or, ribosomal RNA). In brief, the mRNA carries message (code) from DNA to ribosome; the tRNA transfers amino acids from amino-acid-pools to mRNA at ribosome, and the rRNA builds the structure of the ribosome. All three types of nongenetic RNAs are synthesized directly on DNA strand in the nucleus.

The organisms having both DNA and RNA, deploy their DNA as *genetic* material and RNA as nongenetic material. On the otherhand, the organisms having only RNA, deploy their RNA directly in the *genetic* role.

4.8.1 Structure of RNA

We refer to the Table 4.1 to note two important things: (1) The sugar *pentose* are of two types: *ribose* and *deoxyribose,* and (2) Pyrimidine bases are of 3 types: *Cytosine, Thymine* and *Uracil*.

In the structure of RNA, the type of pentose is *ribose* (not deoxyribose) and the pyrimidine bases are *Cytosine* and *Uracil* (not Thymine).

The RNA is *mostly a single stranded* molecule, but often it may have a double stranded structure, which is *nonhelical.*

Thus, chemically, the RNA-single strand is composed of a sugar-phosphate backbone with 4 types of nitrogenous bases, viz., *Adenine, Guanine, Cytosine* and *Uracil*.

Like DNA, the RNA is also a chain of polynucleotide. Here the unit of the polymer is *ribonucleotide* (not deoxyribonucleotide).

The RNA acts as a genetic material in some viruses (Tobacco Mosaic Virus, influenza virus etc. and bacteriophages). But the RNA is mostly nongenetic in prokaryotes and eukaryotes (where DNA is the genetic material).

RNA may be single stranded or sometimes double stranded. The genetic material of most viruses are single stranded RNA (in reovirus, it is non-helical double stranded). Except tRNA (transfer RNA, see 4.8) the nongenetic RNA of all prokaryotes and eukaryotes is single stranded.

4.8.2. Messenger RNA (mRNA)

This is a single stranded RNA molecule synthesized on a single strand of DNA by a process called *transcription* (See chapter 10). The base sequence of mRNA is complementary to the base sequence of DNA from which it is formed. The mRNA carries the message or genetic code (see chapter 10) from DNA to ribosome.

4.8.3 Transfer RNA (tRNA) or Soluble RNA

This single or double stranded RNA *transfers* amino acids from the amino-acid-pool of cytoplasm to the ribosome for protein synthesis. It is an intermediate molecule between the triplet code (see chapter 10) of mRNA and the amino acid sequence of the polypeptide chain (protein). All tRNA molecules have the same basic structure. One basic structure is found in the *clover leaf model* (Fig. 4.26).

In this structure (it looks like a leaf of clover, hence, this name)—the complementary base-pairing produces 4 "stems" which are separated by 4 loops. The 4 loops are I, II, III and IV. The loop II contains the 3-nucleotide *anticodon*. The cloverleaf structure of *yeast alanine tRNA* contains some modified bases: inosine (I), ribothymidine (T), pseudouridine (ψ), dihydrouridine (D), methylguanosine (GMe), dimethylguanosine (GMe_2) and methylinosine (IMe). The loop I is also called *D-loop* as it contains dihydrouridine, the loop II is the *anticodon-loop* which contains the bases I, G and C, the loop III is an extra growth containing D, and the loop IV is called *ψ loop* as it contains pseudouridine. The 3′ end contains free A–C–C–A nucleotides.

4.8.4 Ribosomal RNA (rRNA)

Approximately 80% of the total RNA of a cell is ribosomal RNA (rRNA). The base sequence of rRNA is similar in all organisms. It is associated with the protein molecules in the cytoplasm and together form the ribosomes. The ribosomes are the seats for protein synthesis.

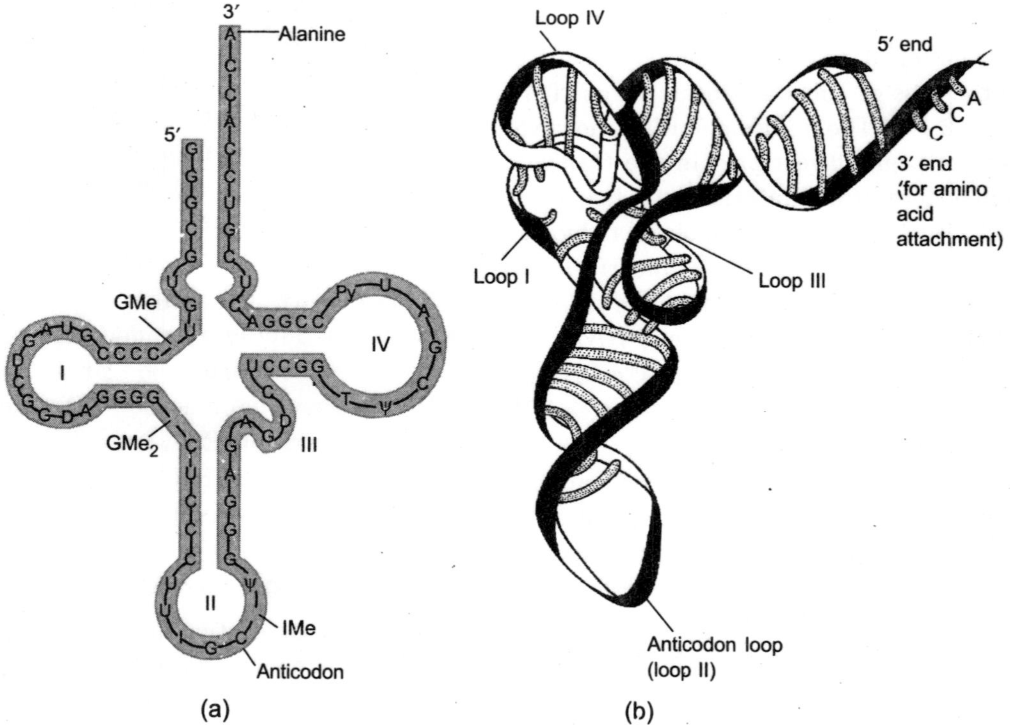

FIGURE 4.26

Transfer RNA: (a) Cloverleaf structure of yeast alanine tRNA. Py = pyrimidine. Modified bases, I = inosine, T = ribothymidine, Ψ = pseudouridine. D = dihydrouridine, GMe = methylguanosine, GMe_2 = dimethylguanosine, IMe = methylinosine (b) Schematic of the three dimensional structure of yeast phenylalanine tRNA, as determined by X-ray diffraction of tRNA crystals, Note the characteristic L-shaped structure.

4.9 Comparison between DNA and RNA

DNA	*RNA*
1. Polynucleotides with *deoxyribose* sugar.	1. Polynucleotides with *ribose* sugar.
2. The nitrogenous bases include adenine, guanine, cytosine and *thymine*.	2. The nitrogenous bases include adenine, guanine, cytosine and *uracil*.
3. Always double stranded.	3. Usually single stranded.
4. Always genetic material.	4. May be genetic or nongenetic.
5. There are 3 forms of DNA: A-DNA, B-DNA and Z-DNA.	5. There are 3 types of RNA: mRNA, tRNA and rRNA.
6. Directly involved in RNA synthesis.	6. Directly involved in protein synthesis.
7. Mainly concentrated in the nucleus of the cell.	7. Scattered throughout the cell.

CHAPTER 5

CHROMOSOME AND GENE

Chromosome Theory, Gene, Gene and Allele, Genome, Chromosome, DNA and Gene, Gene Control, Mechanism of Lac Operon.

Let us look back to the chapter 2 and 3 once more. There we have discussed the term 'Mendelian Factor' and described the structure, number and types of chromosomes in detail. We have understood that what Mendel called *'factor'* was the *'gene'* of today.

5.1 Chromosome Theory

After the rediscovery of Mendelism in 1900, It was soon revealed that there was a distinct correlationship between **chromosome** and **'factors' or genes**. In 1902, *W. Sutton* and *T. Boveri* realized that the Mendelian factors and the chromosomes behave parallelly at the time of gamete production in garden pea. They recognized that—

1. The genes (factors) and chromosomes both remain in pairs.
2. Both the members of paired chromosomes and the paired genes segregate from each other during meiosis and each half enters in each gamete.
3. Both the chromosome pair and gene-pair have distinct role in heredity.

These observations led Sutton and Boveri to conclude that *the genes* (factors) were *located on chromosomes*. This is called *Chromosome theory of inheritance* (Fig. 5.1).

The genes present in the chromosomes constitute the physical basis of heredity. The chromosome are usually present in pair, con-

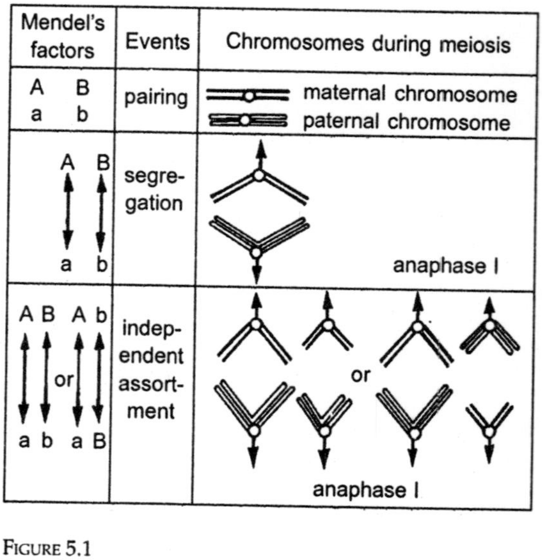

FIGURE 5.1

Parallelism between Mendel's hypothetical factors (or genes) and chromosomes during meiosis (after Suzuki et. al. 1986).

stituting a *diploid* condition (2n). They separate from each other during gamete formation and produce *haploid* (n) cells. When the basic number of chromosome of an organism is more than two, it is called *polyploid* (3n, 4n etc.).

5.2 The Classical Gene and Mole-cular Gene

Mendel referred to *unit factors*. Now we call them *genes*. In classical genetics a gene is—

1. A determiner and controller of a single character of an organism.
2. A unit of inheritance that segregates independently.
3. A specific part of a chromosome.

After a series of investigation in *Drosophila* for *linkage* and *Crossing over* (See chapter 6), Morgan defined gene as the **shortest segment of a chromosome which could be separated from adjacent segments by crossing over**.

Genes can also be defined in terms of function. It is the **shortest segment of a chromosome responsible for the production of a specific polypeptide (protein)**.

It is a part of DNA and can be inherited through generations. Thus, **a gene is the basic unit of inheritance for a given characteristic**.

The term *gene* was coined by Johannsen in 1909.

In molecular genetics, one of the definitions of a gene is—**a part of DNA molecule that carries the information specifying one protein**.

The gene can be expressed as a sequence of nucleotide bases or base-pairs. The sequence is the 'code' that directs the production of one particular sequence of amino acids, characteristic of the specific protein. The gene is both a structural and functional unit of inheritance.

5.3 Gene and Allele (or, Allelomorph)

Alleles are alternative forms of a gene that occur in a given locus on a chromosome.

There is often confusion between the meaning and proper usage of gene and allele.

A **gene** is characterized by—

- what it controls,
- how it behaves during inheritance
- where (on a chromosome) it is found.

While an **allele** is a specific version of a particular gene. A gene may have two, three or more allelic forms or **allelomorphs**.

An analogy can be cited to understand the relationship between the *gene* and *allele*: If the post of principal of a college is analogous to gene, then allele of that gene would be the particular principal at one time. If a new principal is appointed (a different allele) the post (the gene) does not change though the job might be done in a different way and with different results.

5.4 Genome

The term *genome* is used to describe both the number of chromosome and the number of genes. In one definition, **the genome is a complete haploid (n) set of chromosomes inherited as a unit from one parent in sexual reproduction**. In another definition, **the genome is the total number of genes present in a haploid set of chromosomes in a gamete**.

The size of the genome of an organism is determined by the number of nucleotide-pairs per haploid cell. Thus, the size of mammalian genome is calculated between 10^9 and 10^{10}.

Even in another definition of genome it is defined as the **"array of genes carried by an organism"**. However, this definition does not cover the total meaning.

5.5 Chromosome, DNA and Gene

Each species has a definite number of chromosome in each cell. The chromosomes remain in pairs. These are called *homologous pair*. As for example, in human, the specific and characteristic number of chromosomes is 46 in each cell. They remain as 23 homologous pairs.

It is known that each chromosome has two **identical chromatids** (chapter 3) and each of the chromatids contain **one DNA molecule**. On the other hand, we know that one chromosome (or chromatid) contains thousands of **genes**, which are the units of inheritance. Thus, it is evident that each chromosome is made up of a DNA molecule and it harbours thousands of genes. In fact, the genes are very small fractions of a chromosome (or the DNA). In molecular aspect, the role of gene is interpreted in the light of molecular structure of the DNA.

5.6 Gene Control

It is a fundamental question that although all somatic cells of an organism possess the same number of chromosome and that the chromosomes carry the same genes, why the organism show a wide variation in structure and function during development or in a certain phase of life?

This question was answered by the work of Jacob and Monod in 1961. From their work on *E.coli* it was established that **the genes are regulated** and certain protein molecules are synthesized according to circumstances and demand.

The enzymes of *E.coli* are of two types—*constitutive* and *inducible*. The latter is produced only in the presence of an *inducer*. One of the inducers in *E.coli* is *B-galacto-sidase*. Another enzyme is *Lactose permease,* which induces the cell to take up lactose. The genes for the synthesis of those enzymes (proteins) are **not always active**, but are sometimes "switched on" (operating) and sometimes "switched off" (not operating). Jacob and Monod explained these phenomena as 'induction' and 'repression' on the basis of their experiment.

5.6.1 The Experiment and the Jacob-Monod Hypothesis of Gene-control: The Operon Concept: Experiment done on *E.coli*.

The genes that determine the sequences of specific proteins are called *structural genes.* The gene that controls the activity of these genes is called *regulator gene.* Regulator gene prevents the activity of the structural gene. The structural genes for *β-galactosidase* (inducer) and *lactose permease* are present on the same chromosome and are closely linked (see chapter 6.2). The regulator gene is situated some distance apart from the structural gene. The regulator gene is responsible for the production of a *repressor molecule* which prevents the activity of the structural gene **indirectly**. It is thought to influence another gene called the *operator gene* which is located adjacent to the structural gene. **The structural genes and the operator gene together are called the Operon.**

From the experiment on *E.coli,* the control of protein synthesis (or control of genes) could be explained in 6 steps: (Fig. 5.2).

1. The *regulator gene* in a DNA double helix transcribes a specific mRNA.
2. Then, the transcribed mRNA is translated into a protein called *repressor molecule.* It is an allosteric protein.
3. In the third step, the repressor molecule either **binds** with the *operator gene* present in the same DNA and "switch it off" (i.e. suppresses its activity), **or does not bind** with the operator gene and keeps the "switch on" (i.e. the gene remains active).

4. When the operator gene is "switched on", the structural genes on the same DNA remain active.
5. Active structural genes carry on transcription.
6. The transcription is followed by translation to produce protein molecules (enzymes).

On the other hand, if on the fourth step, the repressor gene **binds** with the operator gene to make the latter "switched off" (or inactive) no transcription and translation take place, resulting no protein (enzyme) production.

5.6.2 Mechanism of the Lac Operon

The *lac operon* is the abbreviation of the lactose operon of *E. coli*. It provides an example of regulation of gene control. This hypothesis was postulated by Jacob and Monod.

The *lac operon* of *E.coli*. contains 3 structural genes (z, y and a) which produce enzymes (proteins) for the breakdown of lactose. The gene **z** produces *β-galactosidase*. The gene **y** is responsible for production of *galactoside permease* and the gene **a** produces *thiogalactoside acetylase*.

Usually, in the absence of lactose these enzymes are not produced in *E.coli*; these are produced in the presence of an *inducer*; thus, these are inducible enzymes. Lactose (lac) acts as an inducer. When the inducer is present, the entire set of genes along with an *operator gene* (o) act as a unit, comprising the *operon*. Thus the *operator* is the controlling element.

Soon it was discovered that the controlling mechanism of the lac operon was also regulated by another gene, called *inducer* (i) or *regulatory gene*. The inducer (i) is located close to the operon but separated from the operator; the **i** is responsible for the production of **repressor**. The whole mechanism of the coordination

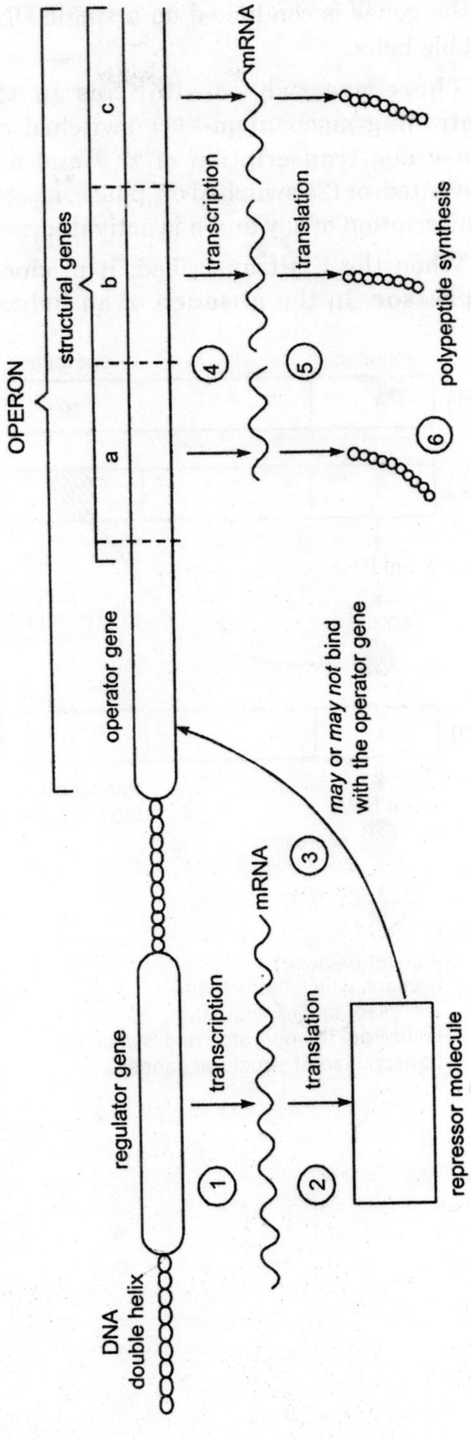

FIGURE 5.2

The basic structures and processes involved in the 'operon concept' according to the hypothesis proposed by Jacob and Monod. The numbers indicate the sequence of events.

of the genes is conducted on a single DNA double helix.

There are two possibilities in the controlling mechanism—(1) 'switched off' phase i.e., transcription of z, y and a is prevented, or (2) 'switched on' phase, i.e., the transcription of z, y and a is activated.

When the **i** is transcribed, it produces **repressor**. In the **absence** of an inducer (lactose), the repressor binds to the operator (o) and prevents the transcription of structural genes (z, y and a).

On the other hand, when the inducer (lactose) is **present**, it binds to the repressor and dissociate the latter from the operator (o). The operator then allows transcription of structural genes; as a result, necessary enzyme synthesis occurs (Fig. 5.3).

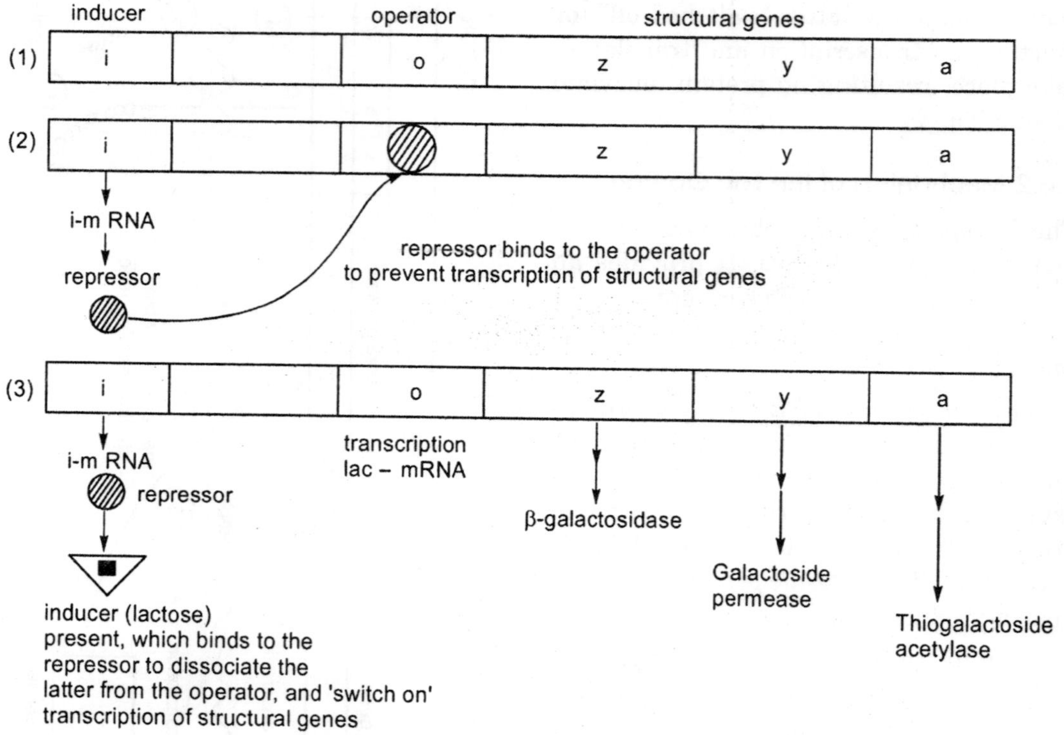

FIGURE 5.3

Lac Operon.

CHAPTER 6

LINKAGE, CROSSING OVER AND CHROMOSOME MAPPING

Mendel's Law of independent assortment is not universal, Coupling and repulsion, Linkage, Linkage-group, Complete and incomplete linkage, Crossing over, Mechanism of crossing over, Crossing over in *Neurospora*, Theories of mechanism of chromosomal crossing over, Crossing over and crossover values, Chromosome mapping, Three-factor cross, Interference, Coefficient of coincidence, Example of chromosome mapping in *Drosophila*, Problems of chromosome mapping worked out, Cytological proof of crossing over, Stern's experiment, Significance of crossing over.

Some facts are now already established—that genes are hereditary units; they are located on chromosomes; number of genes are always greater than the number of chromosomes; the chromosomes along with their genes 'assort independently' during gamete formation (meiosis), and thus Mendel could postulate his 'law of independent assortment' in his dihybrid cross experiment with garden peas.

6.1 The 'Law of Independent Assortment' is not Universally Applicable

The first exception of Mendel's second law (law of independent assortment) was noticed by Bateson in 1905, and it was revealed that the phenomenon of independent assortment was not universal.

6.1.1 Work of Bateson and Punnett: Mendel's Second Law is not Universal

In a dihybrid experiment in sweet pea, Bateson and Punnett crossed a **purple long** (purple flower and long pollen, dominant) plant with a **red-round** (red flower and round pollen, recessive) plant. The F_1 plants were all purple-long. The F_2 result (Table 6.1) showed a discrepancy from the expected 9 : 3 : 3 : 1 ratio. The parental combinations (purple-long and red-round) were produced more in number than were expected (6169 instead of 4345), and the new combinations (purple-round and red-long) occurred only in 783 plants instead of expected 2607.

In explanation of the result, Bateson and punnett opined that the dominant genes for **long-pollen** have a tendency to remain associated with the dominant gene for **purple flowers**, while the recessive gene

TABLE 6.1: Bateson and Punnett's experiment of coupling and repulsion

P: Purple-Long (PPLL) × red-round (ppll)

F_1: Purple-Long (PpLl)

F_2: Type	Number Observed	Expected	Expected ratio
Purple-Long	4381	3910.5	9/16
Purple-round	390	1303.5	3/16
red-long	393	1303.5	3/16
red-round	1338	434.5	1/16

for **round pollen** tends to stay together with the recessive gene for **red flowers**.

When they performed the cross in another way i.e., **purple-round × red-long**, the F_1 plants were all **purple-long**. By self-crossing these F_1 plants (Table 6.2), the result again showed a deviation from the expected 9 : 3 : 3 : 1 ratio. The parental types (purple-round, and red-long) were much more (192 instead of 157) and the new combinations were much less (227 instead of 262) than expected.

Apparently the F_1 plants produce more parental gametes (**PL** and pl in cross 1, Table 6.1, and Pl and pL in cross 2, Table 6.2) instead of equal frequency of 4 kinds of gametes that should have resulted if the genes are assorted independently.

The second cross (Table 6.2) also indicated that there was no tendency for specific genes to stay together (as it was thought earlier by Bateson and Punnett), but only the parental genes tend to stay together.

When the F_1plants were test crossed in each experiment (i.e., crossed to a completely recessive plant), the result (Table 6.3) showed that the F_1 plants did not produce 4 types of gametes in equal frequency. The result deviated from the result of 'law of independent assortment'. Thus it was realized that the *law of independent assortment is not universal* ! The genes are not assorting independently !

TABLE 6.2: Second cross of Bateson and Punnett for coupling and repulsion

P: Purple-round (PPll) × Red-long (ppLL)

F_1: Purple long (PpLl)

F_2: Type	Number Observed	Expected	Expected ratio
Purple-long	226	235.8	9/16
Purple-round	95	78.5	3/16
Red-long	97	78.5	3/16
Red-round	1	26.2	1/16

Table 6.3: Results of the test cross of the F_1 plants of the two previous crosses

	Number of gametes		
Type	Cross 1	Cross 2	Expected
Purple-long	7	1	1
Purple-round	1	7	1
Red-long	1	7	1
Red-round	7	1	1

6.1.2 Coupling and Repulsion

Bateson and Punnett used the term '*coupling*' for the associated dominant genes in one parent and for the associated recessive genes in the other parent, in their first cross. They used the term '*repulsion*' for the situation in which one parent brings in one dominant gene and one recessive gene, and the other parent brings in the other dominant and the other recessive in their second cross.

- **Cis and Trans Arrangement of Genes**

Bateson and Punnett could not explain the reason of their '**coupling** and **repulsion** hypothesis'. In 1910-1911, T.H. Morgan observed '**linkage**' to the segregation of homologous chromosomes and the occurrence of crossing over between homologous chromosomes during meiosis. After interpretation of the phenomenon of 'linkage' in *Drosophila melanogaster*, Morgan was able to explain the arrangement of genes in the 'coupling' state as *cis*-**configuration**; and the alternative arrangement of the 'repulsion' state as the '**trans-configuration**' (Fig. 6.1).

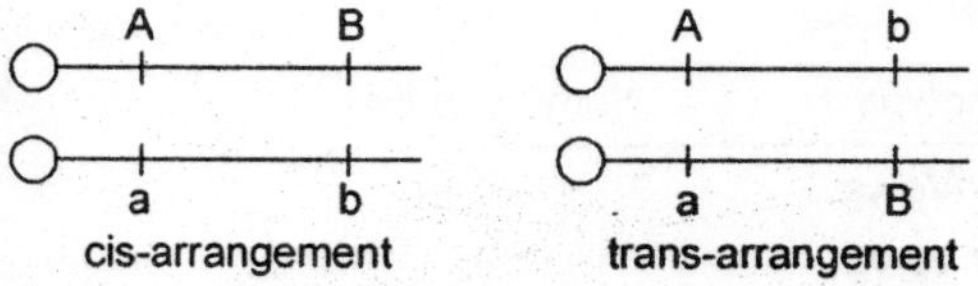

Figure 6.1

Interpretation of coupling and repulsion as cis and trans configuration.

6.2 Linkage

Morgan and his associates (C.B. Bridges, A.H. Sturtevant and H.J. Muller) found that the 'coupling' and 'repulsion' are two aspects of a single phenomenon, which they called **linkage**. Morgan established that the *linked genes tend to remain in their original combination* because they are on the same chromosome.

The alleles of each of the segregating genes are called **genetic markers**. The genetic markers located on the same chromosome *tend to remain together and tend to enter in the same gamete during meiosis*. That is, they do not follow independent assortment. Such genetic markers are said to be *linked* and the inheritance or transmission pattern of the linked genes is called *linkage*. When the genes are closely located they show more strong linkage and when they are distantly located they show less linkage.

6.2.1 Definition

Linkage can be defined as *the tendency of two or more nonallelic genes to remain together and to be inherited together*.

6.2.2 Characteristics of Linkage

1. Phenomenon of linkage proves that the law of independent assortment is not universal.
2. Linked genes are present on the same chromosome.

3. Closely located genes show strong linkage than the widely located genes.
4. Linked genes tend to be transmitted together and to enter in the same gamete.
5. If there is linkage between two genes, there is no crossing over between them, and if there is crossing over between genes, there is no linkage between them. Thus linkage is a phenomenon which is opposite to the phenomenon of crossing over.

6.2.3 Linkage Group

All the genes that are linked together on a chromosome, i.e., the *group of genes* which show **linkage** are called **linkage group**. All genes have their loci on the same chromosome. Linkage group of a homologous pair of chromosomes is considered as a single group. Thus, the linkage group is considered as the *haploid number* of chromosomes of a species. As for example the linkage group of *Drosophila melanogaster* is 4 (because this species contains 4 pairs of chromosome), that of *Pisum sativum* is 7 (it contains 7 pairs of chromosomes) etc.

6.2.4 Complete and Incomplete Linkage

Bridges performed a series of experiments with *Drosophila*, and considering sexes of the insect, found amazing results—as follows:

1. In the first experiment Bridges crossed *wild type* (Red eyed-long winged, PrVg) by flies with *recessive* trait (purple eyed and vestigial winged, prvg). According to Bateson's terminology, this is a cross in *coupling*. The F_1 flies were all wild type. Bridges selected the F_1 **males**

TABLE 6.4

(1) Wild-type (Red-Long) × Purple vestigial (coupling)

P: Pr Pr Vg Vg × pr pr vg vg

F_1: Pr Vg pr vg (Wild)

F_1 Pr Vg pr vg ♂ × pr vg pr vg ♀ (Test Cross)

Result:	Type	Genotype	Number of flies
	Wild (Red long)	Pr pr Vg vg	519
	Purple vestigial	pr pr vg vg	552
	Red vestigial	Pr pr vg vg	0
	Purple long	pr pr Vg vg	0
			1071

TABLE 6.5

(2) P: Purple long (pr pr Vg Vg) × Red vestigial (Pr Pr vg vg) (repulsion)

F_1: Wild (Pr pr Vg vg)

F_1: Pr pr Vg vg ♀ × pr pr vg vg ♂ (Test Cross)

Result:	Type	Genotype	Number of flies
	Wild (Red long)	Pr pr Vg vg	0
	Purple vestigial	pr pr vg vg	0
	Red vestigial	Pr pr vg vg	358
	Purple long	pr pr Vg vg	346
			704

and made a test cross with the completely *recessive purple vestigial* female (Table 6.4). The result showed only the parental combination. It is a case of **complete linkage**.

2. In the second experiment, Bridges mated *Red vestigial* with *purple long* (a cross in repulsion), and again he found that all F_1 flies were *wild- type*. He selected the F_1 *males* and made a test cross with *recessive purple vestigial* female. The result (Table 6.5) showed that the parental genes remained associated and *only parental combinations* were obtained. It is also a case of **complete linkage**. Instances of complete linkage in nature is very rare.

3. In his third cross, Bridges repeated the first cross, as a result, the F_1 flies were all *wild type*. But, now he selected the *females* instead of males from the F_1 flies and *test crossed* them with completely recessive males. (Table 6.6) This time Bridges got a different result—he found both the *parental types* (indicating parental genes stay together) and some new combination (indicating the genes had dissociated in some flies). It is a case of **incomplete linkage**.

4. He made another experiment, similar to the second one (Red vestigial × purple long) and obtained wild flies. But in this case he testcrossed the F_1-wild **females** to the completely

TABLE 6.6

(3) P: Wild type (Pr Pr Vg Vg) × Purple vestigial (pr pr vg vg)

F_1: Wild type (Pr pr Vg vg)

F_1 Pr pr Vg vg ♀ × pr pr vg vg ♂ (Test cross)

Result:	Type	Gene combination	Number of flies
	Wild type	Pr pr Vg vg	2876
	Purple vestigial	pr pr vg vg	2433
	Red vestigial	Pr pr vg vg	270
	Purple long	pr pr Vg vg	270
			5849

TABLE 6.7

(4) P: Purple long (pr pr Vg Vg) × Red vestigial (vg vg Pr Pr)

F_1: Wild-type (Pr pr Vg vg)

F_1 Pr pr Vg vg ♀ × pr pr vg vg ♂ (Test Cross)

Result:	Type	Gene combination	Number of flies
	Wild-type	Pr pr Vg vg	265
	Purple vestigial	pr pr vg vg	234
	Red vestigial	Pr pr vg vg	2038
	Purple long	pr pr Vg vg	2203
			4740

recessive male (Table 6.7). Again he noted that the *parental genes tend to stay together* but some *new combinations* also occurred. It is also a case of **incomplete linkage**.

From the results of the above experiments Morgan deduced that coupling and repulsion are but two aspects of a single phenomenon, the linkage. Also, the questions—why is there no new combination in cross 1 and 2, and why there were some new combinations in cross 3 and 4—led Morgan to discover the phenomenon of *crossing over*.

6.3 Crossing Over

The question of arising of *new combination* in the experiment of Linkage (6.2.4, incomplete linkage) done by Bridges, was solved by Morgan in the light of 'absence of linkage'—or, *the phenomenon of exchange of genes between homologous chromosomes during meiosis*, a phenomenon called *crossing over*. It is the process by which *genetic recombination* takes place, and as a result *new combinations* arise.

6.3.1 Definition

It is a process by which homologous chromosomes exchange genes by breakage and reunion during meiosis. It results in recombination of genes on chromosome. It is effected by *breaks* in the homologous chromosomes, **transhift** of genes between the two homologues, and **reunion** of the transferred parts of the chromosomes (Fig. 6.2).

6.3.2 Mechanism of C.O.

(Please refer to the chapter 3.4.1.3. Meiotic Maturation Phase, Pachynema and Fig. 3.12 and 6.2).

6.3.3 Crossing Over in *Neurospora*

Neurospora is a fungus. Its life cycle includes asexual as well as sexual reproduction. It can grow by ordinary mitosis from a thread-like mass of hyphae, called *mycelium*. The nuclei of the mycelium are haploid and contain 7 chromosomes. Asexual reproduction occurs by the formation of spores called *conidia*.

Sexual reproduction includes a fusion of nuclei from *mycelia of opposite mating types* (A and a). The fusion of the nuclei results in a *diploid zygote* containing 14 chromosomes. The cell containing the zygote form the *ascus*. The zygote undergoes *meiosis*, and the 4 haploid products are arranged in tandems (Fig. 6.3). All divisions occur in the plane of the long axis of the ascus, and the nuclei remain in their primary order. As a result of this regularity of nuclear divisions, the 2 nuclei at one end of the ascus are derived from one of the first-division nuclei, while the 2 nuclei at other end are derived from the other of the first-division nuclei.

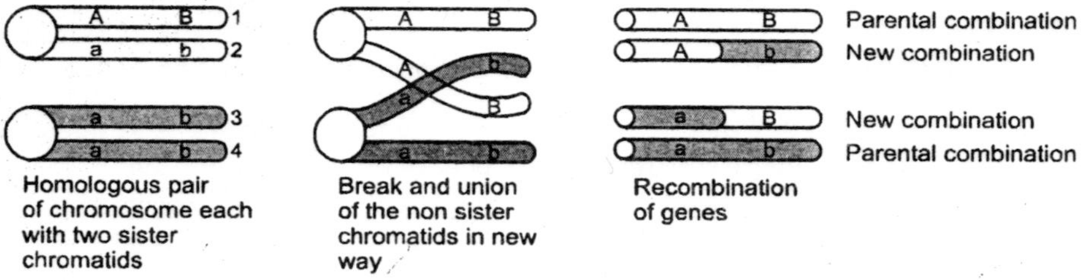

FIGURE 6.2

Process of crossing over during meiosis.

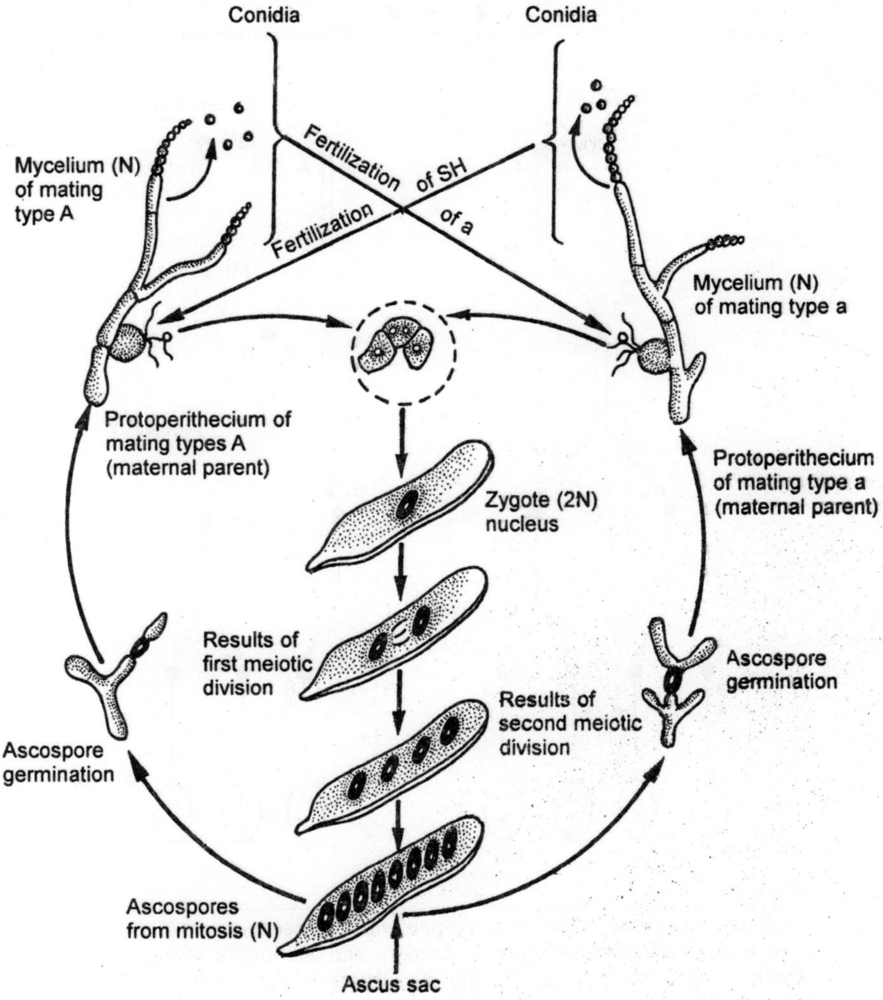

FIGURE 6.3

The life cycle of bread mold, *Neurospora.*

Subsequently, each haploid nucleus divides once mitotically, again in tandem, so that each meiotic product is represented twice within the ascus. Each nucleus develops a hard case around itself and is called **ascospore**. The ascus then be opened and the ascospores are removed individually and cultured in separate tubes to determine their genotypes.

Regarding the question of phase of occurrence of **crossing over** we shall analyse the ascus with its 4 products of meiosis. This is called *tetrad analysis* (Fig. 6.4).

The (Fig. 6.4) shows the meiosis of *Neurospora* zygote without crossing over. In this example, the gene for **a** mating type is linked with the allele '+', while the gene for **A** mating type is linked with '–'. The '+' and the '–' are responsible for coloured and non-coloured mycelium respectively. When separate pairs of alleles are present at different points of the homologous chromosomes, it is possible to detect C.O. and to determine whether the process occurs at the **two-strand** or the **four-strand** stage. If C.O. occurs at the two-strand stage, the

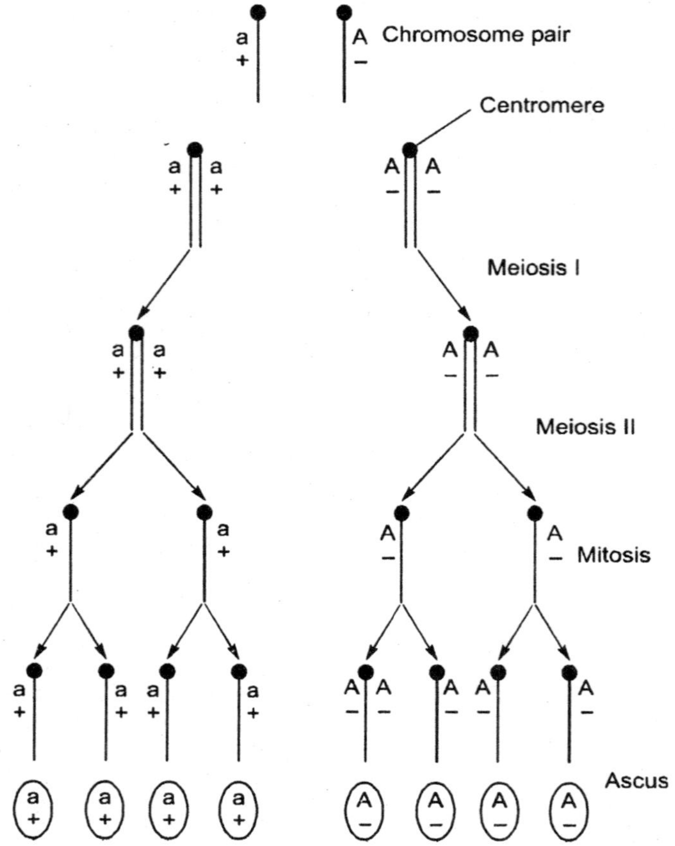

FIGURE 6.4

Diagram illustrating a chromosome pair with the genotype a+/A–, at successive stages during the two meiotic divisions and one mitotic division in *Neurospora*. Due to absence of crossing over, there is a 4 : 4 distribution in an ascus.

asci with C.O. will contain only recombinant types of ascospores. And, if the C.O. occurs between two of the chromatids at the four-strand stage, the asci with C.O. will contain both parental and recombinant ascospores.

In actual cross, the result (Fig. 6.5) shows that C.O. occurs at the *four-strand stage* but only 2 of the 4 chromatids are involved in a single cross-over.

The analysis of the result (Fig. 6.5) shows that if a single cross-over occurs between the mycelium-colour gene and the sex-type gene, 2 of the resultant ascospores will be parental and 2 will be recombinant. If such a cross-over occurs in every ascus, the total number of recombination will be 50%.

It is important to note that in all cases the *maximum frequency of recombination* that can result from crossing over between linked genes is 50%. The recombination frequencies never exceed 50%.

6.3.4 Features of C.O.

1. It is a process in which parts of homologous chromosomes are interchanged.
2. C.O. and independent assortment produce **new combination** of genes.

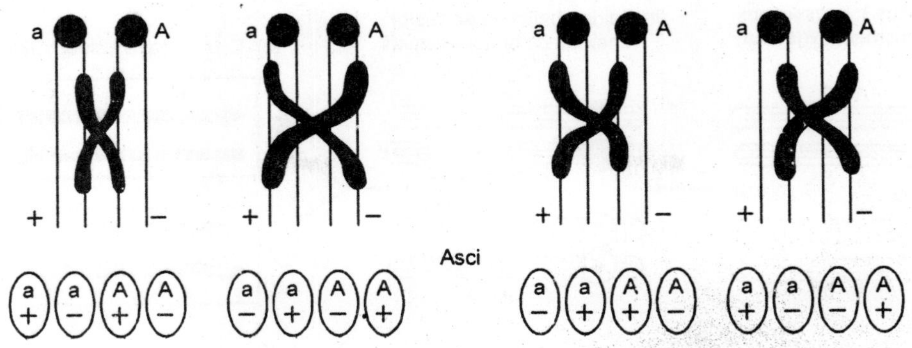

FIGURE 6.5

Diagram illustrating various cross-overs in 4-strand stage in *Neurospora.*

3. It occurs at the **pachytene** stage after the synapsis of the homologous chromosome has occurred in prophase. It occurs at 4-strand stage.
4. C.O. involves the breakage of two non-sister chromatids and the exchange of parts. (C.O. between sister chromatids may occur but genetically not detectable).
5. Recombinant combinations of chromosomes are formed due to transhift of genes.
6. Increased distance between two loci on a chromosome increases the probability of C.O.
7. C.O. takes important part in the process of evolution.
8. C.O. is always less than 50%.

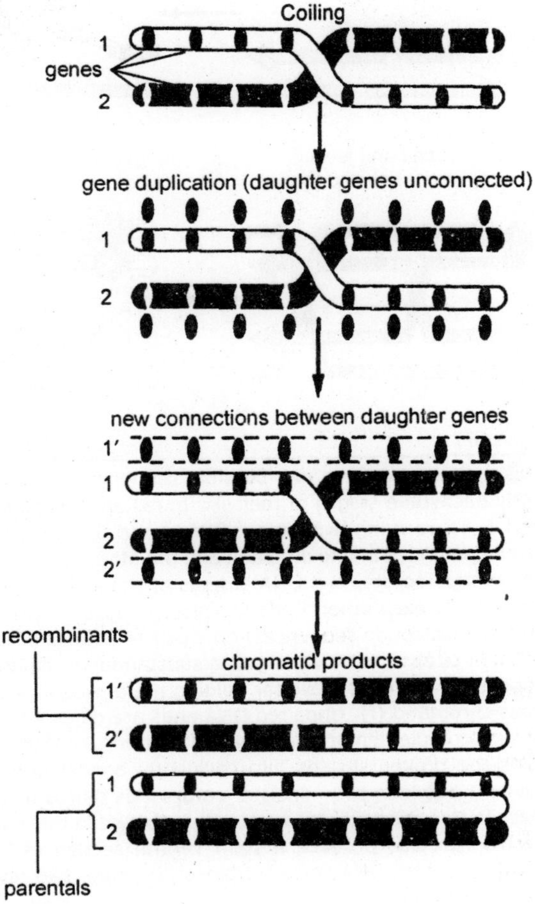

FIGURE 6.6

Simplified diagram of Belling's theory of crossing over as a result of new connections made between duplicated genes on two separate chromosomes that have temporarily intercrossed. The recombinant chromatids therefore derive only from newly synthesized genes, according to this theory (After Strickberger).

6.3.5 Theories of Mechanism of Chromosomal Crossing Over

Following theories have been postulated to explain the mechanism of crossing over.

1. **Chiasma type theory by Darlington**. Darlington proposed that **attractive forces** between homologues during early meiotic stages cause pairing and coiling of them. Then chromatid replication occurs. Next, **repulsion** between them causes breakage and exchange between chromatids.

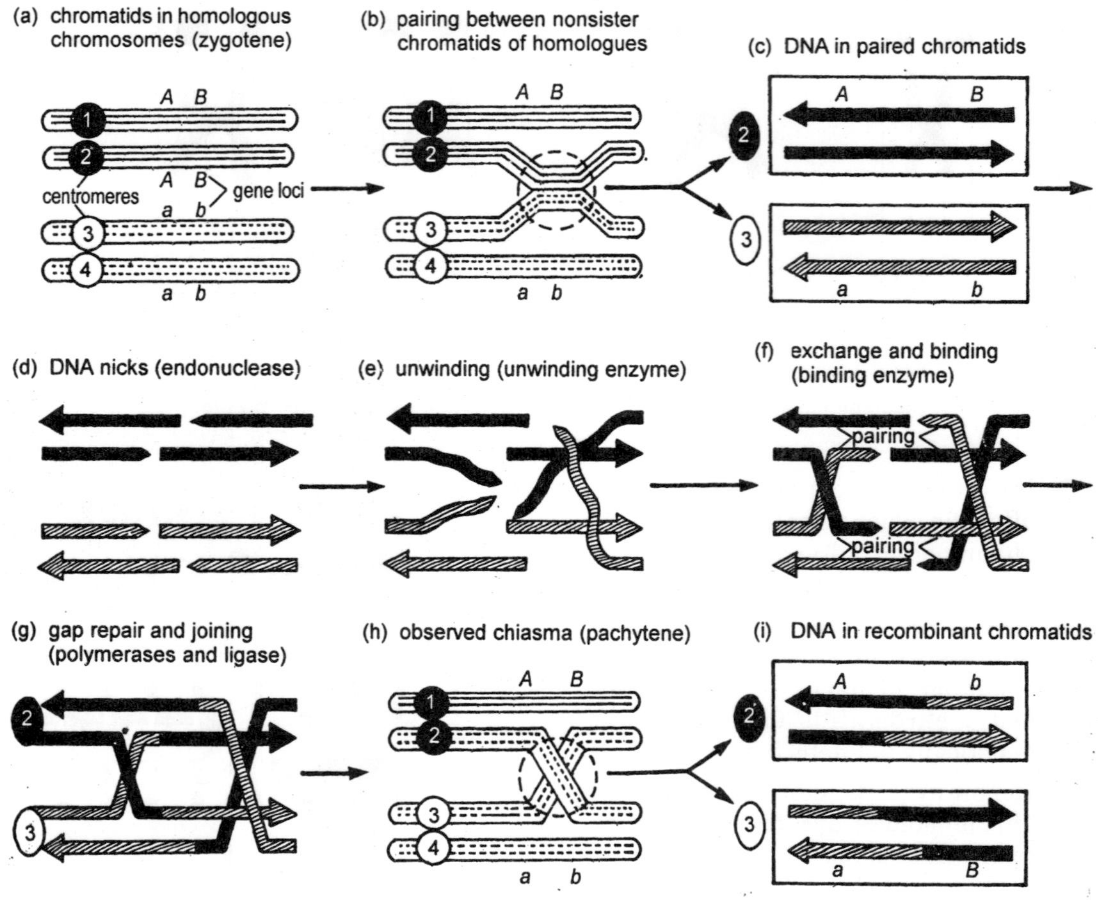

FIGURE 6.7

This model represents one possible sequence of events leading to recombination during the late zygotene—early pachytene stages of meiosis. In (a) are two homologous chromosomes, each with two chromatids (1, 2 and 3, 4), carrying two gene loci, *A, a* and *B, b*. Each chromatid consists of a DNA double helix with one DNA strand polarized in one direction and the other oppositely. In (b) and (c) pairing between homologous nonsister chromatids occurs, and DNA strands of the same polarity (marked by arrows pointing in the same direction) lie close to each other. Endonuclease-produced nicks then occur at similar positions (perhaps recognized by specific nucleotide sequences) on a pair of strands of the same polarity (d). These events allow unwinding and rotation of each helix leading to a substantial exchange of strands of the same polarity (e). A binding enzyme appears to be responsible for holding the unwound DNA strands to the complementary strand of the homologous chromatid (f). Unpaired DNA ends are digested by an exonuclease enzyme, and unpaired gaps are filled in with complementary base-paired sections synthesized by a polymerase enzyme. A ligase enzyme (g) then joins the broken strands into covalently bonded nucleotide chains. These new connections thus produce a chiasma between two nonsister chromatids that is usually first observed during the pachytene stage (h), and lead, in this example, to crossing over between the *A* and *B* loci (i). As can be seen in (i), the recombinational events can also produce a short central section of "hybrid" DNA in a single chromatid, composed of one complementary DNA section from each of the two homologous chromosomes, that is perhaps 100 or more nucleotides long.

2. **Duplication theory of Belling** Belling proposed that during duplication of chromosomes, (a) the chromomeres (genes) duplicate and remain just opposite to the old ones; (b) Later, when the chromomeres are joined by a newly synthesized interchromomeric part, they may switch over from one homologue to the other resulting *crossing over* and *recombination*. This view is not valid today (Fig. 6.6).
3. **Molecular theory** (Fig. 6.7): There are two models under this interpretation.

A. Breakage and Reunion:

(a) The genetic C.O. is associated with *breakage* of 'parental' DNA strands from homologous chromosomes and their subsequent *reunion* into recombinant products.

(b) Howell and Stern have demonstrated that an enzyme system is responsible for breakage and reunion of DNA in zygotene and pachytene stage.

(c) An endonuclease produces nicks in single strands of DNA and a ligase rejoin broken DNA strands.

(d) Single strands of DNA from different homologous chromatids are aligned into new recombinant double-stranded molecule.

The Fig. 6.7 and explanation are based on the breakage of 'parental' DNA strands and their subsequent pairing to form a 'hybrid' or 'heteroduplex' DNA molecules containing one strand from each 'parent'. Here, the recombination is reciprocal.

There are other various ways too by which this type of recombination can be resulted.

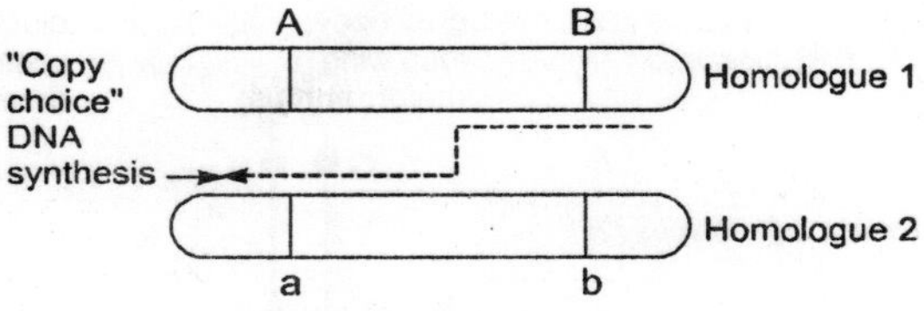

FIGURE 6.8
"Copy choice".

B. Copy Choice: This model was based on the assumption that during synthesis, the DNA molecules could switch from one homologue to the other. They can use DNA of one homologue as template in one region and switch by using the DNA of the other homologue as template in another region (Fig. 6.8).

This theory has been discarded as it was soon discovered that the DNA replicates semiconservatively.

6.3.6 Crossing-over and Cross-over Values (COV)

During the process of crossing over, the cross-over of genetic material occurs between all homologous chromosomes during meiosis. The alleles of parental **linkage group** separate and **new association** of alleles are formed in gametes, a process called **genetic recombination**. Offspring produced from these gametes show 'new' combinations of characteristics and are called '**recombinants**'.

- **Experiment on *Drosophila***

The principle of crossing over is illustrated by a cross between a **male** heterozygous grey-bodied long-winged (GL/gl) *Drosophila* and a **female** homozygous black-bodied vestigial-winged (gl/gl) *Drosophila*.

This cross produced heterozygous F_1 offspring with grey body and long wing (Fig. 6.9).

Test cross of these flies with homozygous double recessive flies (gl/gl) produced the following types of offspring (Fig. 6.9):

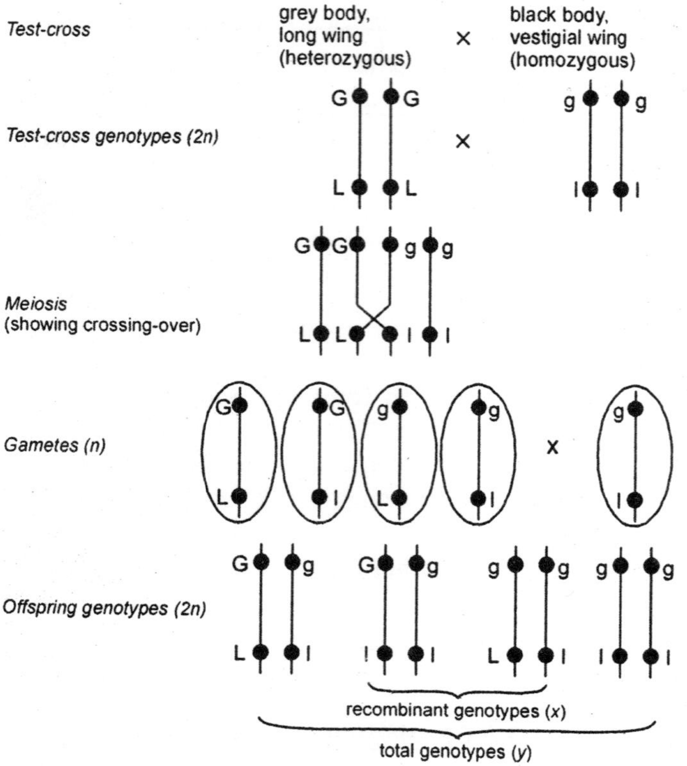

FIGURE 6.9

Genetic explanation of crossing-over and the reappearance of recombinant genotypes. The recombination frequency can be calculated by counting the number of individuals showing recombination and the total number of individuals and applying the following formula: recombination frequency (%) = x/y × 100.

Parental phenotypes
1. Grey body-Long wing—965
2. Black body-vestigial wing—944

Recombinant phynotypes
3. Black body-Long wing—206
4. Grey body-vestigial wing—185

From the result it is evident that the genes for body-colour and wing-length are **linked** (Otherwise, the result would show 1 : 1 : 1 : 1 ratio). Using number of the flies from this cross it is possible to calculate the recombination frequency of the genes for the two phenotypes (body colour and wing-length).

The *recombination frequency* is calculated on the basis of the following formula—

Thus, the *recombination frequency* of the above cross would be—

$$= \frac{206 + 185}{(965 + 944) + (206 + 185)} \times 100$$

$$= \frac{391}{2300} \times 100 = 17\%$$

This value reflects the number of cross-overs which have occurred during gamete formation. Sturtevant posulated that the **recombinant frequency** or **cross-over**

$$\frac{\text{Number of individuals showing recombination}}{\text{Number of offspring (Number of Test Cross progeny)}} \times 100$$

frequency (Cross-over value = COV) reflect the relative positions of the genes on chromosome. He emphasised that distantly located genes show more crossing over (and thus greater cross-over frequency) than that of the closely located genes.

6.4 Chromosome Mapping or Gene Mapping

Most of the concept of chromosome mapping has come from the works of Morgan, Bridges, Muller and Sturtevant.

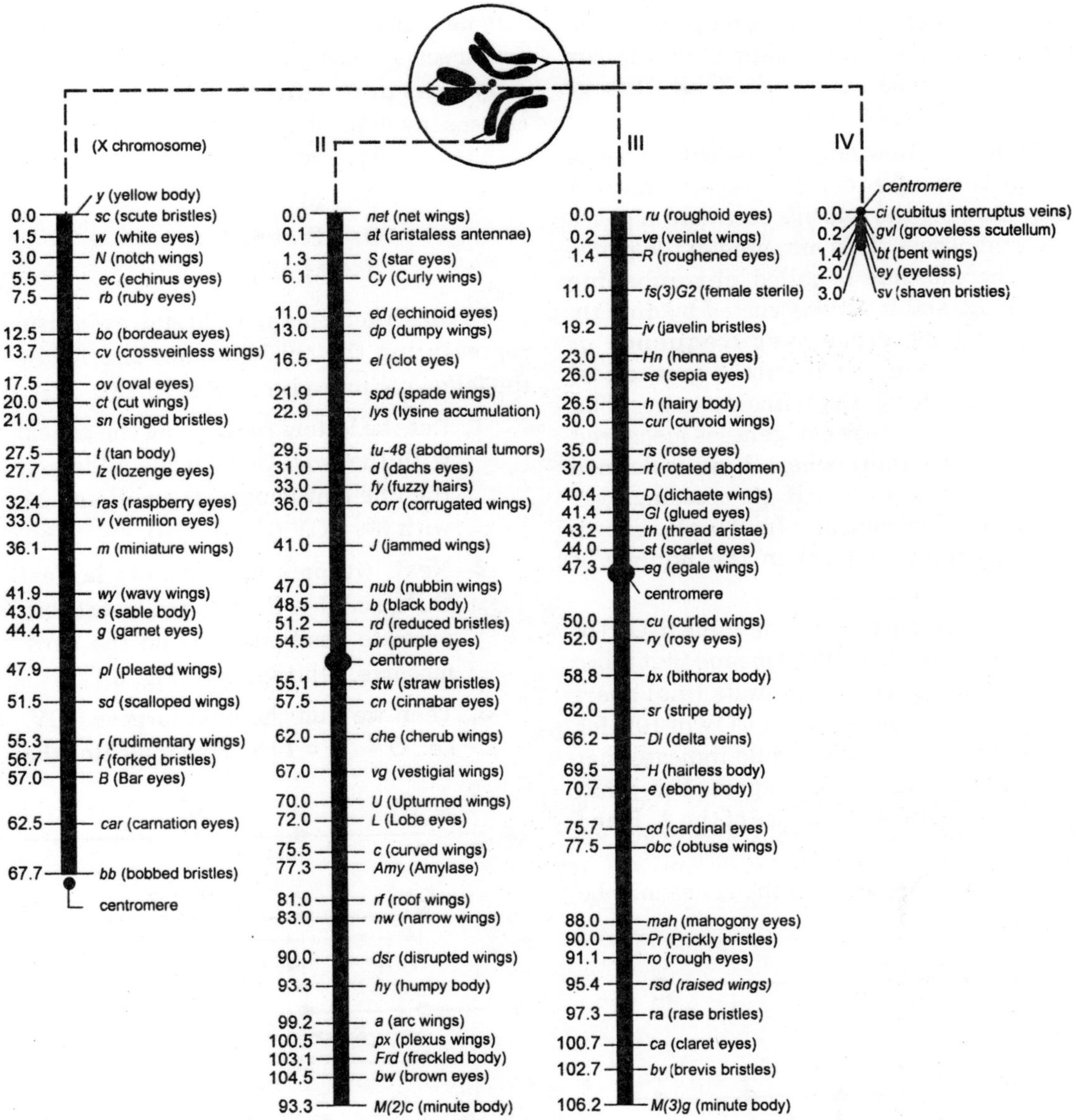

FIGURE 6.10

Abbreviated linkage map of *Drosophila melanogaster.* The map positions are those given by D.L. Lindsley and E.H. Grell, *Genetic Variations of Drosophila melanogaster,* Publication 627, Carnegie Institution of Washington, Washington, D.C., 1968, except where updated by R.C. King, pp. 625-652, in *Handbook of Genetics,* Vol. 3, *Invertebrates of Genetic Interest,* Plenum Press, New York, 1975.

From the knowledge of linkage, it is known that genes occur in a distinct set of **linkage groups** (See 6.2.3), and each homologous chromosomes in the genome. As for example, *Drosophila melanogaster* has 4 pairs of chromosomes and thus, 4 linkage groups (Fig. 6.10). The location of gene-loci on each linkage group is called **linkage map**. All genes in a linkage group shows *linearity*, i.e., they can be mapped in a linear fashion, and has no branch. Thus, *linkage maps are always linear*.

From the knowledge of linearity of genes and cross-over frequencies we can construct **map** showing the *relative positions* of genes on chromosome. This mapping of the genes on a chromosome is called **chromosome mapping**, and it is constructed by directly converting the **cross-over frequency** *or* **value** between genes into hypothetical distances along the chromosome. The genetic distance between genes is measured in **map units** (mu), where 1 mu is defined as the interval in which 1% crossing over takes place. The map unit (mu) is also known as **centimorgan** (cM) in the honour of Morgan.

A cross-over frequency or value (COV) of 5% between genes A and B means that those genes are situated 5 map units (mu) apart on the same chromosome. A COV of 10% for a pair of genes A and C would indicate that they were 10 map units apart. But what would be the linear sequence of the A, B and C? ABC or CAB?

The linear sequence of the genes may be (Fig. 6.11)—

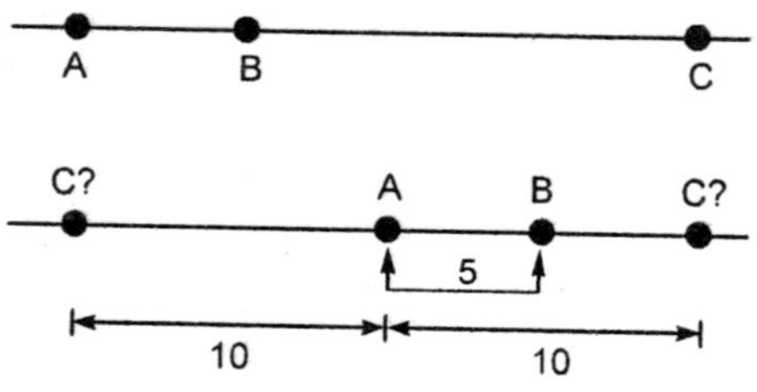

FIGURE 6.11

Possible gene loci of A, B and C on the basis of the data.

In an instance of 3 genes on a chromosome we can construct the map both with *sequence* of the genes and with *distance* between the genes.

6.4.1 Procedure for Constructing a Chromosome Map

Let us consider an imaginary problem to construct a chromosome map. The imaginary cross-over value (COV) in a series of crosses involving 4 genes—M, N, O, P are—

M — N = 24%
O — M = 14%
O — P = 8%
P — M = 6%

To construct a map, we first draw a **line** representing the chromosome, and follow the following procedure—(Fig. 6.12).

1. Put the letters (indicating the genes) with the *least* COV in the middle of the line (chromosome), i.e. P and M, with 6% COV (Fig. 6.12a).
2. Next, we pick up the next largest COV, i.e., O — P = 8%, and insert both possible positions of O on the chromosome, relative to P (Fig. 6.12b).
3. Then we find the next largest COV, i.e., O — M = 14%. This indicates the

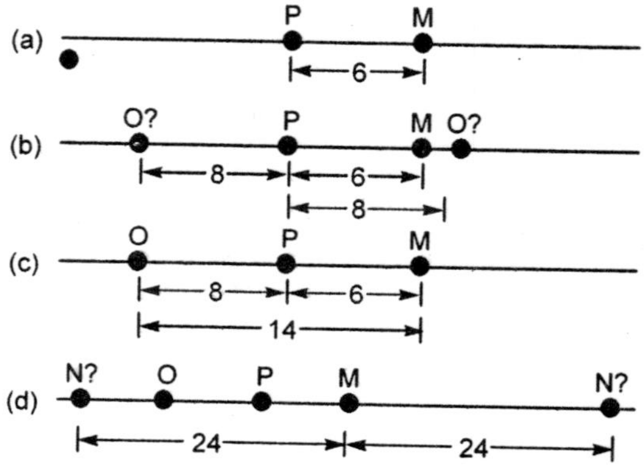

FIGURE 6.12

Method of constructing a chromosome map (follow the text).

right-hand position of O of the (Fig. 6.12b) is *incorrect* (Fig. 6.12c).

4. We repeat the procedure for the COV between M — N = 24% (Fig. 6.12d). The position of N remains uncertain until additional information is received.

6.4.2 Three-factor Cross

A three-factor cross means a cross in which 3 pairs of alleles are segregating. The three-factor crosses allow us *to order the genes (markers) involved* on a chromosome. It is an important tool in **chromosome mapping**.

In diploids, the 3 factor cross is used in detecting linkage. Here, the homozygotes are crossed to produce a *trihybrid* (e.g. ABC/abc), and the trihybrids (triply heterozygotes) are then *test crossed* to triply recessive homozygotes (abc/abc). The types and frequencies of the F_1 gametes can be determined directly from the phenotypes of the test cross-progeny.

In this cross, the trihybrids produce 8 different kinds of gametes. If the 3 genes in question assort independently the 8 types of gametes will occur in equal (1 : 1 : 1 : 1 :

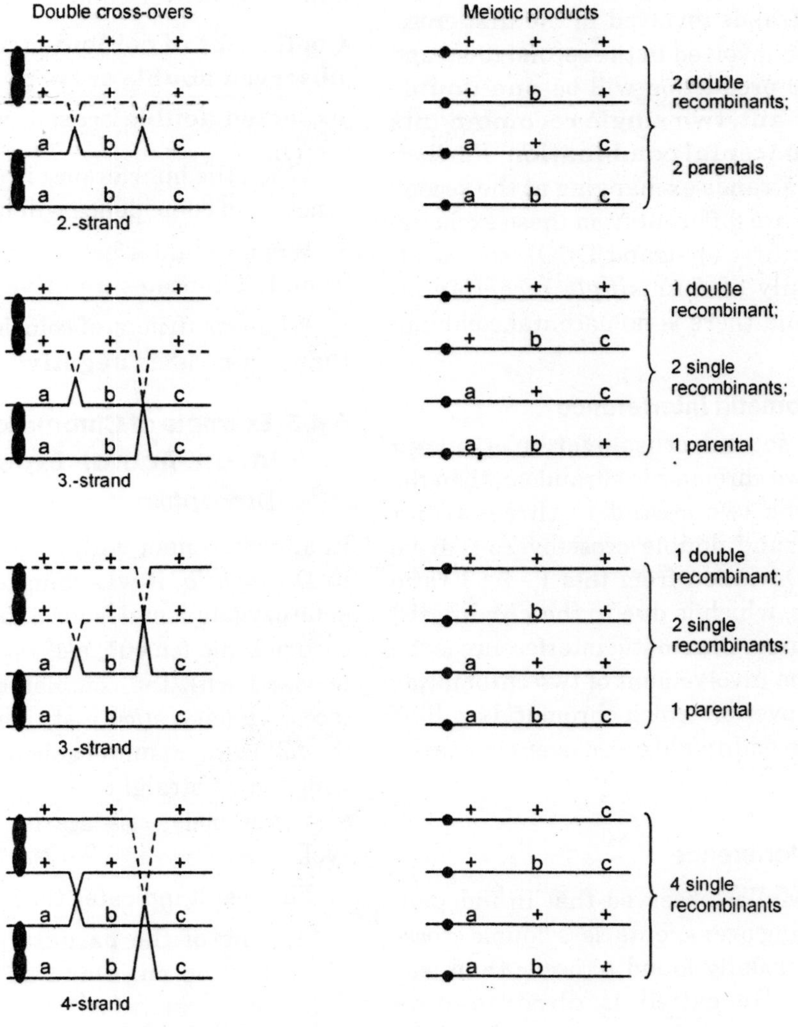

FIGURE 6.13

Double cross-overs and their genetic consequences.

1 : 1 : 1 : 1) ratio. This ratio will deviate if 2 or more of the genes are in linked condition. The exact ratio will be determined by the degree of linkage between them.

In the 3-factor cross, different results are obtained when double cross over (DCO) occur involving two, three or four strands in the tetrad. There are 4 possible types of DCO, which are shown in the (Fig. 6.13).

The figure shows that double cross overs (DCO) between the same two non-sister strands result into **two doubly recombinants** and **two parental combination**. The figure further shows that if only one of the two strands involved in the first cross-over is also involved in the second (3-strand DCO), the production will be **one double recombinant, two single recombinants and one parental combination**. Further, if the two strands exchanging at the second cross over are different from those exchanging at the first (4-strand DCO), the result will be only of *four single recombinant strands*, and there is no parental combination.

- **Chromatid Interference**

If the second cross-over in a tetrad involves two chromatids at random, then the ratio of the two-strand to three-strand to four-strand double cross-overs will be 1 : 2 : 1. Deviation from this 1 : 2 : 1 ratio may occur, which is due to the **chromatid interference**. Chromatid interference is the effect of the involvement of two chromatids in a cross-over on which chromatids will be involved in additional cross-overs in nearby regions.

6.4.2.1 Interference

In 1916, Muller observed that in independent crossing over events, less double cross-overs are usually found to occur than was expected. He called it **chromosome interference or chiasma interference**. It is observed at least in three-factor crosses in eukaryotes. This phenomenon indicates that the occurrence of one crossing over **interferes** with another crossing over nearby. The degree of interference is measured by the **coefficient of coincidence**.

- **Definition**

Interference is the tendency of one crossing over to interfere the occurrence of another C.O. in its adjacent loci.

6.4.2.2 Coefficient of Coincidence

Coefficient of coincidence is the ratio of the observed frequency to the expected one of double cross-over gametes.

$$\textbf{Coefficient of coincidence} = \frac{\textbf{observed double cross-over frequency}}{\textbf{expected double cross-over frequency}}$$

When the interference is **absent**, the coefficient of coincidence will be 1.

When the interference value is between 0 and 1, it indicates **positive interference**.

When coefficient of coincidence is larger than 1, it is called **negative interference**.

6.4.3 Example of Chromosome Mapping in an Acutal Experiment with *Drosophila*

In an experiment with 3-point linkage test in *Drosophila*, it was found that when the heterozygote females for 3 loci on the third chromosome (cu/cu^+, e/e^+, st/st^+) were **test crossed** with the completely homozygous recessive (cu e st/cu e st) males, it showed the following result (Table 6.8) [cu = curled wing, cu^+ = straight wing; e = ebony body, e^+ = grey body; st = scarlet eye, st^+ = red eye].

The result indicates that—

1. One of the parental chromosomes carried cu, e and st^+ whereas the other carried cu^+, e^+ and st. It means the alleles at the **curled** and **ebony** loci were in coupling (cis) condition,

TABLE 6.8

Genotype	Number of progeny	Nature
cu e st^+/cu e st	366	Parental
cu^+ e^+ st/cu e st	380	
cu e st/cu e st	24	Single cross-over in the interval I
cu^+ e^+ st^+/cu e st	30	
cu^+ e st/cu e st	89	Single cross-over in the interval II
cu e^+ st^+/cu e st	105	
cu e^+ st/cu e st	2	Double cross-over one in each interval
cu^+ e st^+/cu e st	4	

Total 1000

and the alleles in the **scarlet** locus were in repulsion (trans) condition.

2. Since the most frequent parental combinations are cu e st^+ and cu^+ e^+ st, and the least frequent double cross-over (DCO) combinations are cu e^+ st and cu^+ e st^+, the cu locus must be present in between the st and e.
3. The result also indicates that a DCO (double cross-over) transhifts the middle gene on a chromosome with the two outside markers from the other parent. On the other hand, the two outside genes on each parental chromosome remain together.
4. In this example, the e and st^+ lie together on one of the parental chromosomes and on one of the DCO chromosomes. Similarly, e^+st lie together on one of the parental chromosomes, and on one of the DCO chromosomes. This indicates these genes are located at outside loci.
5. The alleles of the **curled** locus are present on the DCO chromosomes with the alleles of the **ebony** and **scarlet** loci inherited from the other parent, indicating, the cu must lie in the middle.

Thus, the order of the genes is—

After determination of the order of genes, their relative distances are calculated. For the **st – cu** interval the recombinant genotypes are **st cu** and **st^+ cu^+**, and their distance must be $\left(\frac{24+30+2+4}{1000}\right)\times 100 = 6$ map units. Similarly, for the **cu – e** interval, the recombinant genotypes are **cu e^+** and **cu^+ e**. In this case, the distance must be $\left(\frac{89+105+2+4}{1000}\right)\times 100 = 20$ map units.

Based on this calculation the chromosome map can be drawn as—

6.4.4 Working out Problems on Chromosome Mapping

PROBLEM 1:

In a 3-point test cross $\left(\frac{ABC}{abc}\times\frac{abc}{abc}\right)$ the following results were obtained.

ABC	abc	aBc	AbC	ABc	abC	aBC	Abc	Total
240	250	95	105	140	145	15	10	=1000

Find out the linear order of the genes. Calculate the map distance between the genes and calculate the coefficient coincidence.

Answer:

1. In a 3-point test cross 8 phenotypic classes are expected, and actually 8 types are obtained.
2. The most frequent pairs of classes are the result of non cross-overs and are the parental types. In this problem ABC and abc are the parental classes. That is— 240 + 250 = 490.
3. The order of the genes is calculated by observing the least frequent offspring classes. It represents the result of DCO.

   The aBC and Abc are the least number indicating DCO. So, DCO classes are 15 + 10 = 25, and the frequency of DCO is—

   $\frac{25}{1000} \times 100 = 2.5\%$

   The middle gene pair change their position in relation to the other two pairs due to crossing over. Here this gene pair is A/a. So the order of the genes is BAC or CAB.
4. So, the F_1 cross is $\frac{BAC}{bac} \times \frac{abc}{abc}$
5. The single cross-over recombinants between B and A produces the recombinant Bac and bAC. The number of these recombinants is 95 and 105 respectively. The total is (95 + 105) = 200. Thus the single C.O. frequency between genes B and A is

   $\frac{200}{1000} \times 100 = 20\%.$
6. In this way, the single cross-over frequency between A and C genes produces the recombinants BAc and baC (140 + 145) = 285. And the single cross-over frequency between A and C is $\frac{285}{1000} \times 100 = 28.5\%$
7. The distance between B and A is all single cross-overs + double cross-overs. That is, 20 + 2.5 = 22.5%.

   In the same way, the distance between A and C is single C.O. + double C.O = 28.5 + 2.5 = 31%.
8. Thus the chromosome map is

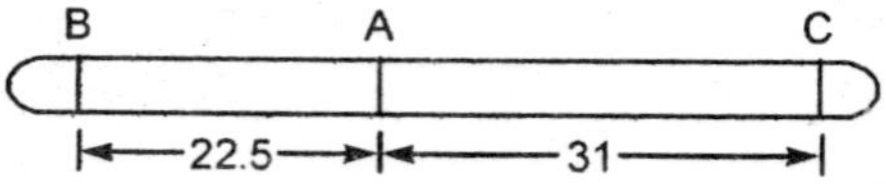

9. The coefficient of coincidence

   $= \frac{\text{Frequency of observed DCO}}{\text{Frequency of expected DCO}}$

   Observed DCO $= \frac{2.5}{100} = 0.025$

   Expected DCO $= \frac{22.5}{100} \times \frac{31}{100}$

   $= 0.225 \times 0.31$

   $= 0.069$

   Thus, the calculated coefficient of coincidence is $\frac{0.025}{0.069} = 0.362 = 36\%$

PROBLEM 2:

In *Drosophila* Lyra (Ly) and stubble (Sb) are dominant mutations at loci 40 and 58 respectively on chromosome 3. A recessive mutation with bright red eyes (br) is also present on that chromosome. By crossing a female heterozygous for all three mutations to a male homozygous for the bright red eye mutation (br), the following results were obtained. Determine the location of br gene on that chromosome.

Phenotype			*Number of flies*
Ly	Sb	br	404
+	+	+	422
Ly	+	+	18
+	Sb	br	16
Ly	+	br	75
+	Sb	+	59
Ly	Sb	+	4
+	+	br	2
			Total 1000

Answer:

1. The most frequent reciprocal pairs of phenotypic classes among the produced 8 types arise from non-cross-over gametes, i.e., the original parental types. Those are Ly Sb br/ +++ and the number is 404 and 422 respectively.
2. The order of genes is determined by observing the least frequent reciprocal pairs of classes representing the double cross-over (DCO). These are Ly Sb + = 4 and + + br = 2.
3. By calculating the frequency of DCO, we get
$$\frac{4+2}{1000}\times 100 = \frac{6}{1000}\times 100 = 0.6\%.$$
4. The genes which are transhifted in relation to the other two pairs is the middle gene pairs (br/+). This indicates the order of genes on the 3rd chromosome as:

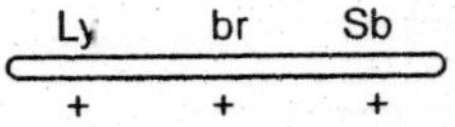

5. Single cross-overs occur in between Ly and br, and br and Sb. The S.C.O between Ly and br produces Ly + + and + br Sb. The single cross-over frequency between them is
$$\frac{18+16}{1000}\times 100 = \frac{34}{1000}\times 100 = 3.4\%.$$
On the other hand, The SCO frequency between br and Sb is
$$\frac{75+59}{1000}\times 100 = \frac{134}{1000}\times 100 = 13.4\%.$$
6. The distance between Ly and br is determined by adding DCO and SCO. That is—
Ly — br = 3.4 + 0.6 = 4 or, *4 map units*. And, the distance between br and Sb is also determined by adding DCO and SCO between them, which is 13.4 + 0.6 = 14 or *14 map units.*
7. Thus, the location of the **br** gene on the 3rd chromosome is:

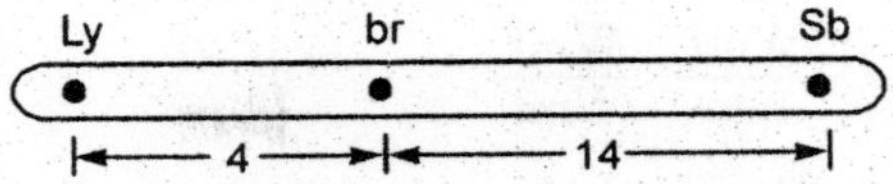

8. From the given data, we know the location of Ly is 40 and that of Sb is 58. So, the location of br (bright red) will be 40 + 4 = 44 or 58 – 14 = 44. That is—

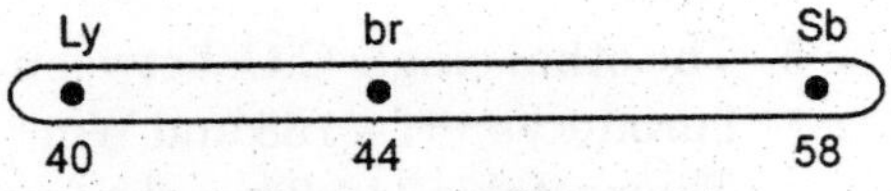

It means, the locus of br is 44.

PROBLEM 3:

The following numbers of *Drosophila* were obtained from a test cross: Construct a linkage map of the genes.

Phenotypes	*Number*	*Phenotypes*	*Number*
+ m +	218	+ m f	95
W + f	236	W + +	101
+ + f	168	+ + +	3
W m +	178	W m f	1

Total no. of progeny obtained 1000.

Answer:

1. The highest number of progeny arises from non-cross-over gametes, and these are the original parental combination. So, the + m + (218) and the W+f (236) are the parental classes.
2. The order of genes is calculated by observing the least member of progeny, which represent the phenomenon of DCO. Those are + + + (3) and W m f (1). So, the frequency of DCO between them is
$$\frac{3+1}{1000}\times 100 = \frac{4}{1000} = 0.4\%.$$
3. To determine the order of the genes, we find out the middle pair of genes which have changed its position in relation to other 2 pairs of alleles due to C.O. This is the +/m. Thus the order

of the genes becomes Wmf or fmW. And the F_1 test cross will be $\frac{Wmf}{+++} \times \frac{Wmf}{Wmf}$.

4. The single cross-over between W and m produces +mf = 95 and W++ = 101. Their total is 196. Thus the frequency of SCO between these two genes is $\frac{196}{1000} \times 100 = 19.6\%$.
5. The other single C.O. between m and f produces ++f = 168 and Wm+ = 178. Their total is 346. Thus the frequency of SCO between m and f is— $\frac{346}{1000} \times 100 = 34.6\%$.
6. The distance between W and m is SCO + DCO = 19.6 + 0.4 = 20% = 20 map unit. Whereas the distance between m and f is SCO + DCO = 34.6 + 0.4 = 35% = 35 map unit.

So, the linkage map of these genes is

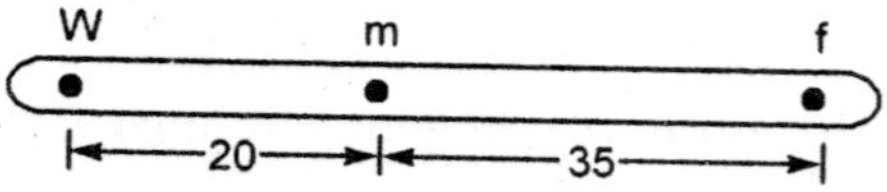

6.5 Cytological Proof of C.O.

So far, the occurrence of crossing over was evidenced only by *genetical recombinations* of the cross-over products. In 1931, however, C. Stern obtained direct **cytological** evidence of crossing over is *Drosophila*, in which he could demonstrate that homologous chromosomes exchange parts during crossing over. H.B. Creighton and B. McClintock also obtained cytological evidence of C.O., working on maize.

6.5.1 Stern's Experiment

Usually, the chromosomes of a homologous pair are morphologically alike. Stern produced a special type of female *Drosophila* in which the two X-chromosomes differed from each other and from the normal chromosome; one of the two X-chromosomes had a part of a Y-chromosome attached to one end. The other X-chromosome was shorter (a broken part) than normal. Thus this female flies were heterozygous for their morphologically distinguishable X-chromosomes. The morphological features of the chromosomes could be detected under microscope. These flies were also heterozygous for *two genes* located on these special type of X-chromosomes. One of the genes was for eye-shape and the other for eye-colour. A partially dominant mutant allele **B** is responsible for the bar (narrow) shape of the eye; its normal (wild) allele **B⁺** produces normal round eye in homozygous condition. Another mutant recessive allele **car** is responsible for the carnation colour of the eye; its dominant (wild) allele **car⁺** produces normal red-coloured eyes. The two X-chromosomes of this special female *Drosophila* were like (Fig. 6.14).

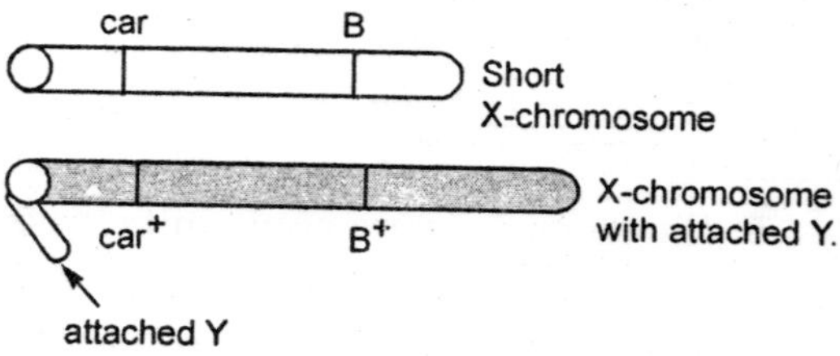

FIGURE 6.14
The two X-chromosomes for Stern's experiment.

The arrangement of the allelic pairs was that the 'short (broken) X-chromosome' carried **car** and **B** gene and the 'X-chromosome with attached Y' carried car^+ and B^+ (Fig. 6.14).

Stern crossed these heterozygous females with males that had carnation coloured round eyes (car B^+/Y males) (Fig. 6.15).

Now, according to theories of crossing over it was expected that—

1. If there occurs no crossing over, the two X-chromosomes will enter to the two gametes unaltered (i.e. different from normal).

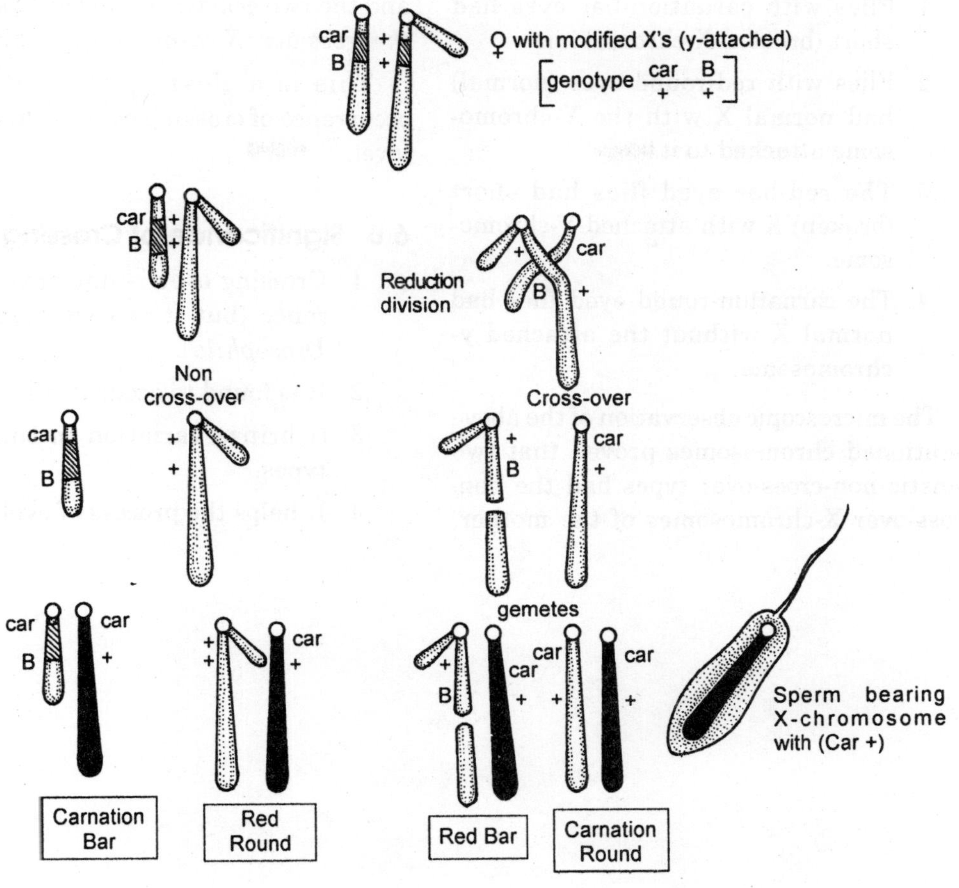

FIGURE 6.15

The cross-over and Non-cross-over products in Stern's experiment exhibiting cytological evidence of C.O.

2. But, if there occurs crossing over between **carnation** and **bar** (car, B), the short (broken) X-chromosome bearing the bar gene (B) will have the Y-chromosome attached to it, the other will lose the Y-chromosome and will have the carnation gene (car).

The Y-chromosome in this experiment is used as a marker so that cross-overs can be distinguished under microscope.

Thus, there would be 4 types of eggs: (Fig. 6.16). If these eggs are fertilized by a sperm with an X-chromosome bearing *car and* B^+, only female offspring will be produced. Stern examined the chromosomes of these flies under microscope which showed the following features (Fig. 6.15).

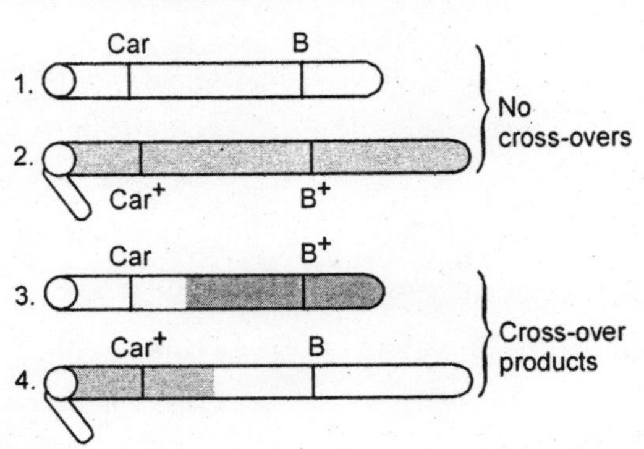

FIGURE 6.16

4 types of eggs of the modified female having two special X-chromosome.

1. Flies with carnation-bar eyes had short (broken) X-chromosome.
2. Flies with red-round eyes (normal) had normal X with the Y-chromosome attached to it.
3. The red-bar eyed flies had short (broken) X with attached Y-chromosome.
4. The carnation-round eyed flies had normal X without the attached y-chromosome.

The microscopic observation of the above mentioned chromosomes proved that two genetic non-cross-over types had the non-cross-over X-chromosomes of the mother, and the two *genetic cross-over types carried the cross-over X-chromosomes* of the mother.

This is a clear cytological proof of occurrence of crossing over at chromosomal level.

6.6 Significance of Crossing Over

1. Crossing over is universal in occurrence (but it is very rare in male *Drosophila*).
2. It is found to occur in all organisms.
3. It brings variation in the gamete-types.
4. It helps the process of evolution.

CHAPTER 7

SEX DETERMINATION AND SEX-LINKAGE

Sex-chromosome, Human chromosome, XX and XY, Y-chromosome, Barr body, Lyon hypothesis and Dosage compensation, Types of Sex-determination, XO and XY type of Sex determination, Sex determination in heterogametic females, ZO-ZZ type of sex determination, ZW-ZZ type of sex determination, Genic Balance or Balance Concept in Sexdetermination, Non-disjunction in *Drosophila*, Gynandromorphs and Mosaics, Sex determination in Drosophila (Molecular approach), Sex determination in humans, Sex-linkage (Sex-linked traits), Criss-cross transmission, Non disjunction, Sex-linked genes and inheritance in humans.

7.0 Sex-Chromosome

To understand the nature of sex-chromo some in a sexually reproducing organism, we first refer to the sub-section 3.1.2.2 of chapter 3. Further, it was noticed long ago that in sexually dimorphic organisms, the males and females not only differ in their morphology and behaviour from each other, but also in their chromosome-constitution.

In 1891, Henking first noticed that in certain bugs, the meiotic nuclei contained **11 pairs of chromosomes and an unpaired 'element'**. This element had no homologue. He called it **the X-body**. He deduced that during spermatogenesis, two kinds of sperms were produced in equal number, one with this 'X-body' and one without it. But he could not explain the nature of the 'X-body'. McClung in 1902, suggested that this 'X-body' was concerned with sex-determination. He called it **accessory chromosome**. He also noted that the **female** grasshopper contained 24 chromosomes, while the **male** contained only 23. He suggested a hypothesis that a particular chromosome is the **sex-determiner**. In 1905, E.Wilson noted that in *Protenor* (a bug, Hemiptera), the females have 14 chromosomes (7 pairs) and males have 13 chromosomes (6 pairs and 1 unpaired). Wilson referred to this unpaired chromosome as **X-chromosome**. He analysed that in many organisms the female had **two X-chromosomes**, and the male had only **one X-chromosome** (thus, the chromosome number became odd). In the same year, Nettie Stevens found that in beetle, both the males and females have the same number of chromosomes but in the males one pair is not identical in shape and size. Stevens showed that in the beetle, the males possessed a homologue to the X, which is dissimilar in shape and size, i.e., smaller than the X. Wilson called it **Y-chromosome.** Thus when a female beetle produces gametes, each gamete (egg) contains **half of the autosomes and an X-chromosome (AX)**, whereas when a male produces gametes (sperms), half of them contain **half of the autosomes and an X-chromosome (AX)**, and the other half **either contain only half of the autosomes** (AO in case of odd number of chromosomes) or contain **half of the autosomes and a Y-chromosome** (AY, in case of even number of chromosome) (Fig. 7.1).

7.1 Human Chromosome

Humans are diploid eukaryotic organisms. Diploid eukaryotic cells have two haploid sets of chromosome, one set from the father and the other set from the mother. The complete set of chromosomes in a cell is called its **Karyotype** ("Carry-o-type", meaning 'nucleus-type'), which is species-specific. Staining with particular dyes (Giemsa dye) show banding patterns in chromosome (called G-bands). The bands help to define chromosome regions and subregions.

Female (♀) — XX —Oogenesis→ All X-eggs (AX)

Male (♂) — XY —Spermatogenesis→ Half X-sperms and half Y-sperms. (AX and AY)

Male (♂) — XO* —Spermatogenesis→ Half X-sperms and half without sex-chromosome (AX and AO)

FIGURE 7.1

Formation of gametes in XX, XY and XO individuals.

It was explained that when an X-sperm fertilizes an X-egg, XX females are produced; when a Y-sperm fertilizes an X-egg, XY males are produced; when sperm without sex-chromosome fertilizes an X-egg, XO males are produced.

Usually, in sexually dimorphic animals, the females have **sets of autosomes** (which determine the somatic characters) **and a pair of identical sex-chromosomes**, i.e., XX (responsible for sex-determination); on the other hand, the males have **sets of autosomes as well as a pair of non-identical sex-chromosomes**, i.e., XY. Thus, both X and Y are regarded as sex-chromosomes. The females are called *homogametic* (as they produce all similar type of gametes) and the males are called *heterogametic* (since, they produce two called *heterogametic* (since, they produce two dissimilar type of gametes, X-containing and Y-containing). Sometimes, a male may lack a Y-chromosome showing only XO constitution. Sex-chromosomes are also called **heterosome or allosome**.

In humans, there are 46 (or 23 pairs) chromosomes (established by Tijo and Levan in 1956), of which 22 pairs are autosomes (A) and one pair is sex-chromosomes (XX or XY) (Fig. 7.2). Chromosomes are usually seen at mitotic metaphase stage in somatic cells, where each chromosome has duplicated to form two chromatids that remain attached to the common centromere until anaphase. Based on size and centromere-position, the 46-chromosomes have been arranged into 7 groups (Table 7.1).

The chromosome group 1-3 is designated as group I or A, which are large in size and are **metacentric or submetacentric**.

The chromosome group 4-5 is called group II or B, which are also large but **submetacentric** in appearance.

The chromosome group 6-12 and X are designated as group III or C, which are medium in size and **submetacentric**.

* O = no chromosome; XO = only one X-chromosome is present.

TABLE 7.1: Grouping of Human Chromosomes

Chromosome Group	Group Name	Size	Centromere Location
1-3	A	Large	Approximately median
4-5	B	Large	Submedian
6-12 and X	C	medium	submedian
13-15	D	medium	Acrocentric
16-18	E	short-medium	Median or submedian
19-20	F	short	Approximately median
21-22 and Y	G	very short	Acrocentric

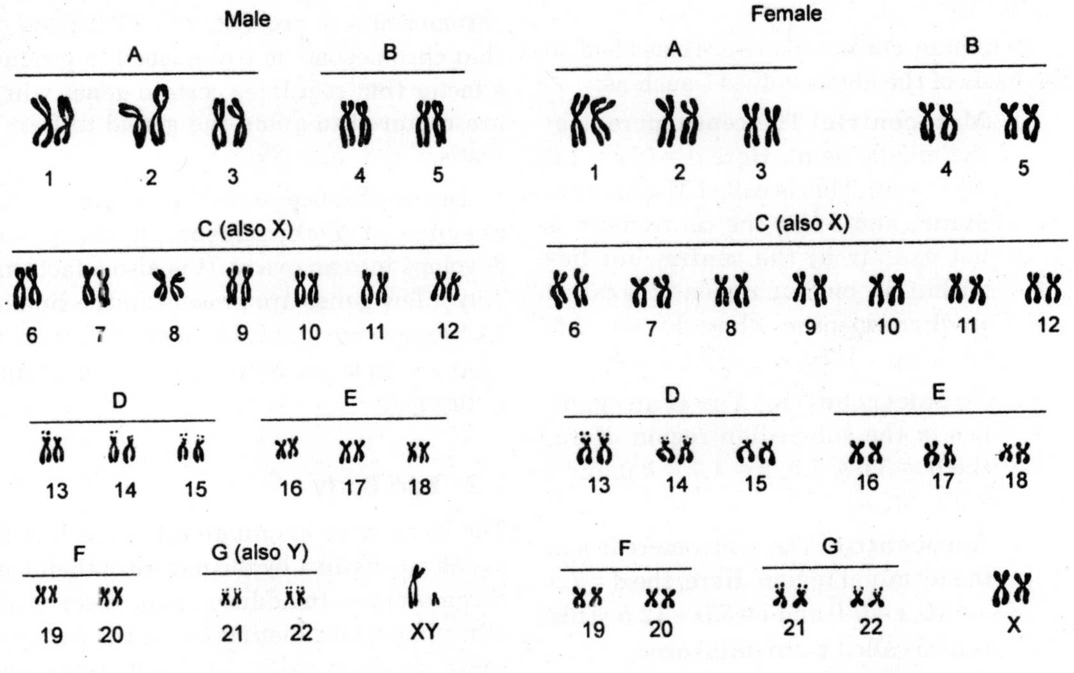

FIGURE 7.2

Metaphase chromosomes of mitosis in a human male and female arranged according to homologous pairs.

The chromosome group 13-15 is called group IV or D, which are medium in size and **acrocentric**.

The chromosome group 16-18 is called group V or E which are shorter than IV, and **metacentric** or **submetacentric**.

The chromosome group 19-20 is designated as group VI or F, which are short in size and approximately metacentric.

The chromosome group 21, 22 and Y are designated as VII or G, which are very short in size and **acrocentric** in appearance.

7.1.1 Classification and Nomenclature of Human Chromosome

In acrocentric and submetacentric chromosomes the short arm is called p and the long arm is called q. Three morphological parameters, d, r and i are used to delimit regions of the chromosomes in which centromeres are considered. The 'd' is the **d-value** denoting the difference in length between the long arm and the short arm, i.e., d = q − p. The 'r' is the **r-value** represented by the ratio of the lengths of the two arms. It means r = q/p. The 'i' is the centromeric index representing the distance from the centromere to the tip of the short arm (p), expressed in percentage of the total length of the chromosome. That is, $i = \frac{P}{(p+q)} \times 100$.

In human, chromosomes are classified on the basis of the above values—such as:

1. **Metacentric**: The centromere is at the middle point. Here d = 0, r = 1.0 and i = 50. This is called **M-chromosome**, and when the centromere is not exactly at the centre, but lies within the median region, it is called **m-chromosome**, where d => 0 – 2.5, r => 1.0 – 1.7 and i =37.5 – < 50.
2. **Submetacentric**: The centromere lies in the submedian region. Here, the d = 5.0 – 7.5, r = 1.7 – 3.0 and i = 25 – 37.5.
3. **Acrocentric**: The centromere lies in the terminal region. Here, the d = 7.5 – < 10, r = 7.0 and i = 7.0 – 12.5. This is also called **t-chromosome**.

7.1.2 XX and XY Chromosome

The mammalian females have two identical (homomorphic) X-chromosomes (XX) in their somatic cells. They are homogametic and produce only one kind of eggs (ova), each with 22 autosomes and a single X (22AX). The mammalian males, on the other hand, possess one X chromosome and one Y-chromosome (XY) in their somatic cells. The sex-chromosomes of males are dissimilar (heteromorphic). The males are heterogametic, i.e., they produce two types sperms; half the sperms possess 22 autosomes and one X-chromosomes (22AX) and the other half possess 22 autosomes and one Y-chromosome (22AY).

When a (22AX)-sperm fertilizes an ovum (22AX), a female (44AXX) is produced and when a (22AY)-sperm fertilizes an ovum (22AX), a male (44AXY) is produced.

7.1.3 The Y-Chromosome

In humans, like all mammals, sex of an individual is determined by the presence or absence of the Y-chromosome. If the Y-chromosome is present, the SRY gene on that chromosome is transcribed to produce a factor that regulates certain genes which are required to guide the gonad to form a testis.

In the absence of SRY gene (i.e., in the absence of Y-chromosome), the gonad develops into an ovary. It is also a fact that only a few genes are present in the human Y-chromosome and it is genetically inert in comparison to the X-chromosome and other autosomes.

7.2 Barr Body

The human (or mammalian) males and females are distinct by the nature of their sex-chromosomes. In addition to it, there is another important distinction in the nuclei of their somatic cells. In 1949, Barr and Bertram discovered that there was a **dark-staining spot** in many female-cell-nuclei, which was absent in the male-cell-nuclei. This dark-spot is a small chromatin body and is called **Barr body** or **sex-chromatin** or **chromatin positive body** (Fig. 7.3).

It was observed that the maximum number of Barr body in the nucleus was always **one less than the number of X-chromosomes**. That is, in normal female (XX), there is one Barr body and in the normal male (XY) there is no Barr body. In the same way, in abnormal females and males (say tetrasomic metafemales XXXX or tetrasomic Klinefelter males XXXY) the number of Barr bodies are 3 and 2 respectively. (See Chapter 8, Chromosomal mutation and chromosomal aberration).

7.2.1 Lyon Hypothesis and Dosage Compensation

The reason behind the 'One less number of Barr-bodies than the number of X-chromosome' was explained by M.F. Lyon and by Russell. Their proposed hypothesis is called **Lyon-hypothesis** which advocates that Barr-bodies are formed from the *inactivation* or *heterochromatinization* of all but one of the X-chromosomes. Lyon explained that there is a **different dosage of genes** in the X-chromosomes in males and females. In mammals, a **dosage compensation** mechanism equates expression of X-linked genes in males and females by **inactivating** one of the two X-chromosomes in a female at early development. Only one X-chromosome remains active, as in the case of males.

This phenomenon of inactivation or heterochromatinization of one of the two X-chromosomes in female mammals, leaving the other to remain metabolically functional, is known as **Lyon hypothesis**, which is the result of **dosage compensation.** Thus, in abnormal females (with more than 2X chromosomes) any additional X-chromosomes are condensed to genetically inactive heterochromatin (or Barr body). In the process of heterochromatinization, which X-chromosome will be 'inactive' is a matter of chance, but once an X-chromosome has become inactivated all cells arising from that cell will maintain the inactivation of the same chromosome.

- **Definition**

H.J. Muller (1932) defined dosage compensation as "the equality of amounts of genes in hemizygous and homozygous

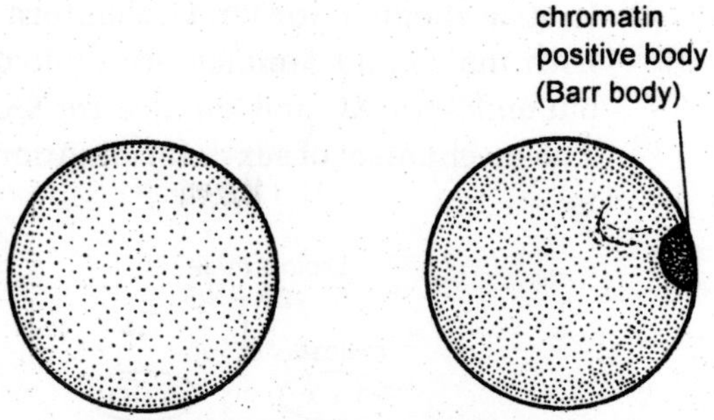

Normal male nucleus with one X-chromosome

Normal female nucleus with two X-chromosome one being Barr body

FIGURE 7.3

Interphase nuclei showing appearance of chromatin-negative (left) in normal males and chromatin positive (right) in normal females.

individuals", which is well applicable in the Lyon hypothesis.

7.3 Types of Sex Determination

There are two types of sex-chromosomes in dioecious organisms; on this basis, the dioecious organisms are of two types: **heterogametic** and **homogametic**.

The heterogametic males (producing two types of sperms) are again of two types: XO* and XY. In some species, the females may be heterogametic (ZO or ZW) and males may be homogametic (ZZ).

7.3.1 XO – XX Type of Sex Determination

A. In honey bee (Order Hymenoptera) the males are haploid and XO, and the females are diploid and XX. The mechanism of sex determination in them is called **haplo-diploidy** and is illustrated in the following diagram:

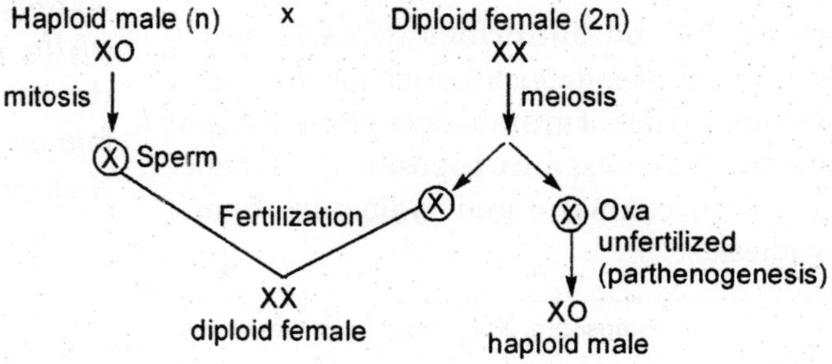

B. In grasshopper (order Orthoptera) both males and females are diploid but males are XO and females are XX. The mechanism of sex determination is—

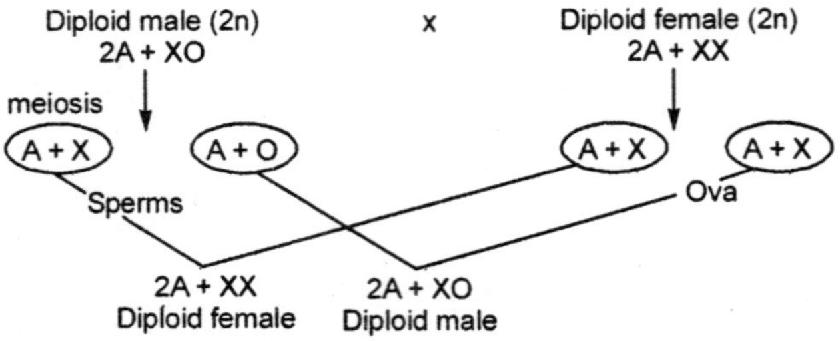

7.3.2 XY – XX Type of Sex Determination

In *Drosophila*, humans and other mammals both males and females are diploid and the males are XY and females are XX. Their sex determination mechanism is—

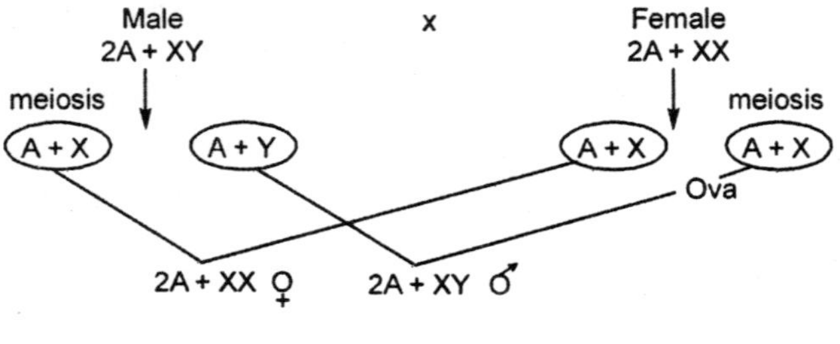

7.3.3 Sex Determination in Heterogametic Females

In moths and butterflies the females are **heterogametic** (ZO) with odd number of chromosome, and in certain fish, reptiles and birds, females are *heterogametic* with even number of chromosome (ZW).

7.3.3.1 ZO – ZZ Type of Sex Determination

In moths and butterflies (♂ 2AZZ and ♀ 2AZO), the sex-determination mechanism is similar to that of grasshopper (7.3.1.{b}), but for, here, the sexes are opposite, i.e., females are heterogametic and males are homogametic.

7.3.3.2 ZW – ZZ Type of Sex Determination

In fish, reptiles and birds the sex-determination mechanism follows that of *Drosophila* (7.3.2), but for, here also the sexes are opposite, i.e., the females are heterogametic (2A + ZW) and the males are homogametic (2A + ZZ).

7.4 Genic Balance or Balance Concept in Sex Determination in *Drosophila*

After the discovery of sex-chromosome it was noticed that the sex-determination in *Drosophila* was not very simple. In the period of classical genetics (before the introduction of molecular genetics), C.B. Bridges (1922) showed that the **Y-chromosome of *Drosophila* plays no significant role in sex-determination** of that insect. Rather, the female-sex-determiners are present on the X-chromosome and the male-sex-determiners on the autosomes. He proposed a **genic**

* O = no chromosome; XO = only X chromosome present.

balance theory to explain sex-determination mechanism in *Drosophila melanogaster*.

- **Bridges' Experiment**

Bridges produced various combinations of X-chromosomes and autosomes in *Drosophila* and deduced that **one X-chromosome and 2 sets of autosomes** (IX, 2A) produced a **normal diploid male.** Their X/A ratio is 0.5. On the otherhand, the combination of 2X chromosomes and 2 sets of autosomes (2X, 2A) produced normal diploid female. There X/A ratio is 1. Thus, Bridges in 1925, hypothesized that **a balance of number of X-chromosome to the number of autosomes** (i.e., X/A ratio) is important in sex determination in *Drosophila* (Table 7.2).

Thus, the **genic balance theory** established that whenever the X/A ratio is 0.5, it is a male; whenever the ratio is 1, it is a female, whenever the ratio is more than 1, it is a metafemale, when the ratio is less than 0.5, it is a metamale, and finally; whenever the ratio is between 0.5 and 1, it is an intersex.

- **Non-disjunction in *Drosophila***

Non-disjunction is the failure of paired X-chromosomes to separate in anaphase in meiosis. Consequently in *Drosophila,* two types of eggs can be produced in females; one with three autosomes and two X-chromosomes (3A 2X), and the other with only three autosomes with no X-chromosome. The nonseparation of X-chromosomes is called **non-disjunction** of X-chromosome (Fig. 7.4).

When these 'abnormal' eggs are fertilized by normal sperms (AX and AY), four types of offspring developed. Those were—(Fig. 7.5)

1. 2A 3X, metafemale, usually died.
2. 2A XXY, female
3. 2AX, sterile male
4. 2AY, all died.

The 2A XXY females and 2AX males were exceptional flies in Bridges' experiment. The ratio of X-chromosomes to sets of autosomes, and their predicted sexes in the genic balance theory was confirmed in the appearance of metafemales (3X/2A = 1.5). These flies did not survive. The XXY flies (2X/2A = 1) were normal females (although a Y was present !). XO flies (1X/2A = 0.5) were males, but sterile (as it lacks a Y). The flies having autosomes, Y-chromosome, but no X-chromosome (AAY) did not survive at all. This experiment and results indicated that in *Drosophila,* the **Y-chromosome has**

Table 7.2: Genic balance / (X : A) and sex in *Drosophila*

X-chromosome (X) and sets of autosomes (A)	Ratio of X and A (X/A)	Sex
1 X 2A	0.5	Male (diploid)
2 X 2A	1.0	Female (diploid)
3 X 2A	1.5	Super female (metafemale)
4 X 3A	1.33	Super female (metafemale)
4 X 4A	1.0	Tetraploid female
3 X 3A	1.0	Triploid female
3 X 4A	0.75	Intersex
2 X 3A	0.67	Intersex
2 X 4A	0.5	Tetraploid male
1 X 3A	0.33	Super male (metamale)

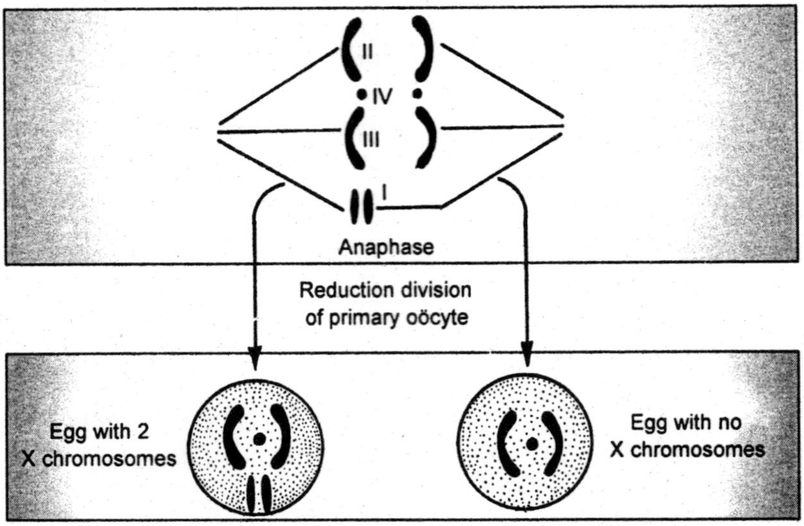

FIGURE 7.4

Nondisjunction, and production of two types of eggs.

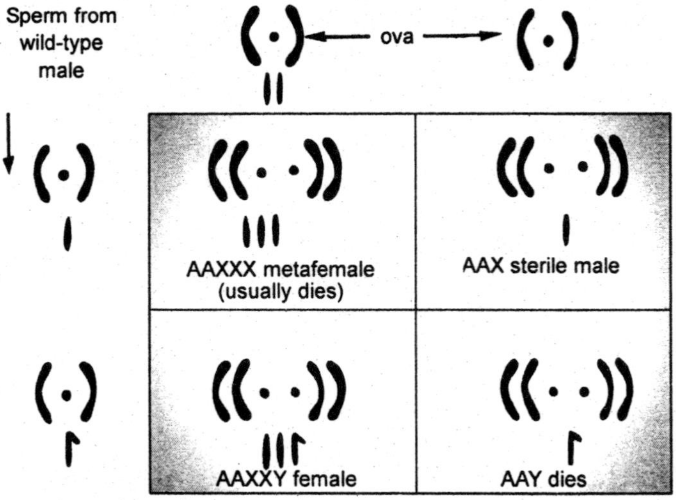

FIGURE 7.5

Ova due to Nondisjunction in *Drosophila,* and the zygotes that result from fertilization by wild-type males (AAXY). The AAXXY females and AAX males were exceptional flies in Bridges experiment (After Bridges, 1925).

no role in sex-determination. But the Y-chromosome is necessary for fertility in male (hence, XO is a sterile male).

With similar experiments, flies with chromosome combination 3X4A ($^3/_4 = 0.75$) and 2X3A ($^2/_3 = 0.67$) were obtained. These flies were called "**intersexes**", as they exhibited intermediate characteristics between males and females. Other Karyotypes of flies were 2X4A and 1X3A. their X:A ratio were 0.5 and 0.3 respectively. The 2X4A were males and 1X3A were metamales justifying the genic balance concept of Bridges.

7.4.1 Gynandromorphs and Mosaics

Occasionally a chromosomal change occurring immediately after the formation of a zygote may lead to a condition called **mosaics**. That is, two parts of tissue of the zygote may contain different chromosomal constitution and grow side by side. The individuals with such different chromosomal make up in different tissues are called **mosaics or chimeras**. It may develop either spontaneously or by induction.

A striking instance of mosaic may be observed in *Drosophila* which develops in the following way: A female embryo (XX) may lose one of its X-chromosomes during cell division in a very early stage of its development, and thus, its one half of the body may bear XX, and the other half XO. As a result half of the body shows female sexual phenotype and the other half shows male sexual phenotype. Such sexual mosaics are called **gynandromorphs** (gynae = female, andros = male) or **gynanders** (Fig. 7.6).

The two sex-phenotypes are exhibited in bilateral symmetry; it is called bilateral gynandromorph.

Mosaics may occur in various other organisms. It can be produced artificially too, in laboratories.

Gynandromorphs may also result from "double fertilization" by two sperms of a binucleate egg containing one egg nucleus and one polar body.

Gynandromorphs and the **intersexes** are not the same. Gynandromorphs differ from the intersexes, since the gynandromorphs are obviously mosaic having different chromosome constitution in different tissues, but the intersexes have the same chromosome constitution in all cells.

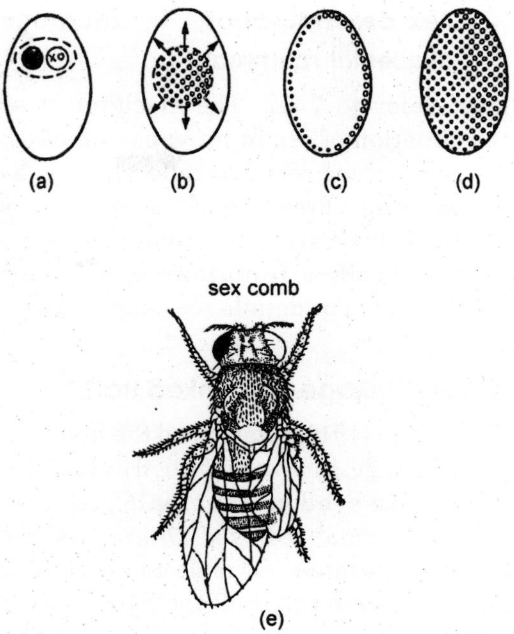

FIGURE 7.6

Formation of a gynandromorph in *Drosophila melanogaster* mosaic for cells carrying two X chromosomes (left side) and one X chromosome (right side). (a) The loss of one X chromosome from one of the two diploid nuclei formed soon after the fertilization of an X-bearing egg by an X-bearing sperm. In (b-d) the XX "female" nuclei are shown as small black circles. The open circles represent "male" XO nuclei in which the wild-type X-chromosome has been lost and which therefore bear only the mutant X chromosome. (e) Adult fly showing the XX heterozygous wild-type phenotype on the left, and the XO hemizygous mutant phenotype (mutant eye, miniature wing) on the right. Note that the male sex comb is only present on the right foreleg (After Strickberger).

7.4.2 Sex Determination in *Drosophila*: Molecular Approach

In the *genic balance concept* it was established that in *Drosophila melanogaster* the sex is determined by the X-chromosome and autosome ratio (X : A). In molecular level (genic level), this has been explained in full details. The principle is that the ratio of chromosomes results **in different amounts of proteins encoded by numerator genes on the X-chromosomes versus a denomenator gene on an autosome**. In female *Drosophila* embryos, there is an excess of numerator proteins that form transcription-factors which activate a master regulatory gene for sex determination, **Sxl**. In male embryos, numerator proteins are bound to denominator proteins, so **Sxl** does not work. This regulatory events in transcription employ a **cascade of regulated alternative RNA-splicing events** that finally leads to differentiation of sex (For details, see chapter 10).

7.4.3 Sex-determination in Humans (Placental mammals)

Please refer to 7.3.2. The mechanism of determination of sex in humans and other mammals is entirely controlled by the **Y-chromosome**. Presence of the Y specifies **maleness** by testis development and its absence specifies **femaleness** by ovary development (For details see chapter 10).

7.5 Sex-Linkage (Sex-linked traits)

T.H. Morgan (1910) established the first case of sex-linkage in *Drosophila*. He found a single **white-eyed mutant male** in the culture of a normal red-eyed *Drosophila*. He mated that white-eyed male to a red-eyed female and found that all the F_1-flies were red-eyed. The males and females from the F_1 were mated to one another to obtain the F_2 offspring. The result of F_2 was unusual; all the females were red-eyed while half of the males were white-eyed and half were red-eyed (Fig. 7.7).

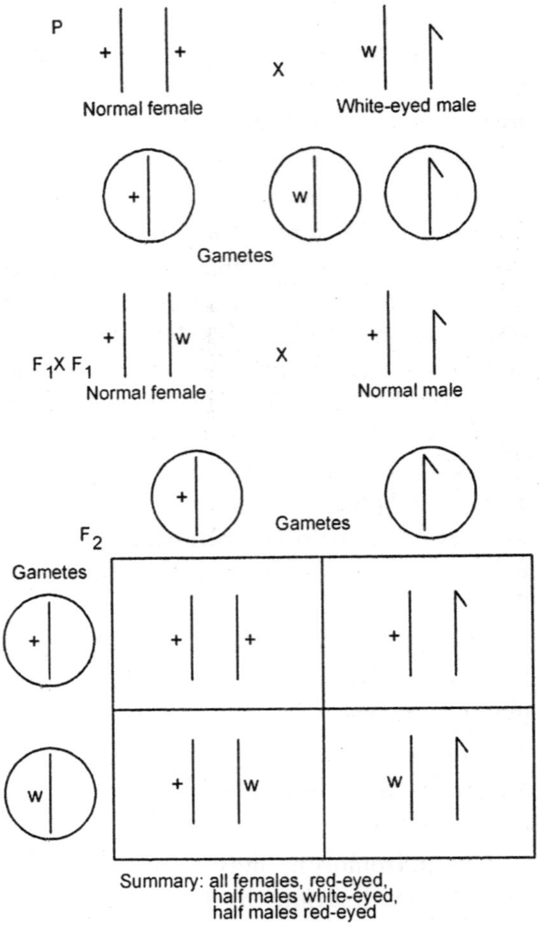

FIGURE 7.7

Diagram of a cross between a red-eyed female *Drosophila* and a white-eyed male, illustrating sex linkage.

Morgan interpreted the results by postulating that **the gene for white eyes was located on the X-chromosome**, and, although it was a recessive gene, it was capable of expression in the male. This expression of the white-eyed gene in the male was a case of **pseudodominance**. The fly was **hemizygous for the gene** because it had only one copy of the gene for the said characteristic (only in the X, not in the Y). The gene is exclusively X-linked.

7.5.1 Criss-Cross Transmission

The inheritance of X-linked characteristic is interesting. It is never transmitted from male-to-male, i.e., from father to son. The reason behind this is that the father transmits its X only to his daughters, never to his sons. The X-chromosome follows a "**Criss cross**" **transmission**, i.e., for example, **beginning with a P male, passing to the F_1 daughter, and thence to the F_2 sons** (From father to daughter to grandson).

A male affected by an X-linked trait will have his sons **unaffected** (as there is no X-chromosome from the father) and his daughters **carriers** (bearing the recessive gene in one of its two X-chromosomes, coming from the father). The 'carrier' daughters have two X-chromosomes, one received from the affected father (carrying the gene in question) and the other from the normal mother.

If the daughters also receive the X-chromosome from a 'carrier' mother (i.e., one of their X bears the gene in question), the daughters will be homozygous for the allele and will exhibit the characteristic (Fig. 7.8).

Thus, wherever the X-chromosome (sex-chromosome) is transmitted, it carries the X-linked gene and the offspring become either carrier or affected. This type of transmission of X-linked gene is called "sex-linkage". It is also true in case of the Y-linked gene, which will be discussed later.

underwent two types of fertilization: When the 'egg without an X' was fertilized by an 'X-sperm' from a red-eyed male, the red-eyed male offspring were produced. These males were sterile due to absence of a Y. On the other hand, when the 'eggs with XX chromosome' were fertilized by 'Y-sperms' from a red-eyed male some exceptional females were produced which were white-eyed, and carried 2X and one Y (Fig. 7.9).

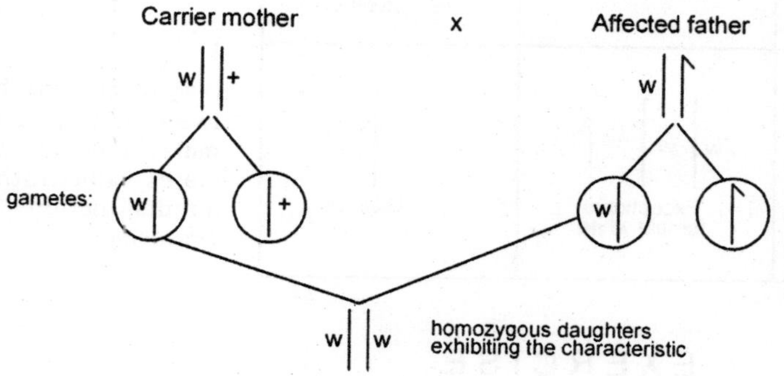

FIGURE 7.8

Transmission of X-linked character.

7.5.2 Non-disjunction

C.B. Bridges in 1916 reported a variation in sex-linkage in *Drosophila*. He mated a white-eyed female to a red-eyed male. It was expected that this female would produce such male offspring, which would inherit the mother's X-chromosome (carrying the white-eye gene) and would show white eye.

But in actual experiment, some red-eyed males were produced (Fig. 7.9). To explain such result, Bridges hypothesized that the two X-chromosomes of the mother *did not separate* during gamete formation and produced two special types of eggs— one with two X-chromosomes (XX) and the other with no X-chromosome. (Both the types, however, contained the haploid set of autosomes). These two types of eggs

The production of 'XX-egg' was due to *non separation* of homologous chromosomes in meiosis. This phenomenon of non-separation of homologous chromosomes during meiosis is called **non-disjuncton**.

7.6 Sex-linked Genes and their Inheritance in Humans

The recessive X-linked traits are easily identified than the recessive autosomal traits in human. Two most well-known X-linked traits in human are **haemophilia** and **colour blindness**. Haemophilia, is a blood-clotting disorder (bleeder's disease) and colour-blindness is a vision-disorder.

The inheritance of these disorders strictly follow the 'sex-linkage'. Their inheritance pattern has been discussed in chapter 17, Human Genetics.

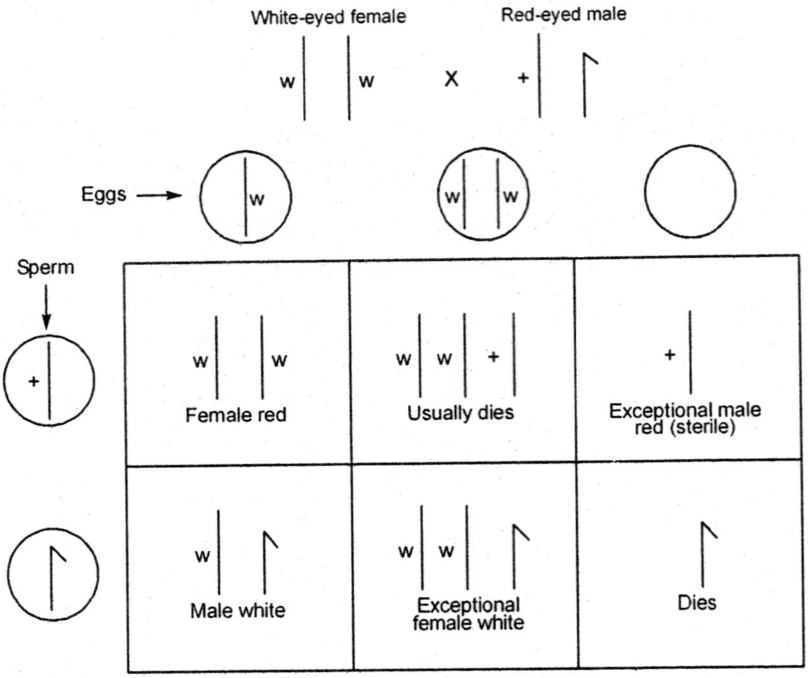

FIGURE 7.9
Diagram of a cross between white-eyed females and red-eyed males, illustrating nondisjunction in *Drosophila*.

EXERCISE

PROBLEM 1:

The wild type body colour in *Drosophila* is grey and is dominant to yellow. When a yellow female fly was mated to a wild type male, all the F_1 male progeny were yellow and all the F_1 female progeny were wild type. The F_1 progeny were allowed to interbreed. Show the two crosses diagrammatically and state the expected ratio of wild type to yellow in the male and female F_2 progeny.

• **Solution**

Yellow female × Wildtype male

$X^c X^c$ $X^C Y$

Gametes ♀   Gametes ♂

↓   ↓

all X^c $\frac{1}{2}X^C$ $\frac{1}{2}Y$

Cross:

	$\frac{1}{2}X^C$	$\frac{1}{2}Y$
all X^c	$\frac{1}{2}X^C X^c$	$\frac{1}{2}X^c Y$
	half wild type F_1 females	half Yellow F_1 males

$X^C X^c$ $X^c Y$

F_1 wild type female × F_1 yellow male

gametes ↓   ↓ gametes

	$\frac{1}{2}X^c$	$\frac{1}{2}Y$
$\frac{1}{2}X^C$	$\frac{1}{4}X^C X^c$	$\frac{1}{4}X^C Y$
$\frac{1}{2}X^c$	$\frac{1}{4}X^c X^c$	$\frac{1}{4}X^c Y$

F_2 female progeny are in the ratio 1 yellow ($X^c X^c$) : 1 wild type ($X^C X^c$). F_2 male progeny are in the ratio 1 yellow ($X^c Y$) : 1 wild type ($X^C Y$).

PROBLEM 2:

The phenotypes and genotypes of certain coat-colours in cats are shown in the table below:

Phenotype	Genotype	
	Male	Female
Black	X^gY	X^gX^g
Ginger	X^GY	X^GX^G
Tortoise-shell	—	X^GX^g

A breeder has a number of black, ginger and tortoise shell cats.

(a) If she wants all the female kittens to be tortoise-shell, what are the genotypes and phenotypes of the parents she should use? List all possible answers. (b) What will be the phenotypes of the male kittens in the crosses you have suggested. (c) Suggest a reason to account for the appearance of an infertile black female in the litter of a cross between a ginger tom and a black female.

• **Solution**

(a) i. X^GX^G (ginger female) × X^gY (black male)

ii. X^gX^g (black female) × X^GY (ginger male)

(b) i. ginger

ii. black

(c) The infertile female is X^gO. The sperm which fused with the ovum did not carry a sex-chromosome.

Problem 3:

A female *Drosophila* with the mutant phenotype 'notched wings' was mated to a wild type male. Two-thirds of the progeny were female. Half of the females had 'notched wings' and the other half had wild type wings and all the male progeny had wild type wings. When the notched progeny were mated to their wild type brothers, the same results were obtained. Explain.

• **Solution**

The male and female phenotypic ratios are different in the progeny suggesting that the gene is sex-linked. Male constitute only one-third instead of one-half of the progeny and they are all wild type. This suggests that 'notched' is lethal in males. Notched females must be heterozygous because their female progeny are either wild type or notched. The cross made was X^NX^n (notched female) × X^nY (wild type male).

		Gametes from the male	
		$\frac{1}{2}X^n$	$\frac{1}{2}Y$
Gametes from female	$\frac{1}{2}X^N$	$\frac{1}{4}X^NX^n$	$\frac{1}{4}X^NY$
	$\frac{1}{2}X^n$	$\frac{1}{4}X^nX^n$	$\frac{1}{4}X^nY$

F_1 progeny are

$\frac{1}{4}X^NX^n$ notched females
$\frac{1}{4}X^nX^n$ wild type females } $\frac{2}{3}$ females

$\frac{1}{4}X^nY$ wild type males
$\frac{1}{4}X^NY$ die } $\frac{1}{3}$ males

Problem 4:

(In birds and moths the female is the heterogametic sex). The wing markings of budgerigars (bird) are normally black. The sex-linked gene **cinnamon** causes brown colouration and is recessive to the allele for black. A breeder has a number of birds, none of which carries the **cinnamon** allele. He purchases a **cinnamon** female. Illustrate by means of diagrams the breeding programme he should carry out in order to obtain, in as few generations as possible, birds which have only cinnamon progeny of both sexes. Bear in mind that the budgies live for several years and can breed every year.

• **Solution**

P_1: X^cY Cinnamon female × X^CX^C Black male

X^CY X^CX^c

F_1: Black female Black male

P_2: X^CX^c Black males × X^cY Cinnamon female

F_2: X^CX^c Black males; X^CY Black females; X^cX^c Cinnamon males; X^cY Cinnamon females

X^cX^c Cinnamon males × X^cY Cinnamon females

F_3: All Cinnamon offspring

PROBLEM 5:

There is a rare black form of the Pale Brindled Benty moth. All the moths of this type are female and all their female offspring are black while their sons are normal. With the help of diagrams explain these observations. In moth, the female is heterogametic.

- **Solution**

The allele for the black, is on the Y-chromosome.

Black female XY^B × Wild-type male XX

	Gametes from the female: $\frac{1}{2}X$	$\frac{1}{2}Y^B$
Gametes from male: All X	$\frac{1}{2}XX$ Wild type males	$\frac{1}{2}XY^B$ Black females

CHAPTER 8

CHROMOSOMAL VARIATION: ITS NATURE AND CONSEQUENCES

(Chromosomal Aberration, Chromosomal Mutation)

Change in the structure of chromosome and change in the number of chromosome, Deletion, Duplication, Inversion, Translocation, Euploidy, Monoploidy, Polyploidy, Aneuploidy.

The constancy of a chromosome as a structural entity lies in its capacity to reproduce itself at each cell division with extraordinary precision. Still, occasionally chromosome can undergo spontaneous structural and numerical change. However, the change in the structure or chromosomal constitution can be induced by X-rays and certain chemicals too.

- **Definition**

The changes in the genome, involving chromosome structure and number are called chromosomal aberrations, chromosomal variations or chromosomal mutations.

- **Types**

Chromosomal aberrations are of two types: **Structural changes in chromosome** and **numerical changes in chromosome**. The structural changes are again of two types: **Changes in the number of genes in a chromosome and changes in the arrangement of genes** in a chromosome. Changes in the number of genes are of two categories: Loss of gene or **deletion** or **deficiency** and gain or addition of gene or **duplication**. Change in the gene arrangement are of two kinds: Rotation of a group of genes in 180° on the same chromosome, called **inversion** and transfer of a gene or a group of genes from one chromosome to a non-homologous chromosome, called **translocation** (Table 8.1).

8.0 Types of Structural Changes

8.1 Deletion or Deficiency

Deletion or deficiency involves the detachment and loss of a block of genes (chromatin) from the remainder of the chromosome. If the detached portion is acentric, it does not survive; and the portion of the chromosome carrying a centromere functions as a genetically deficient chromosome. This is called deletion or deficiency.

Deletion may be of two types—**terminal** and **interstitial or intercalary**. The **terminal deletion** is the loss of a terminal part due to a single break in a chromosome fol-

TABLE 8.1: Types of chromosomal aberration in relation to structure

Structural Change in a Chromosome

- Changes in number of genes
  - Loss of genes (Deletion or Deficiency)
  - Addition of genes (Duplication)
- Changes in arrangement of genes
  - Rotation of a group of genes in 180° (Inversion)
  - Transfer of a set of genes to a nonhomologous chromosome (Translocation)

lowed by a 'healing' of the broken end. The **interstitial** *or* **intercalary deletion**, on the other hand, is the loss of an intermediate part due to two breaks followed by the reunion of the broken parts of the chromosome (Fig. 8.1).

The terminal deletion is less frequent than is the interstitial type.

8.1.1 Consequences of Deletion

1. It results a quantitive loss of genetic material.
2. It produces deleterious effect; it may be lethal in homozygous condition.
3. In many higher plants it acts as gametophytic lethals, causing pollen abortion.
4. In *Drosophila*, homozygous losses of the tip of the X-chromosomes are noticed. Cytologically visible homozygous loss of bands 3C2, 3 and 5 in X-chromosome are also noticed; inspite of these losses, the flies may be viable.
5. In *D. melanogaster*, a loss of more than 80 bands is severely detrimental, if not lethal.
6. In *Drosophila*, the characters like *Blond, Pale Beaded, Carved, Snipped,* and *Plexate* are associated with deletions and all are dominant.
7. The *notch-wing* mutation in *Drosophila* is due to a small deletion. In the presence of a deletion, a recessive allele of the normal homologue will

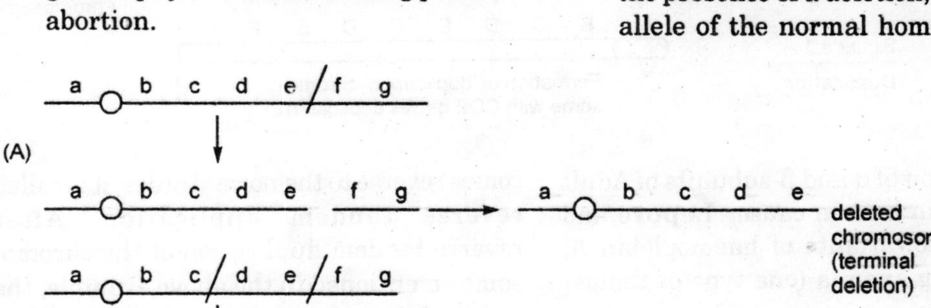

FIGURE 8.1
The mechanism of formation of a terminal (A) and an interstitial (B) deletion.

behave like a dominant allele and the character is phenotypically expressed. This phenomenon is called **pseudodominance**.

8. In humans, the **cri du chat** (cat-cry) syndrome results from the loss of the short arm of chromosome-5; ($5P^-$). See chapter 17.
9. In humans, loss of the short arm of chromosome 4 leads to abnormalities somewhat similar to cri du chat, but without the characteristic cry.
10. Loss of a part of chromosome-13 results in **retinoblastoma**.
11. Deletion of short arm of X-chromosome results in **Turner Syndrome**. See chapter 17, human genetics.
12. The presumed deletion of the Y-chromosome has reduced its male-determining effect.

Duplication may be of 5 types: These are **tandem duplication, Reverse tandem duplication, Displaced duplication, Transposed duplication and Extra-chromosomal duplication**. The Displaced duplication is of two types: **Homobrachial** displaced duplication and **Heterobrachial** displaced duplication. The Transposed duplication may also be of two types: **Interstitial** and **Terminal** (Table 8.2).

- **Tandem duplication**: The duplicated (extra) genes lie serially by the side of the normal part of the chromosome, keeping the genic sequence unaltered and forming a tandem duplication (Fig. 8.3).

In the Fig. 8.3, the CDE part has been inserted tandemly to the original chromosome forming a duplication of CDE.

- **Reverse tandem duplication**: When the gene order in the duplicated part becomes reverse to the normal order, it is called reverse tandem duplication. After reverse tandem duplication of the chromosome, mentioned in the above example, the order of the genes will be AB **CDE EDC** FG.

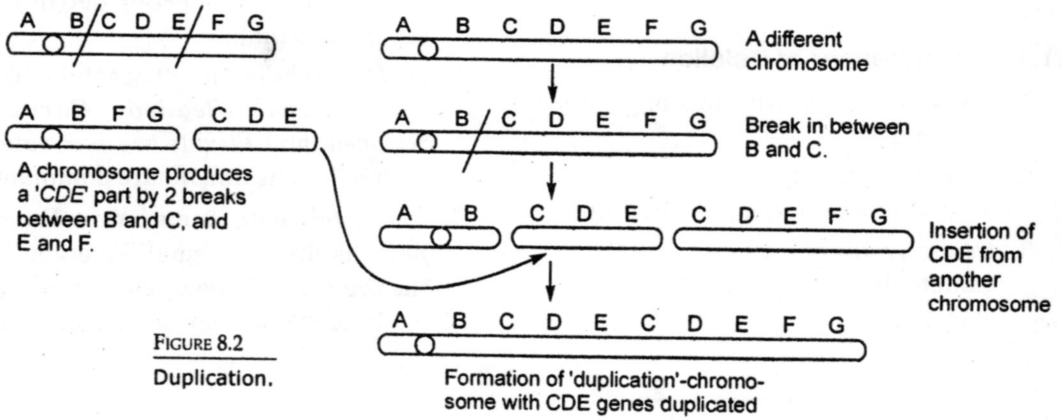

FIGURE 8.2
Duplication.

13. Deletion of σ and β-subunits of Adult Hb A in human causes **Lepore** and **Kenya** variants of haemoglobin A, causing anemia (one type of thalassemia).

8.2 Duplication

Addition of a part of a chromosome to the normal one is called **duplication**. The diagram 8.2 illustrates how a duplication may occur:

- **Displaced duplication**: When the

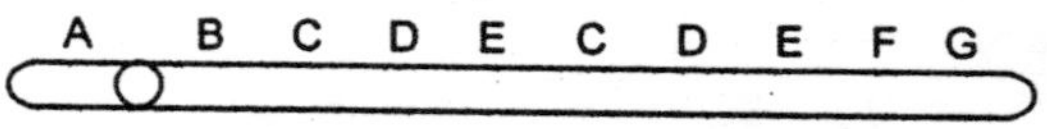

FIGURE 8.3
Tandem duplication.

TABLE 8.2

Duplication

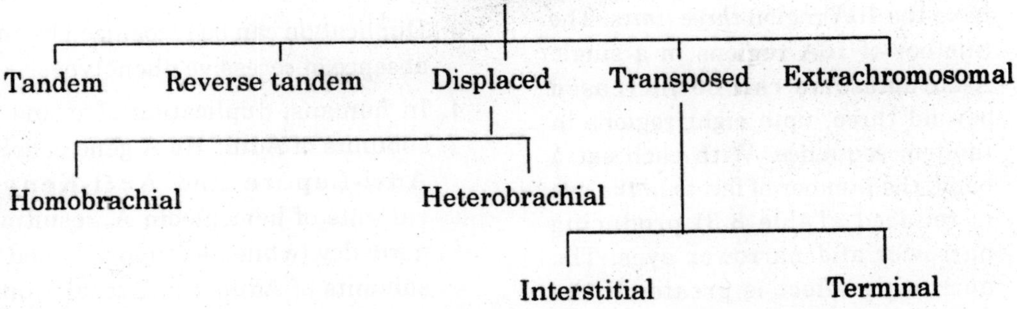

duplicated part is displaced from one arm of the chromosome to the other arm, it is called *displaced duplication*. It may be *homobrachial* (duplicated portion remaining on the same arm as the original section), or *heterobrachial* (duplicated portion is shifted on the other arm of the chromosome) (Fig. 8.4).

• **Transposed duplication**: This type of duplication involves a transfer of the duplicated portion to a nonhomologous chromosome. It is illustrated in (Fig. 8.5).

• **Extrachromosomal duplication**: Sometimes, the duplicated part may bear a centromere and considered an extrachromosome with may act independently.

The adjacent gene repetition due to duplication is more frequent in *Drosophila* and is called **repeats**.

8.2.1 Consequences of Duplication

1. The well known *Bar-eye* phenotype in *Drosophila melanogaster* is a result of **tandem duplication** (Fig. 8.6). Four or five bands in 16A region of the salivary gland chromosome are involved in producing a *bar-eye* (narrow eye) phenotype. It has been evidenced that the *normal* males bear

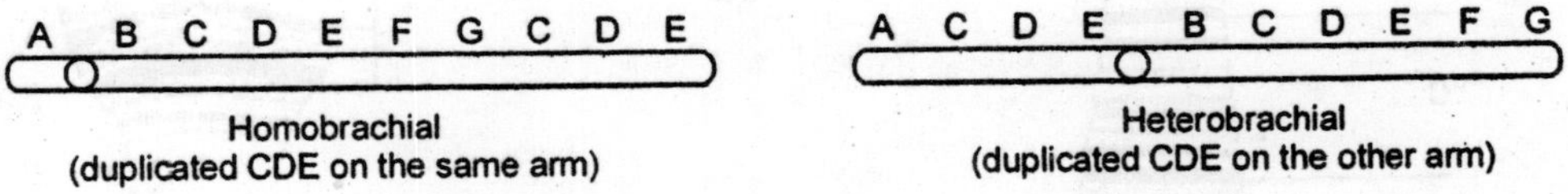

FIGURE 8.4
Displaced homobrachial and heterobrachial duplication.

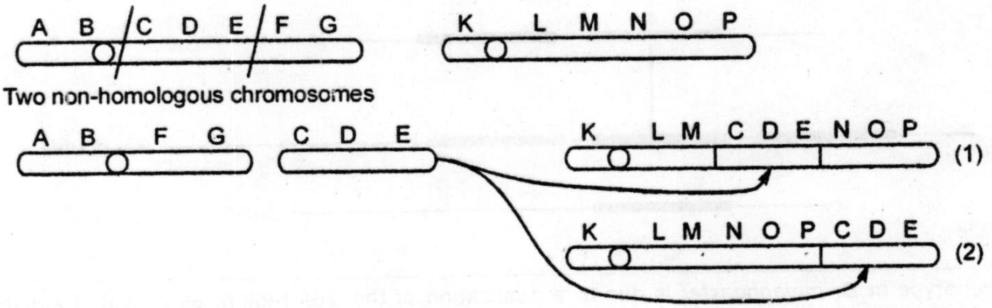

FIGURE 8.5
Transposed duplication. 'CDE' is transposed to the nonhomologous chromosome either (1) interstitially or (2) terminally.

the 16A region *once*, *Bar*-males bear the region *twice*, *Bar-double* males have the 16A region *three times*. The number of 16A regions in a single X-chromosome can be increased beyond three, upto eight regions in tandem sequence. With each extra band, the number of facets in the eye is reduced (Table 8.3) producing narrower and narrower eyes. The narrowing effect is greater if the duplication occurs in the same chromosome. This is called **position effect**.

2. Duplications are less lethal to the organism than in the case of deletion.
3. Duplication can be recognized by the absence of recessive phenotypes.
4. In humans, duplication of σ and β subunits of Adult Hb A gene causes **Anti-Lepore** and **Anti-Kenya** variants of hemoglobin A, resulting normalcy (while, deletion of σ and β subunits of Adult Hb A results into Lepore and Kenya variants of Hb A causing severe anemia or a kind of thalassemia).

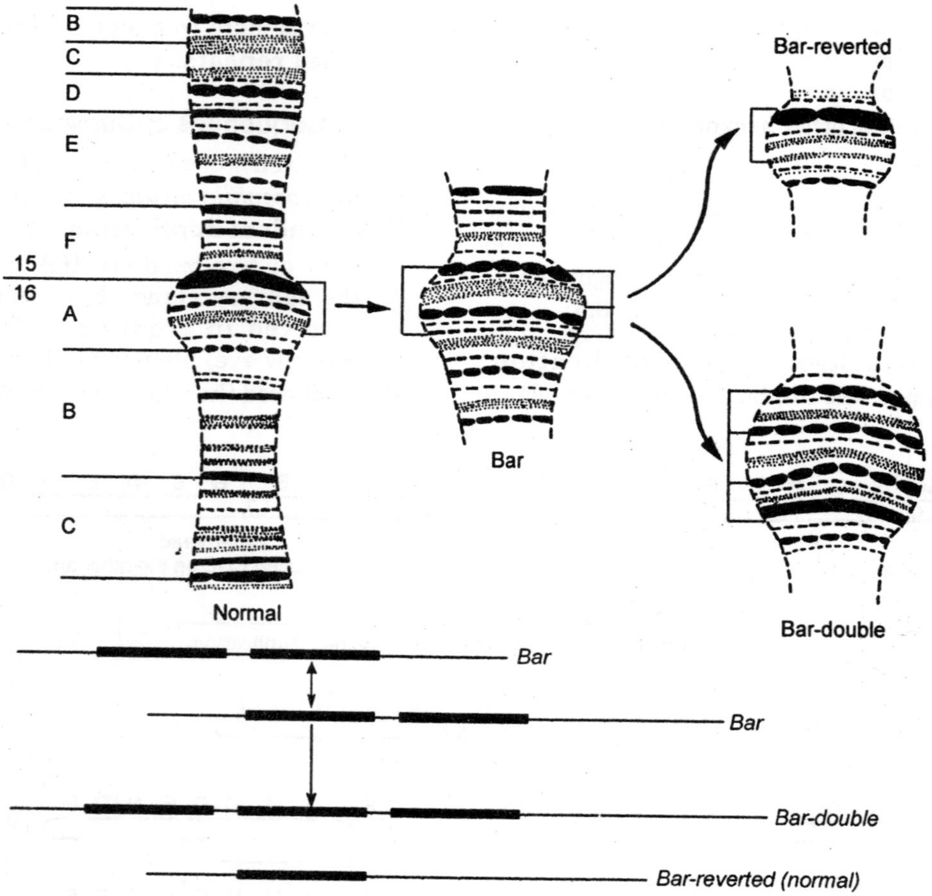

FIGURE 8.6

The *Bar* phenotype in *D. melanogaster* is due to a duplication of the *16A* region, as indicated above, while *Bar-double* is due to triplication of the same region of the X chromosome. Below, diagram to illustrate how unequal crossing over in *Bar* females leads to *Bar-double* and to normal phenotypes. (After Swanson).

TABLE 8.3: Comparison of genotypes and phenotypes for bar eye in *Drosophila* females showing position effects of 16A segment (Source: **Burns and Bottino, 1989**)

X chromosomes	Phenotype		Mean number of facets
1. 16A/16A	Normal		779
2. 16A, 16A/16A	Heterozygous bar eye		358
3. 16A, 16A/16A, 16A	Homozygous bar eye		68
4. 16A, 16A, 16A/16A	Heterozygous ultrabar (=double bar)	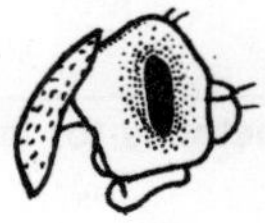	45
5. 16A, 16A, 16A/16A, 16A, 16A	Homozygous ultrabar		25

8.3 Inversion (Rotation of a group of genes in 180° on the same chromosome)

After the discovery of linkage and recombination (chapter 6) in *Drosophila* it was noticed that some factors ('C-factors') reduce or suppress the C.O. in certain part of a chromosome in females, heterozygous for C-factor. This is called *cross-over suppressor factor*. In 1926, Sturtevant found that it was due to "an *inversion* of a section of chromosome".

The **inversion** involves a rotation of a part of chromosome (bearing a set of genes) by 180° so that the gene sequence is arranged in reverse order. In *inversion*—

1. Breakage occurs between two pairs of genes.
2. The broken part rotates in 180°.
3. Reunion of broken part occurs in reverse order.
4. There is no loss or gain of genes (Fig. 8.7).

Types: Inversion may be of two types according to the *inclusion* or *non-inclusion* of centromere in the process. When the inversion **does not include the centromere** it is called **paracentric** inversion which is more common. The other type is called **pericentric** inversion which **includes the centromere**.

8.3.1 Paracentric Inversion

If, for example, the sequence of a normal chromosome is represented as A B C D E F G H, an *inverted* homologue might be A B *F E D C* G H, with the 180° rotation of the *C D E F* section (Fig. 8.7). This example is of a paracentric inversion since it **excludes** the centromere.

8.3.2 Consequences of Paracentric Inversion

1. The rearrangement of gene order in inversion shows important consequences. One of them is the change in gene functions. The alterations of

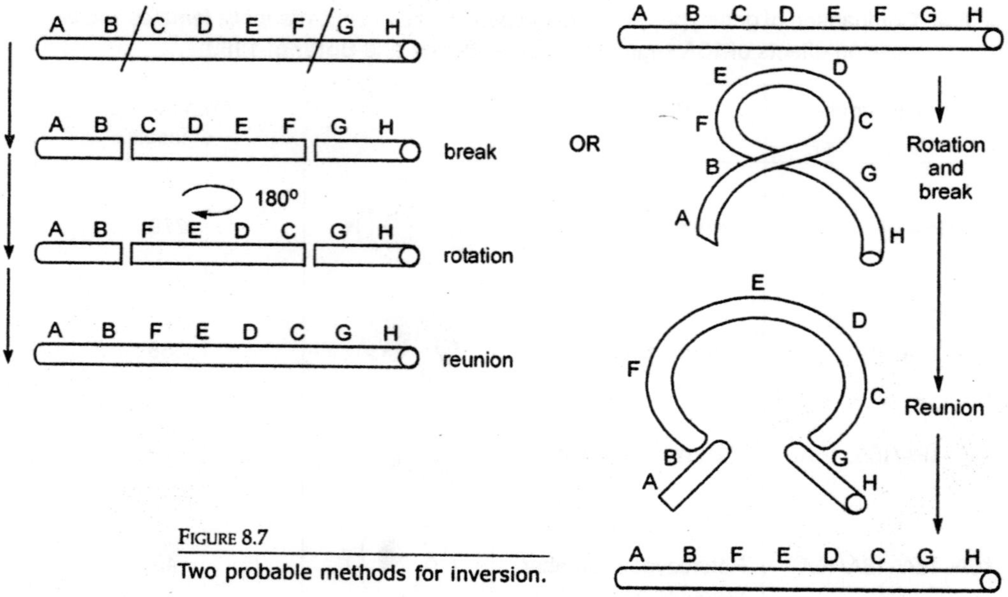

FIGURE 8.7
Two probable methods for inversion.

gene function due to a change in location of the genes, i.e., inversion, are called **position effects**. Position effects appear as mutations which may be dominant or recessive.

2. Another important consequence of paracentric inversion is noticed in a *paracentric heterozygote* (i.e., one homologue is normal and the other is with a paracentric inversion). In these organisms, during gamete formation, **synapsis and crossing over** produce unusual structure. It is known that the synapsis is a gene-by-gene pairing; but since in this paracentric heterozygote, one homologue is inverted, there will be a twisting and **loop formation** in chromosome in diplotene stage (Fig. 8.8).

Now, if a crossing over occurs within the inverted portion, then, upon separation of the homologous chromosomes, four types of chromosomes will be produced: (a) One non-cross-over chromosome with **normal gene** complement. (b) Another non-cross-over chromosome with **inverted gene part**. (c) One **dicentric** chromosome, and (d) One **acentric** chromosome (Fig. 8.8). Under this condition, gametes will be produced with **deficiencies** and **duplication** of chromosomal materials. These gametes then give rise to **chromosomally imbalanced** organisms that usually die. Since only gametes with chromosomes in which no crossing over occurred, result in offspring that survive, inversion appear as **a domi-nant mutation for cross-over suppressor**.

8.3.3 Pericentric Inversion

Inversions that include the centromere within the inverted region are called pericentric inversions.

The result of single crossing over within the inverted region in a pericentric inversion is the production of **duplications** and **deficiencies** (Fig. 8.9).

Pericentric inversions are less frequent in nature.

Sometimes, pericentric inversion may produce morphological changes in a chromosome. For example, a **metacentric** chromosome with unequal pericentric breaks can change into an **acrocentric** chromosome (Fig. 8.10).

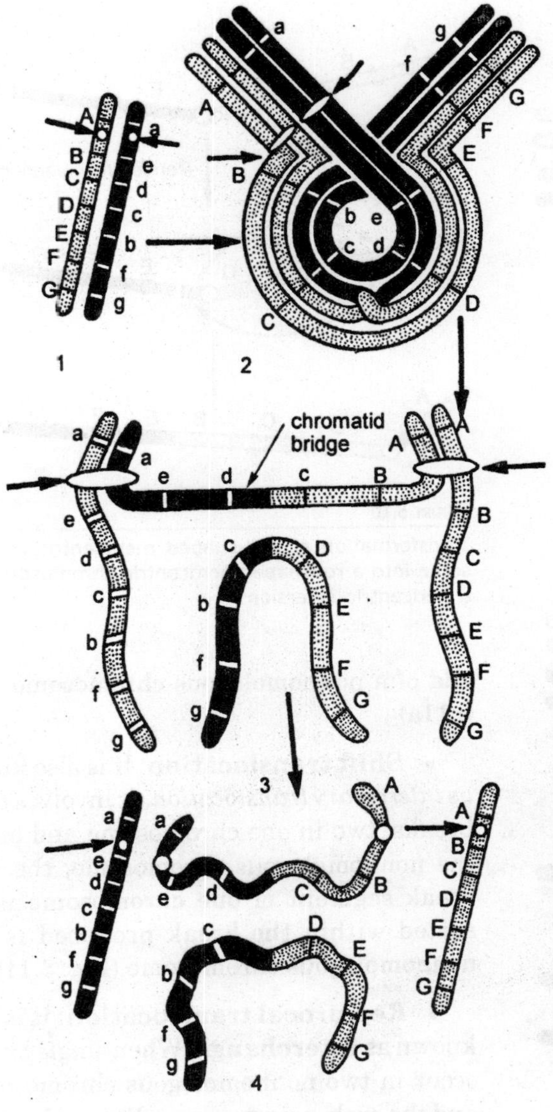

FIGURE 8.8

Crossing-over in a heterozygote for a paracentric inversion: (1) Two chromosomes differing in an inversion, (2) diplotene stage of meiosis showing a chiasma inside the inverted section, (3) anaphase of the first meiotic division, showing a chromatid bridge and an acentric fragment, (4) outcome of meiosis—2 non-cross-over chromosomes with normal gene complements and (in the middle) decentric and acentric chromosomes. After Sinnott, Dunn and Dobzhansky.

8.3.4 Consequences of Pericentric Inversion

1. Crossing over in heterozygotes for pericentric inversions may produce unbalanced gametes and inviable zygotes.
2. Acts as *crossing over suppressor.*
3. Fertility of inversion homozygotes and sterility of inversion heterozygotes lead to establishment of two varieties—which are mutually fertile but do not breed well with rest of the species.
4. Both of the above breed evolve in different directions and later become reproductively isolated species (sibling species). Cytological evidences for such evolutionary mechanisms have been found in *Drosophila* and mammals.

8.4 Translocations

Transfer of a segment of a chromosome (or a set of genes) to the nonhomologous chromosome is called Translocation.

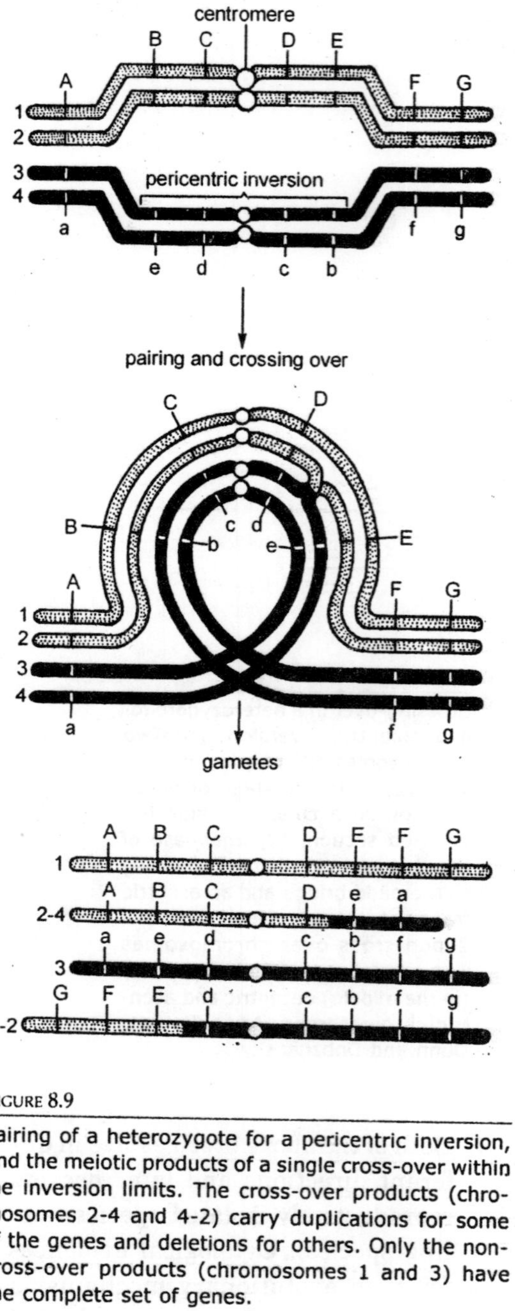

FIGURE 8.9

Pairing of a heterozygote for a pericentric inversion, and the meiotic products of a single cross-over within the inversion limits. The cross-over products (chromosomes 2-4 and 4-2) carry duplications for some of the genes and deletions for others. Only the non-cross-over products (chromosomes 1 and 3) have the complete set of genes.

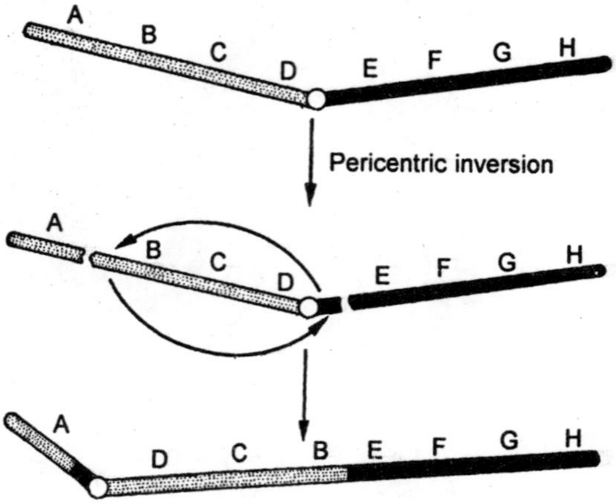

FIGURE 8.10

Transformation of a V-shaped metacentric chromosome into a rodshaped acrocentric chromosomes by a pericentric inversion.

Translocation may be of 3 types:
(a) **simple**, (b) **shift** and (c) **reciprocal**.

- **Simple translocation**: It involves a single break in the chromosome and the transfer of the broken part directly at the end of a nonhomologous chromosome (Fig. 8.11a)

- **Shift translocation**: It is also known as *intercalary translocation*. It involves three breaks, two in one chromosome and one in the nonhomologous chromosome; the two-break segment of one chromosome is inserted within the break produced in the nonhomologous chromosome (Fig. 8.11b).

- **Reciprocal translocation**: It is also known as **interchange**. When single breaks occur in two nonhomologous chromosomes and the broken parts are exchanged between them, it is called *reciprocal translocation* (Fig. 8.11c).

8.4.1 Consequences of Translocation

As a result of translocation three kinds of cytological changes may occur—

1. New linkage relationship is found.
2. Size of chromosome and position of centromere change.
3. Variation in chromosome morphology is found. Development of *isochromosome* (both arms are equal),

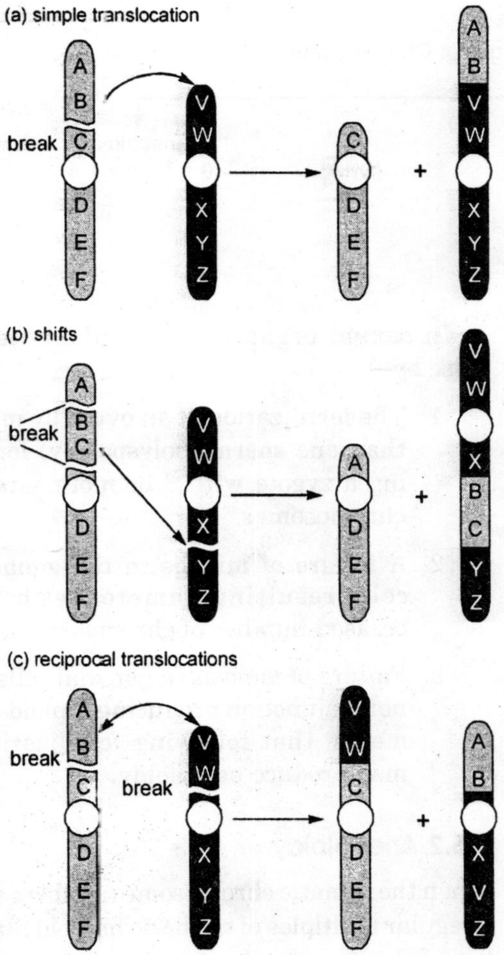

FIGURE 8.11

Three different types of translocations that may occur between two nonhomologous chromosomes, *A B C D E F* and *V W X Y Z*.

ring chromosome and **Robertsonian chromosome** may occur. In Robertsonian chromosome, acrocentric chromosomes produce metacentric chromosomes as a result of translocation (Fig. 8.12).

The *genetic consequences* of translocation are—

1. Mutation. (The first case of *mutation* was observed by De Vries in *Oenothera* which was the result of a translocation).
2. Lethality may develop.
3. In human, the chronic myelocytic leukemia may develop. In this disease, a distal part of the long arm of the **Philadelphia chromosome** (Ph'), i.e., the 22nd chromosome is translocated to a larger chromosome (usually to the 9th), and a deletion ($22q^-$) is resulted. (See chapter 17, human genetics).
4. **Robertsonian translocation** may have a role in human evolution. Humans have 46 chromosomes, while great apes (*Gorilla* etc.) have 48. It is thought that a common human-ape-ancestor evolved due to a centromeric fusion of **two acrocentric chromosomes** of great apes to produce a **single metacentric chromosome**.

8.5 Types of Numerical Changes

The second type of chromosomal aberration, i.e., the **numerical changes** in chromosome, or variation in chromosome number is of two types: **Euploidy** and **Aneuploidy**.

• **Euploidy**: When somatic complements are exact multiples of the basic haploid (n) number of the species, it is called **euploidy**. The euploidy may be of two types:

1. with higher multiple numbers of chromosome above the diploid (2n) or **polyploidy**, and
2. with half the number of the diploid by the loss of an entire set of chromosome or **monoploidy** (n).

• **Aneuploidy**: When the somatic complements are irregular multiples of the basic haploid number (n) it is called **aneuploidy** (Table 8.5).

8.5.1 Euploidy

The word "ploidy" means the number of chromosome sets. When the somatic karyotypes (chromosome numbers) are **exact multiples of the basic haploid number** of the

TABLE 8.4

Numerical Changes in Chromosome

- Euploidy
  - Monoploidy (n)
  - Polyploidy (3n, 4n, 5n etc.)
- Aneuploidy (2n + 1, 2n – 1 etc.)

species, it is called **euploidy**. Here, the variation involves entire sets of chromosome. Organisms with chromosome complements n, 3n, 4n, 5n etc. all are euploids. Euploidy is of two types: (1) Monoploidy or haploidy and (2) Polyploidy.

8.5.1.1 Monoploidy (haploidy)

Various fungi and algae show haploidy or monoploidy throughout the majority of their life cycles. Certain species of insects, mites and rotifers produce *haploid males*, through parthenogenesis. The drone (male) of honey bee is an example of monoploid organism. Monoploidy is denoted as 'n'.

8.5.1.2 Polyploidy

Polyploidy is the state where the number of complete chromosome set is more than that of the diploid. Polyploid organisms are denoted as 3n (where the number of chromosome is three times of its haploid number), 4n, 5n etc. And, accordingly, these organisms are called triploid (3n), tetraploid (4n), pentaploid (5n) etc. respectively. Thus, a tetraploid *Drosophila* means it has 16 chromosomes, which is a multiple of 4, its basic haploid number. It is denoted as 4n.

Polyploidy may be of two types:

(a) **Autopolyploidy**:
Variation arises within a species. (auto = self).

(b) **Allopolyploidy**:
Multiplication of chromosome sets is derived from two different species. (allo = different).

In sexual organisms, polyploidy may arise by—

1. The fertilization of an ovum by more than one sperm (polyspermy) forming a zygote with 3 or more sets of chromosomes.
2. A failure of mitosis in the gonadal cells resulting gametes with increased number of chromosomes.
3. Failure of meiosis in gonadal cells or nondisjunction producing diploid gametes, that following fertilization, may produce polyploidy.

8.5.2 Aneuploidy

When the somatic chromosome numbers are irregular multiples of the basic haploid number of the species, they are termed *aneuploidy* (aneu = uneven, ploid = set). A normal diploid organism is called *disomic* (di = two). It may *gain* or *lose* one or two extra chromosome to or from its set, and may be 2n + 1 or 2n + 2 and 2n – 1 or 2n – 2 respectively. Both the karyotypes are aneuploids. Such aneuploids have been classified variously as follows:

Hypoploidy: It involves *loss* of chromosome from a set.

Nullisomy: The loss of two homologous chromosomes from a set (2n – 2) is called **nullisomy**, and the organisms are called **nullisomics**. (loss of a set of homologues).

Monosomy: Condition, where one chromosome is lost from a set (2n – 1) is called *monosomy*. The organisms are called *monosomics* because they have only one homologue of a particular chromosome.

• **Hyperploidy**: Aneuploidy due to **gain** of chromosome is called **hyperploidy**, such as—

Trisomy: Organisms with additional chromosome to a set (2n + 1) are called **trisomics,** because they possess three homologues of one chromosome. The condition is referred to as trisomy.

Similarly, **tetrasomy, pentasomy** etc. are the conditions of having 4 and 5 homologues of one chromosome (2n + 2), (2n + 3) respectively. Organisms with two extra chromosomes to the set (2n + 2) are called tetrasomics and those with three extra chromosome to the set (2n + 3) is called **pentasomics.**

The table 8.5 shows the chromosomal complements in hypoploidy and hyperploidy with example.

8.6 Consequences of Euploidy and Aneuploidy

Polyploidy: Polyploidy has great contribution in plant hybridization. Seedless varieties of water-lemons, sugar beet, tomato, grapes etc. can be produced by autopolyploidy. The crop plants like rye, corn, marigolds, snapdragons etc. can be improvised by autotetraploidy. Autopolyploidy produces gigantism of plant tissue increasing quantity of fruit, flower and leaves.

The production of *Raphanobrassica* is an example of allopolyploidy. It has been produced by crossing raddish (*Raphanus sativum,* 2n = 18) and cabbage (*Brassica oleracea,* 2n = 18). The product *Raphanobrassica* is a tetraploid with 36 chromosomes.

The new world cotton (2n = 52) is also a production of allopolyploidy.

1. Polyploidy induces gigantism
2. It reduces fertility.
3. It may play role in evolution.

• **Consequences of Aneuploidy**

A classical example of aneuploidy is found in **Down syndrome** (DS) in humans. It is a **trisomy** of 21st chromosome. (See chapter 17).

Other human syndromes, called **Edward syndrome** and **Patau syndrome** are also abnormal products of chromosomal aneuploidy. The former syndrome is a result of trisomy of 18th (Trisomy-18) chromosome and the latter syndrome is a result of trisomy-13 (See chapter 17, for details).

Table 8.5: Types of Aneuploidy with example

Type	No. of chromosome	Example
Disomic	2n	AA BB CC
Monosomic	2n – 1	AA BB C
Nullisomic	2n – 2	AA BB
Trisomic	2n + 1	AA BB CCC
Double trisomic	2n + 1 + 1	AA BBB CCC
Tetrasomic	2n + 2	AA BB CCCC
Pentasomic	2n + 3	AA BB CCCCC
Hexasomic	2n + 4	AA BB CCCCCC
Septasomic	2n + 5	AA BB CCCCCCC

Trisomy-21 and Trisomy-22 are reported from gorilla and chimpanzee respectively.

In *Datura*, the normal chromosome number is 12, and due to aneuploidy twelve mutant varieties are found in this plant. Each mutant produces different phenotypes due to an extra chromosome. The trisomics in *Datura* may be of primary, secondary or tertiary types.

EXERCISE

1. Define aneuploidy and euploidy.
2. Define deletion, duplication, inversion and translocation.
3. Describe consequences of duplication and inversion.
4. What is cross-over suppressor? Explain.
5. A triploid is formed through fertilization of a diploid gamete containing 12 chromosomes with a haploid gamete containing 6 chromosomes. If the average number of gametes of this triploid is 7, how many of its chromosomes are lost on the average during meiosis?
6. Draw a picture of pairing between two chromosomes, one of which has the sequence 1·23456789 and the other the sequence 1·27843659. (The dot is the centromere).
7. According to the sex-determination theory of Bridges, what are the sexual phenotypes of the following tetraploid *Drosophila*, all with 4 sets of autosomes but carrying different number of X and Y chromosomes: (a) XXYY (b) XXXY (c) YYYY (d) XXY, (e) XXXXY?
8. What is *Cri du Chat*? How does it occur?
9. Write notes on 'Bar-eye' in *Drosophila*, giving possible reasons for it.
10. Mention the consequences of translocation.

CHAPTER 9

GENE EXPRESSION AND GENE FUNCTION

(Transcription and Translation)

Metabolism, Metabolism and genes, Genetic control of metabolism, One-gene-One polypeptide concept, 'Inborn errors', Transcription, RNA polymerase, Transcription in *E.coli,* Transcription in eukaryotes, Eukaryotic RNA-polymerases, Processing of RNA, Splicing, RNA editing, Transcription of rRNA, Transcription of tRNA, Translation, Proteins and amino acids, The Genetic Code, Protein-synthesis, Regulation of gene-expression.

In one definition, **gene** is a segment of DNA that *encodes* a polypeptide. That is, genes encode enzymes (polypeptides). These enzymes are necessary to catalyze the **metabolism** in all living cell. Hence, there is an inseparable and integrated relation between **genes** and **metabolism**.

9.1 Metabolism

Metabolism is the sumtotal of all chemical processes in living cells by which energy is produced and used. The chemical processes or reactions are catalyzed by **enzymes** (which are biological catalysts and protein in nature). In principle, **genes control the metabolism**. Organic molecules synthesized or degraded in cells are called **metabolites**. The synthesis of metabolites is called **anabolism**, and the degradation or breaking of metabolites is called **catabolism**.

9.2 Metabolism and Genes

Metabolism needs **enzymes**. Most enzymes are proteins with catalytic activity. Proteins (= polypeptides) are made up of **amino acids**. Each enzyme **catalyzes** (takes part in the chemical reaction but remains unmodified in the process) a reaction **specifically**. Their *specificity* is due to the unique sequence of the 20 amino acids of which they are made. The sequence of amino acids in each polypeptide (= a single chain of amino acids) is controlled by **one gene**. An enzyme does not catalyze any metabolic reaction, but catalyze a **specific** metabolic reaction. The metabolic process of a particular metabolite usually occurs by *a* **series** *of* **enzyme-catalyzed reactions**, called a **metabolic pathway**. The metabolic pathway includes **several steps** of reactions. Each reactionary step requires an enzyme (= polypeptide). Each polypeptide is coded by a gene. Thus, each of the steps of the metabolic pathway needs the production of a gene. Thus, there is an inseparable correlation between process of metabolism and guidance of one or more genes (Fig. 9.1).

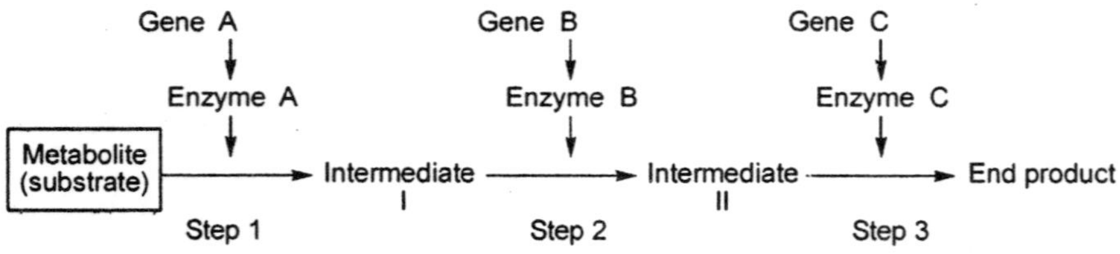

FIGURE 9.1

A metabolic pathway with 3 reactionary steps which are catalyzed by 3 enzymes, produced by 3 genes, Gene A, Gene B and Gene C.

The genetic control of metabolism is further elaborated in Chapter 5.

9.3 Genetic Control of Metabolism

Please refer to 5.6, 5.6.1, 5.6.2 and 9.4 chapters.

9.3.1 One gene-One Polypeptide Concept

The **one gene-one polypeptide concept** is a central tenet in molecular genetics. We owe to Sir Archibald Garrod (1909), G.W. Beadle and E.L. Tatum (1958) for the establishment of this principle.

Sir Archibald E. Garrod was the pioneer worker to study some hereditary metabolic disorders in humans. His studies and observations were published in details in his famous book, "Inborn Errors of Metabolism" in 1909. One of his works was on the inherited disease **alkaptonuria**, in which the urine became black upon exposure to air. This blackening is due to the presence of **alkapton** or **homogentisic** acid. This substance is an intermediate product in the degradation of two amino acids—*tyrosine* and *phenylalanine*. Garrod believed that the presence of this substance in the urine was due to a block in the normal metabolic pathway of it. By observing family pedigree, Garrod advocated that **alkaptonuria** was inherited as a single recessive gene; his insight was that there must be a relationship between gene and metabolic reaction. His concept was the fore-runner to the one gene-one enzyme or one gene-one polypeptide concept.

Garrod's discovery, like that of Mendel's, was unknown for a long time. G.W. Beadle and E.L. Tatum revived his concept in 1941. They worked on the growth of *Neurospora* and proved that the biosynthesis of the growth factors of this fungus were under genetic control. In their experiment on the growth of *Neurospora*, they showed that **one mutation** resulted in a new requirement for **one growth factor**. They correlated their genetic analysis with biochemical studies of the mutant strains and established that **one mutation resulted in the loss of one enzyme activity**. They were awarded Nobel Prize in 1958 for this discovery. Their work led to the discovery of the concept of **one gene-one enzyme**.

Later, it was known that many enzymes and proteins (as for example hemoglobin) consisted of two or more different **polypeptide-chains**, each of which was a product of a separate gene. Thus the concept of **one gene-one enzyme** was changed to **one gene-one polypeptide**.

9.4 "Inborn Errors" and their Genetic Causes: (Genetic control of metabolic pathway)

Out of hundreds of "inborn errors" in humans, the **phenylketonuria**, **alkaptonuria**, **tyrosinosis** and **albinism** are well studied 'inborn errors' or metabolic disorders.

Humans cannot manufacture phenylalanine or tyrosine (two amino acids) and these must be supplied in the diet.

- **Phenylketonuria**

The 'inborn disorder' phenylketonuria (PKU) is a result of defect in the phenylalanine-tyrosine metabolism; the enzyme *phenylalanine (Phe)–hydroxylase*, that converts Phe to Tyr (tyrosine) is not produced or remains inactive due to an autosomal (chromosome no. 12) recessive mutation of the concerned gene for its production. The conversion from Phe to Tyr is the first step in the catabolism of Phenylalanine. Due to the non-conversion of phenylalanine (due to inactivity of *Phenylalanine hydroxylase*) it accumulates in blood of the patient. As a result, some metabolic derivatives of Phe (e.g. Phenylpyruvic acid) are formed and concentrated in the cerebrospinal fluid (CSF) causing toxic effect to the brain-tissue. It causes irreversible brain damage. It results in severe mental retardness and light-pigmentation (Fig. 9.2).

If PKU is detected and diagnosed in an early stage, the children (homozygous for the gene) can be treated with a restricted phenylalanine-diet. Adequate Phe must be included in the diet for protein synthesis, but the amount must be restricted and *insufficient*, so that it cannot be converted into phenylpyruvic acid.

- **Alkaptonuria**

Alkaptonuria is the blackening and hardening of cartilage and blackening of urine when exposed to air. It is also an inborn metabolic disorder due to a defect in the Phenylalanine and Tyrosine metabolism. In the metabolic pathway one of the end product is **homogentisic acid**.

An enzyme, *homogentisic acid oxidase*, present in liver catalyses the break-down of homogentisic acid to CO_2 and H_2O. **When the enzyme is lacking or defective,** large amount of homogentisic acid are excreted in the urine, which turns black upon exposure to air. Some excess homogentisic acid accumulate in the cartilage tissues (e.g. external ear) and turn them black. Their accumulation in the joint of the body may cause arthritis.

The defect in the production of the enzyme *homogentisic acid oxidase* is due to a mutation of an autosomal recessive gene. It is an example of genetic cause for enzyme block (Fig. 9.2).

- **Tyrosinosis**

Tyrosinosis is a metabolic disorder caused due to the lack of *tyrosine transaminase* enzyme. It is a rare disease and the affected persons show increased tyrosine level in their blood and urine. This disorder is caused by a mutation in the genes that encode the enzymes required for the catabolism of Tyrosine (Fig. 9.2).

- **Albinism**

Albinism is the genetic disorder causing absence of melanin in the skin, hair and eyes. It is resulted due to the biochemical block in the Phenylalanine and Tyrosine metabolism at the point between 3, 4-dihydroxy-phenylalanine (DOPA) and melanin (Fig. 9.2). The block is due to the lack or inactivity of the enzyme *tyrosinase*, whose production is controlled by a recessive gene in homozygous condition.

The above examples of 'inborn errors' are focusing on the phenomenon of genetic control of metabolism. The mutation of concerned gene causing production of defective enzymes or inappropriate enzymes hamper the metabolic pathway in humans as well as in other organisms.

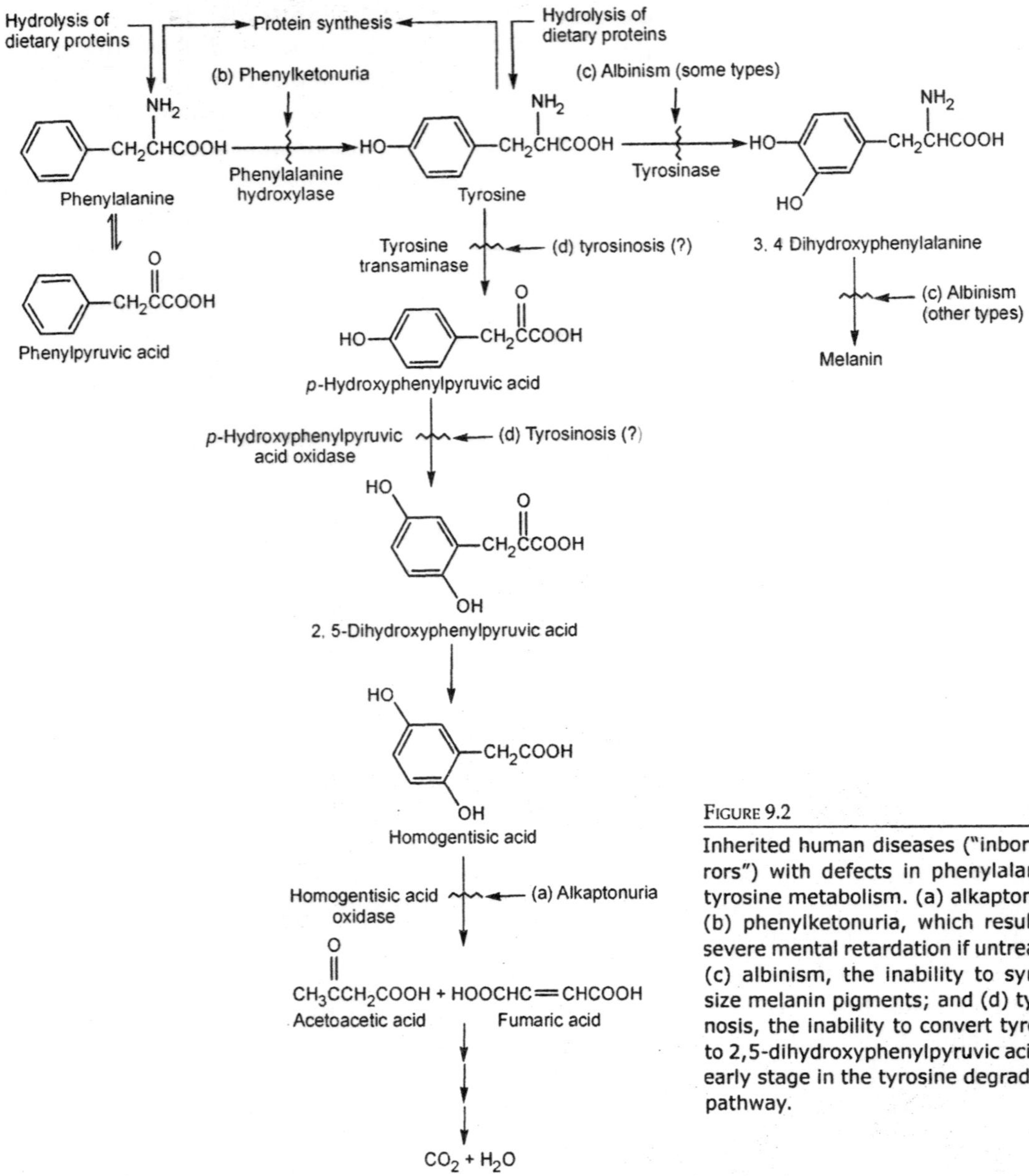

FIGURE 9.2

Inherited human diseases ("inborn errors") with defects in phenylalanine-tyrosine metabolism. (a) alkaptonuria, (b) phenylketonuria, which results in severe mental retardation if untreated; (c) albinism, the inability to synthesize melanin pigments; and (d) tyrosinosis, the inability to convert tyrosine to 2,5-dihydroxyphenylpyruvic acid, an early stage in the tyrosine degradative pathway.

9.5 Transcription

Definition: Transcription is the process of synthesis of single stranded RNA from a complementary strand of a DNA molecule.

9.5.1 The Central Dogma

Francis Crick proposed the name *central dogma* for a two-step biochemical process of protein synthesis: DNA $\rightarrow$ RNA $\rightarrow$ Protein. In this process, the first step (DNA $\rightarrow$ RNA), i.e., production of RNA from DNA is called **transcription**, and the second step (RNA $\rightarrow$ Protein) is called **translation** or **protein synthesis**. Thus, 'Central Dogma' of biochemistry is the process of transcription followed by translation. Wherever DNA is absent, to central dogma tends to be reversed, i.e., RNA $\rightarrow$ DNA $\rightarrow$ RNA $\rightarrow$ Protein.

5 main types of RNAs are transcribed from DNA. Those are—

(a) **mRNA**: It encodes the amino acid sequence of a polypeptide. These are transcripts of protein-coding genes.

(b) **rRNA**: It forms the ribosome in association with ribosomal proteins. Ribosome is the site for protein synthesis.

(c) **tRNA**: It brings amino acids to the ribosomes.

(d) **snRNA** (small nuclear RNA): It is required in the eukaryotic RNA processing.

(e) **MiRNAs** (micro RNAs): These are short 20-22 nucleotide single stranded RNAs.

Although there exists a basic similarity, the transcription process is different in prokaryotes and eukaryotes.

9.6 Major Events in Prokaryotic Transcription

1. Genetic information are copied from DNA to RNA.
2. The DNA unwinds, and an **RNA-polymerase** catalyses the synthesis of a single stranded RNA.
3. **Initiation, elongation and termination** are three steps in RNA synthesis.
4. In *E.coli* (a prokaryote), the *initiation* of transcription of protein coding genes requires a complex of RNA-polymerase and the sigma factor-protein binding to the *promoter*.
5. A single RNA-polymerase synthesizes mRNA, tRNA and rRNA, in *E.coli*.
6. Elongation begins after the dissociation of sigma factor and by involvement of the RNA polymerase core-enzyme.
7. **Termination** is signalled by a specific sequence of the DNA.
8. The transcript of protein-coding genes are linear precursor mRNAs (pre-mRNAs).

9.6.1 RNA Polymerase

The *E.coli* RNA polymerase consists of 5 polypeptide subunits, two of which are identical. The complete molecule of the enzyme is called **holoenzyme** and consists of **2α-subunits, one β-, one β′-**and **one σ-subunits.** The σ-factor or σ-subunit is involved only in the **initiation** process to recognise and bind the holoenzyme to the transcription initiationsite. The two α and ββ′ ($\alpha_2\beta\beta'$) is called **core enzyme** that catalyzes the **elongation** of chain. The β-subunit contains the NTP-binding site, the β′-subunit bears the DNA template binding region, whereas the α-subunits are involved in the assembly of the four factors (without σ), forming the **core enzyme.**

9.6.2 Transcription Process (RNA synthesis) in *E.coli*

There are certain sequential elements associated with genes that regulate the transcription process. These are called *gene regulatory elements.* The enzyme *RNA polymerase* denatures and unwinds the DNA double helix, and catalyzes the synthesis of RNA-chain in 5′ to 3′ direction, using only one strand of DNA (3′ to 5′) as a **template strand** (Fig. 9.3).

The resultant RNA-chain molecule is called **transcript**. The other DNA-strand (5′–3′) is called **non-template** and does not take part in transcription.

Before transcription, the unwinding of DNA for a short region takes place next to the gene. Four ribonucleoside triphosphates ATP, GTP, CTP and UTP are necessary energy suppliers. They are collectively called NTPs (Nucleoside Triphosphates). They help in polymerization reaction. Each time, the RNA polymerase selects the required nucleotide by the help of a specific NTP and adds it to the chain (Fig. 9.3).

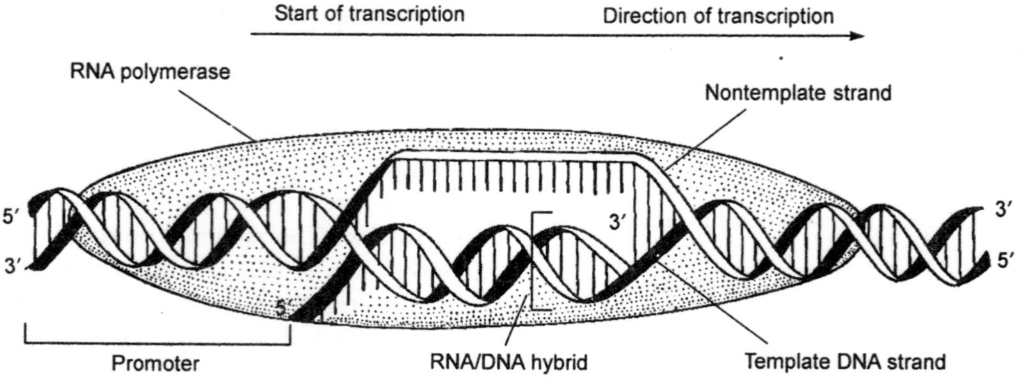

FIGURE 9.3

Transcription process: The DNA double helix is denatured by RNA polymerase in prokaryotes and by other proteins in eukaryotes. RNA polymerase then catalyzes the synthesis of a single-stranded RNA chain. The RNA chain is made in the 5′-3′ direction, with only one strand of the DNA used as a template to determine the base sequence.

The **Adenine** of the template strand is complemented by the **Uracil** in the growing RNA polymer instead of a Thymine. Thus, if a template DNA-strand reads 3′-ATACTGGAC-5′ then the RNA transcript will have the sequence 5′-UAUGACCUG-3′.

9.6.2.1 Initiation

In *E.coli,* there are three sequences of gene that are used in transcription.

1. The first sequence is the **promoter**. It is the **upstream*** of the start of the RNA-coding sequence. The RNA polymerase attaches at this sequence to start initiation.
2. The second is the **RNA-coding sequence,** which begins after the promoter at nucleotide +1 site.
3. The third is the **terminator sequence** which is the terminator site of the transcription (Fig. 9.5).

Base pairs preceding the initiation site are given – signs, and those following the initiation site are given + signs.

* Nucleotide sequences preceding the initiation site are called *upstream* and those following the initiation site are called *downstream*.

The major steps of the initiation process are—

1. The holoenzyme binds to a promoter region mediated by the σ-factor.
2. The two strands of DNA unwinds locally and provides a **template strand**.

There are 2 short sequences within the promoter which are usually *conserved.* The mid points of these two sequences are at about 10 and 35 nucleotide pairs, respectively, before the initiation site. As such, they are termed **–10 sequence** and **–35 sequence**, respectively. The nucleotide sequence present in such conserved genetic elements are called **consensus sequences.** The consensus sequence for –35 region (–35 box) is 5′-TTGACA–3′; The σ subunit of RNA polymerase binds to this site. On the other hand, the consensus sequence for the –10 region (–10 box) is 5′-TATAAT-3′. It was formerly called **Pribnow box.** This A : T rich –10 sequence **facilitates the localized unwinding of DNA**. (Fig. 9.6).

9.6.2.2 Elongation

After the binding of *RNA polymerase* (Fig. 9.7) at the promoter region, the α factor is

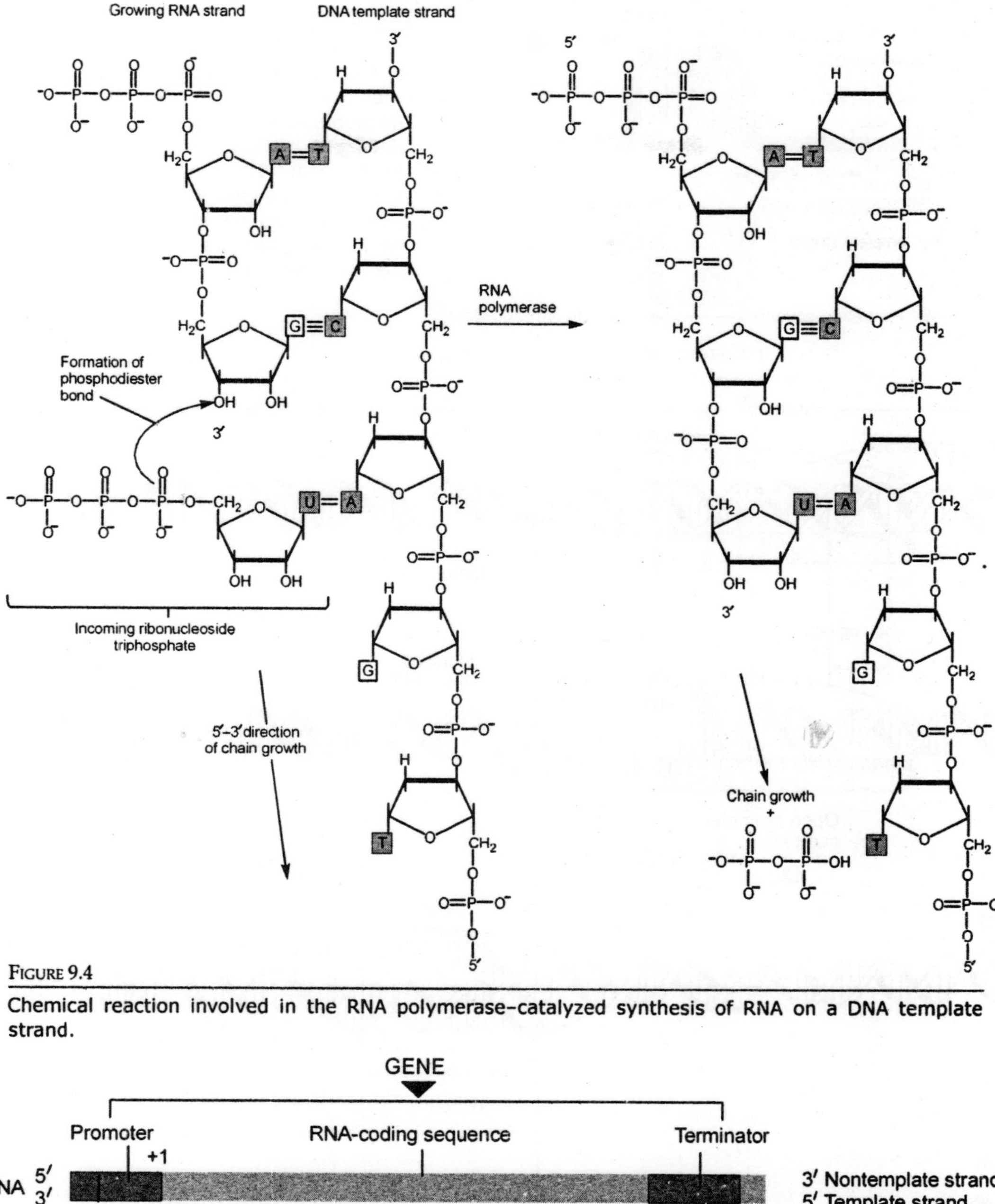

FIGURE 9.4

Chemical reaction involved in the RNA polymerase-catalyzed synthesis of RNA on a DNA template strand.

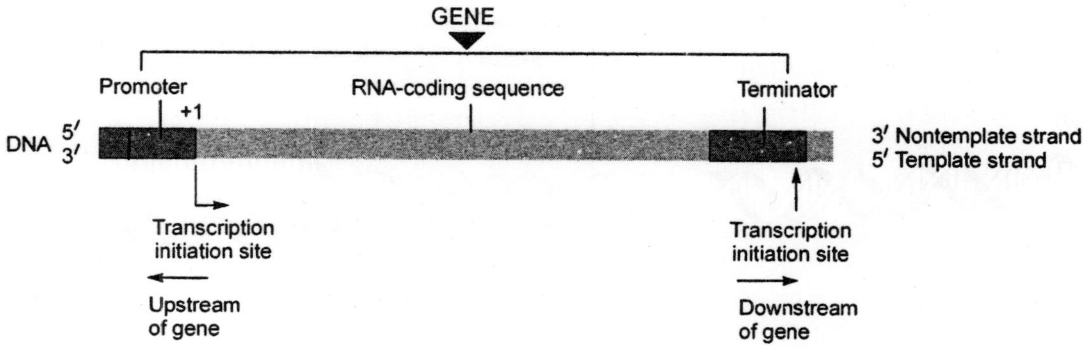

FIGURE 9.5

Promoter, RNA-coding sequence, and terminator regions of a gene. The promoter is upstream of the coding sequence, the terminator downstream. The coding sequence begins at nucleotide +1.

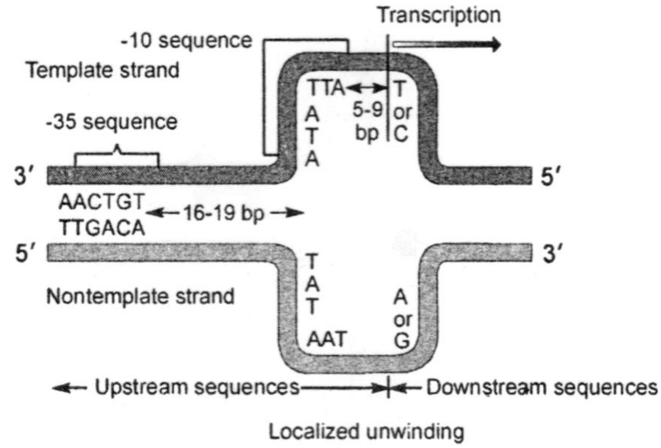

FIGURE 9.6

Structure of a typical promoter in *E.coli.* RNA polymerase binds to the -35 sequence of the promoter and initiates unwinding of the DNA strands at the A:T-rich -10 sequence. Transcription begins within the transcription bubble at a site five to nine base pairs beyond the -10 sequence.

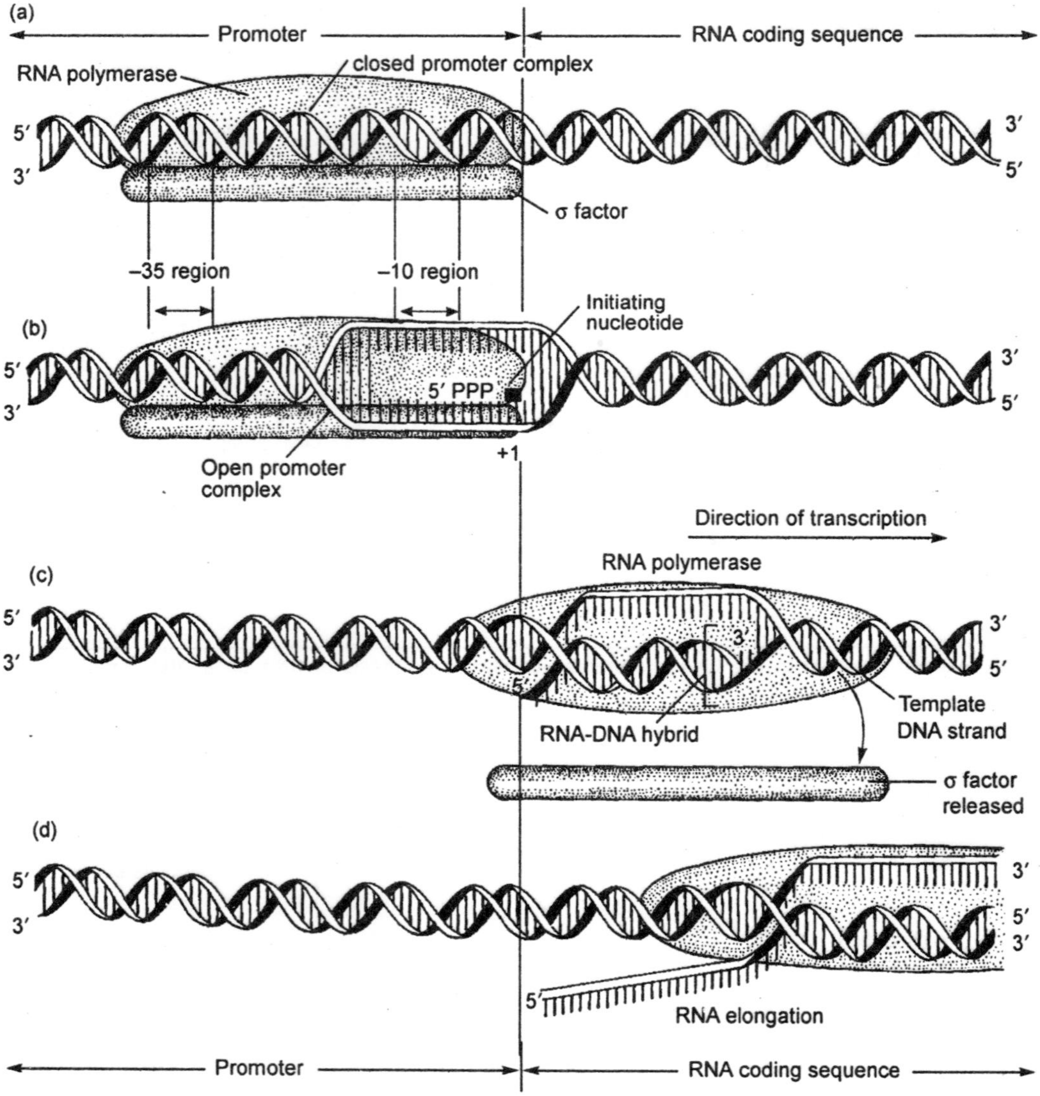

FIGURE 9.7

Action of *E.coli* RNA polymerase in the initiation and elongation stages of transcription.

released and elongation of RNA chain begins. The site of elongation is within the locally unwound segment of the DNA, called the '**transcription bubble**'. The core enzyme (with no σ) moves forward (downstream) along the DNA double helix; during this time—(a) it continuously *unwinds* the DNA strands (Fig. 9.8). (b) elongation of RNA-chain goes on within the 'bubble'. (c) the enzyme *rewinds* the DNA strands behind the elongationsite. (d) the nascent RNA-chain is displaced from the DNA template strand. A little area of DNA-RNA helix hybrid is formed that continuously gives rise to a nascent growing chain behind. The whole area of RNA formation and elongation together is called transcription-complex. The whole process is entirely catalyzed by RNA polymerase in *E.coli*. Recent evidence suggests that RNA

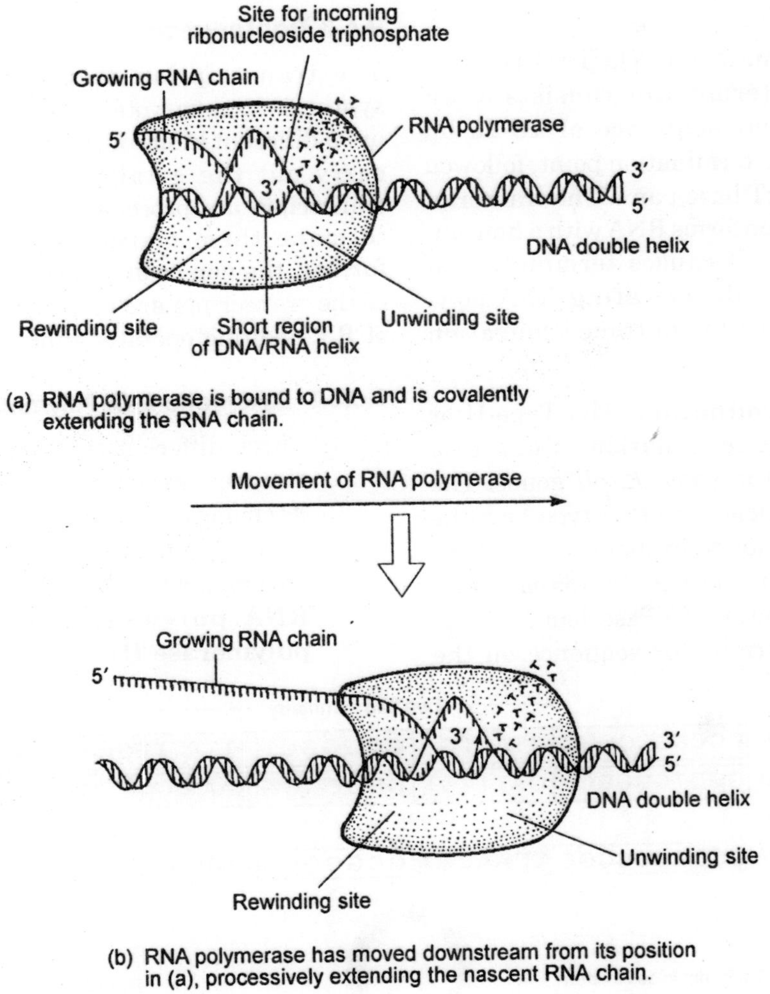

FIGURE 9.8

Elongation of an RNA chain catalyzed by RNA polymerase in *E.coli*.

polymerase comprises two proof-reading activities too.

9.6.2.3 Termination

The termination of transcription may be of two kinds—(1) Rho-independent (Type-I termination), (2) Rho-dependent (Type-II termination).

Certain **terminator sequences** in the template impose signals for stopping the RNA synthesis.

Type-I termination: The Type-I or Rho-independent terminator consists of an inverted repeat sequence of 16-20 bp upstream of the termination point, followed by about 4-8 A:T base pairs. The transcription at this region forms RNA with a hairpin-loop-structure. It causes termination of transcription by **deactivating** RNA polymerase's ability to continue elongation (Fig. 9.9).

Type-II termination: The Type-II or Rho-dependent termination is due to a protein, *rho* (ρ) in some *E.coli* genes. The terminator sequences in this type lack the special A : T bp found in the type-I, and the hairpin-loop is not formed. In this case, Rho has RNA-binding and ATPase domains. Rho binds to the terminator sequence on the RNA as it exits the RNA polymerase. After binding it hydrolyses ATP; as a result, the ATP dissociates the RNA from the RNA polymerase and from the DNA templates, stopping transcription.

Before completion of transcription process, translation or protein synthesis begins in a prokaryote. Hence it is called *coupled transcription and translation* (For translation see 9.7).

9.6.3 Transcription in Eukaryotes

The transcription or process of RNA synthesis in eukaryotes is not simple. There are 3 different kinds of RNA polymerases to carry out the catalysis; moreover the transcripts are processed in a special way to make them mature and functional; further, the non-coding sequences (introns) of the transcripts are cut off and a process of 'RNA-editing' operates after transcription.

9.6.3.1 Eukaryotic RNA Polymerases

1. Three different types of RNA polymerases are present in eukaryotes. Each one catalyses the transcription of a specific class of genes. The enzymes are **RNA polymerase I, RNA polymerase II** and **RNA polymerase III**.

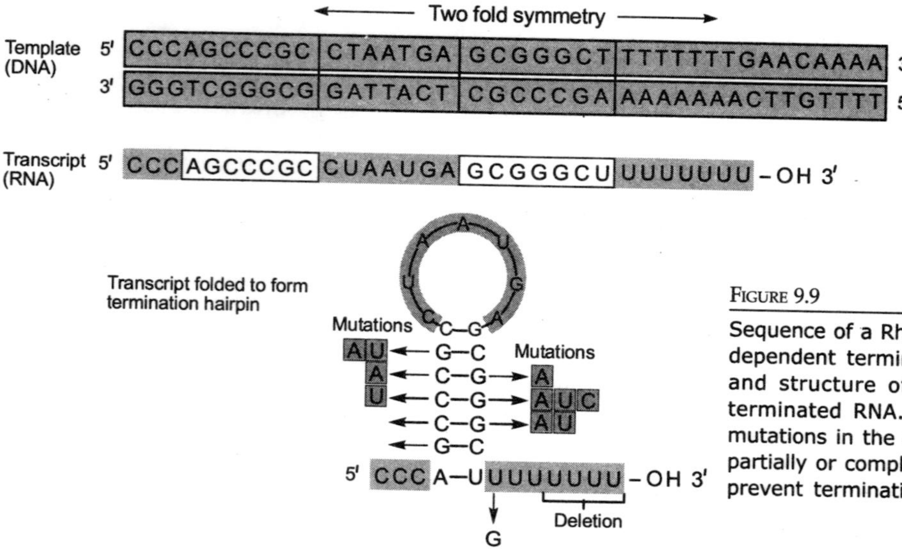

FIGURE 9.9

Sequence of a Rho-independent terminator and structure of the terminated RNA. The mutations in the stem partially or completely prevent termination.

2. All the enzymes need help of proteins called **transcription factors** in initiation process.
3. **RNA polymerase I** is located in the nucleolus and catalyses the production of r RNAs (except small 5 S r RNA).
4. **RNA polymerase II** is located in the nucleoplasm and transcribes nuclear pre-mRNAs and some sn RNAs (small nuclear RNAs).
5. **RNA polymerase III** is also located in the nucleoplasm and transcribes tRNAs, 5S r RNAs and sn RNAs.
6. In eukaryotes, the **RNA polymerase II** transcribes the protein-coding gene, which is a **precursor-mRNA**. This transcript is modified and processed to finally produce the functional mRNA molecule.
7. **Promoters** of the protein-coding gene (pre-mRNA) have about 200 bp upstream (–) of the initiation site.
8. The promoter is divided into 2 parts —(a) the **core promoter** and (b) **promoter proximal elements**. The elements of core promoter are—
   (a) **Initiator** (Inr), spanning from the initiation start site (+1), and
   (b) The **TATA box** (Goldberg-Hogness box), located at about –30 position. The consensus sequence of TATA box is 5′-TATAAAA-3′ on the nontemplate strand.

   The promoter proximal elements are found in the area from 50-200 bp upstream of TATA box. These elements are **CAAT-box, GC-box** and the **octamer box**. The CAAT box is centered at about –75 or –80 (consensus sequence GGCC AATCT) and the GC box at about –90. The consensus of GC box is GGGCGG. The consensus sequence of octamer box is ATTT GCAT.
9. There are other base pair sequences called **enhancers** and **silencers** which are short sequence elements and modulate transcription from a distance.
10. Likewise the prokaryotic transcription, the eukaryotic transcription is also completed in three steps—**initiation, elongation** and **termination**.
11. The process is followed by **RNA processing**, **splicing** and **editing**.

9.6.3.2 Initiation (in eukaryotes)

1. The initiation of pre-mRNA synthesis by RNA polymerase II starts with the local uncoiling of DNA segment, in which one of the DNA strands acts as a template.
2. Several *basal transcription factors* are required to start initiation. The factors interact with the promoters and initiate transcription. Each of the basal factors is called **TFIIX** (T = transcription, F = factor, II = RNA polymerase II, X = the particular factor involved) (Fig. 9.10).
3. The **initiation** process can be summarized in five steps:
   (a) In the **first step**, TFIID factor interacts with the promoter; it contains a TBP (TATA-binding protein) and other small TBP-associated proteins.
   (b) In the **second step**, TFIIA joins the initiation complex.
   (c) In the **third step**, TFIIB joins the complex.
   (d) In the **fourth step**, TFIIF associates with the RNA polymerase II, and then, both of them join the initiation complex together.
   (e) In the **fifth step**, the TFIIE joins the complex, and binds to the DNA-downstream.

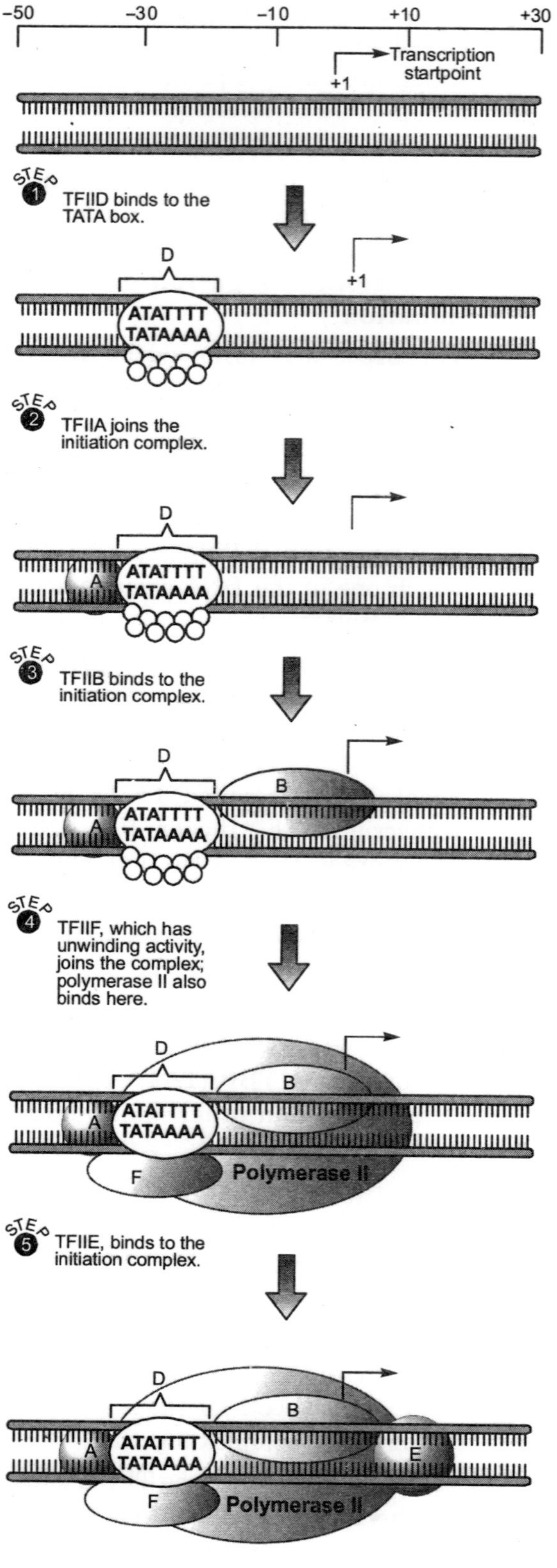

4. The initiation process forms a **basal transcription initiation complex** at the promoter region. The TFIID (factor for first step) binds to the TATA-box by the help of TBP. It is followed by the addition of the other factors successively and unites with polymerase-II, which together unwind DNA and initiate the transcription process.

9.6.3.3 Elongation (and addition of 7-MG)

The RNA chain elongation in eukaryotes proceeds after the release of RNA polymerase. The elongation process is similar to that of prokaryotes (See Fig. 9.7). But in most eukaryotic mRNA, the 5′ cap is a **7-methyl guanosine residue** joined to the initial nucleoside of the transcript by a 5′-5′ triphosphate linkage. The 3′ poly (A) tail is polyadenosine chain of small segment.

In the elongation process of eukaryotic mRNA, the 5′ ends joined with the **7-methyl guanosine caps** are called 7-MG caps. When the growing mRNAs are only of about 30 nucleotide long, these 7MG caps and two or more methyl groups are added. This is also called 5′-capping (Fig. 9.11) and (Fig. 9.12).

9.6.3.4 Termination (and addition of 3′ poly (A) tail)

In fact, the termination in eukaryotic transcription is not typical and there is *no* termination sequence in the DNA to give signal for ending the process. Usually, the last part of RNA synthesis is associated with **5′-capping, 3′-poly (A) tail forming, formation of mature mRNA from pre-mRNA** (splicing) and **editing**.

FIGURE 9.10

The initiation of transcription by RNA polymerase II requires the formation of a basal transcription initiation complex at the promoter region. The assembly of this complex begins when TFIID, which contains the TATA-binding protein (TBP), binds to the TATA box. The other transcription factors and RNA polymerase it join the complex in the sequence shown.

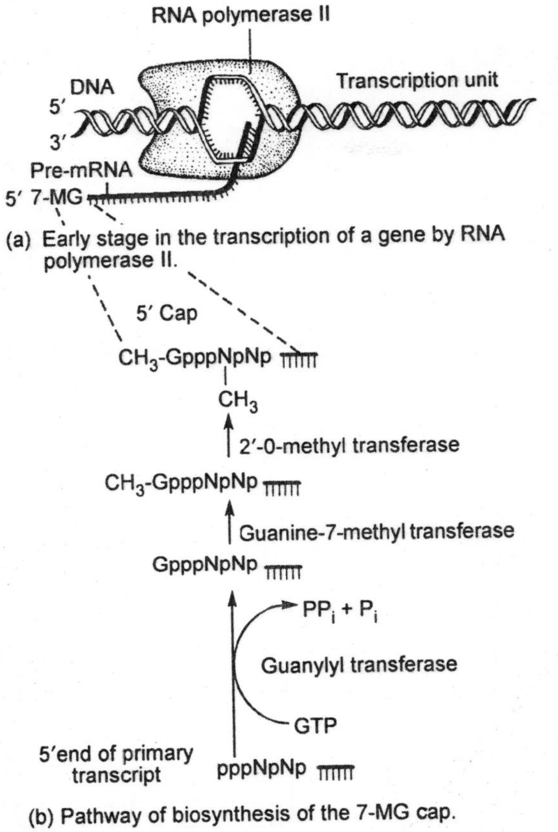

FIGURE 9.11

7-Methyl guanosine (7-MG) caps are added to the 5′ ends of pre-mRNAs shortly after the elongation process begins.

The eukaryotic pre-mRNAs become modified during the end of elongation at their 3′ ends by addition of a sequence of 50-250 *Adenine* nucleotides called **poly (A) tail**. For this, a number of proteins (CPSF, CstF, CFI and CFII) bind to the RNA and cleave it (Fig. 9.13). The enzyme **Poly (A) polymerase** (PAP) catalyses the addition of adenine nucleotides at the 3′ end.

9.6.4 RNA Processing and Maturation of pre-mRNA to mRNA

The eukaryotic pre-mRNAs mostly contain certain parts (genes) that do not specify an amino acid sequence. These parts of the nascent pre-mRNA are called **introns**. The parts in between introns are called **exons**. That is, the introns and exons are alternately present in the pre-mRNA. These introns are now **excised** from the pre-mRNA transcript to produce a **mature** and **functional** mRNA, containing only **exons**. The exons are the amino acid coding sequence, needed for the process of translation.

After capping and tailing, and removing the introns of pre-mRNAs, **the exons are joined together** by a process called **mRNA splicing**. The process occurs in **spliceosomes** (a complex of small nuclear ribonucleoprotein particles, **snRNPs** or **snurps**) (Fig. 9.14).

9.6.4.1 Splicing of mRNA

The pre-mRNA contains coding sequence or **exons** and non-coding sequence or *introns*. **Splicing is the excision of introns from the transcript and rejoining of the exons** [Fig. 9.15 (a), (b)]. Five **sn RNAs** (small nuclear RNAs) called U1, U2, U4, U5 and U6 are involved in splicing. These sn RNAs do not exist as free molecules but remain associated with **snRNPs**. Each of U1, U2 and U5 is present in a specific sn RNP; U4 and U6 are present together in a fourth sn RNP.

The first step in splicing involves a cleavage at the 5′ intron splice site (↓Gu-intron), and a phosphodiester linkage is formed between 5′ carbon of the G at the cleavage site and the 2′ carbon of a conserved A-residue. It forms the branch point in the lariat structure of the spliced intron. Next, the U5 snRNP binds at the 3′ splice site, and U4/U6 is added to the complex forming a complete **spliceosome**. When the 5′ intron splice-site is cleaved in the first step, the U4 snRNP is released. In the second step, the 3′ splice site of the intron is cleaved. Next, the two adjacent exons are joined by a phosphodiester bond. The resultant spliced mRNA is then released to the cytoplasm for translation (protein synthesis).

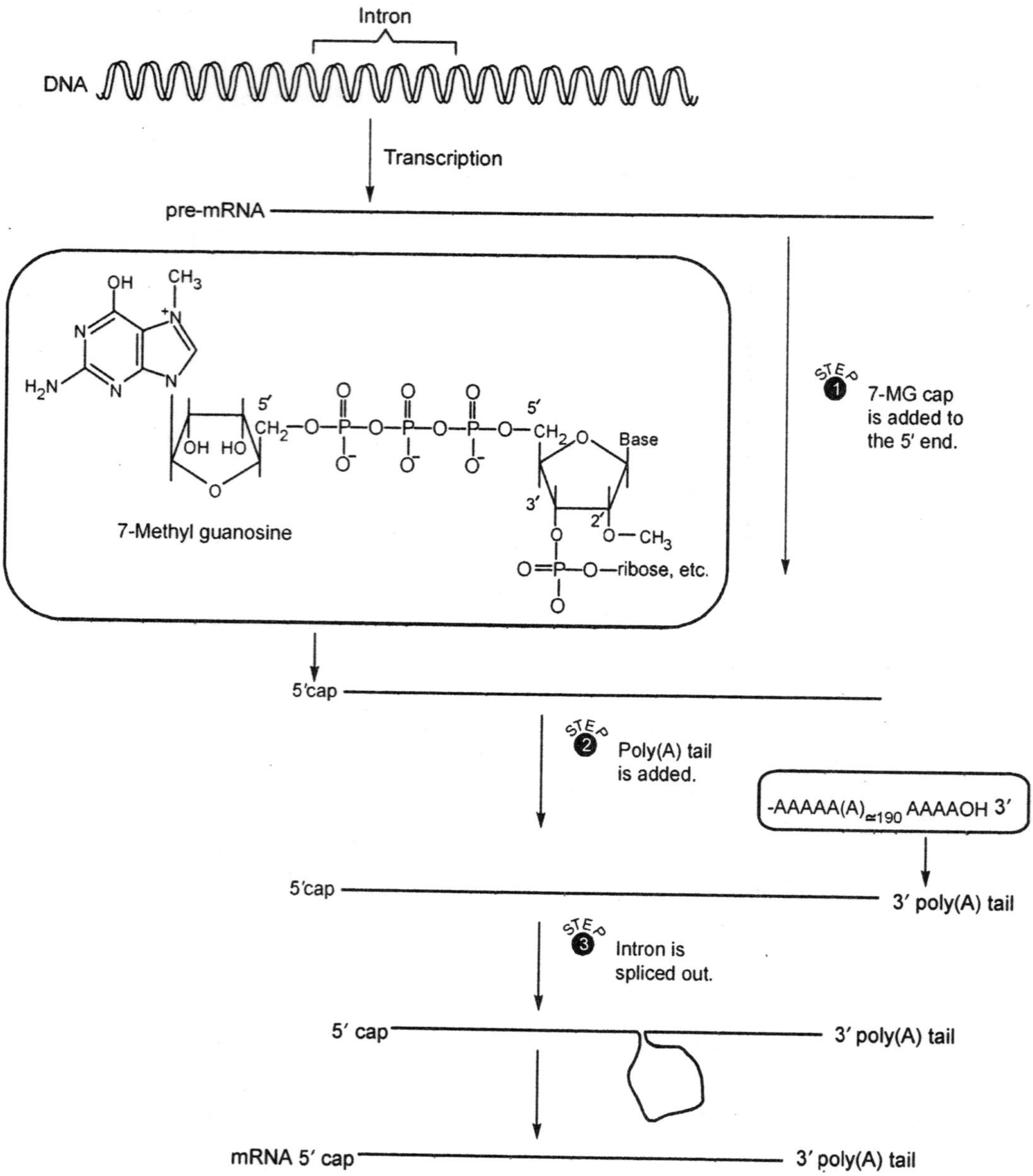

FIGURE 9.12

Post transcriptional process in eukaryotes.

The excision of introns is very precised and accurate, so that *codons* in exons are read correctly at the time of protein synthesis (translation).

9.6.4.2 RNA Editing

After transcription, RNAs are edited by adding or deleting nucleotides; sometimes, a base is converted into another. The process of editing produces a functional RNA that bears a little different base sequence from its original DNA template. The RNA editing must be accurate is order to get appropriate codons for translation into desired proteins. In mammals, the editing involves the change of Cytosine (C) to Uracil (U) in the nuclear

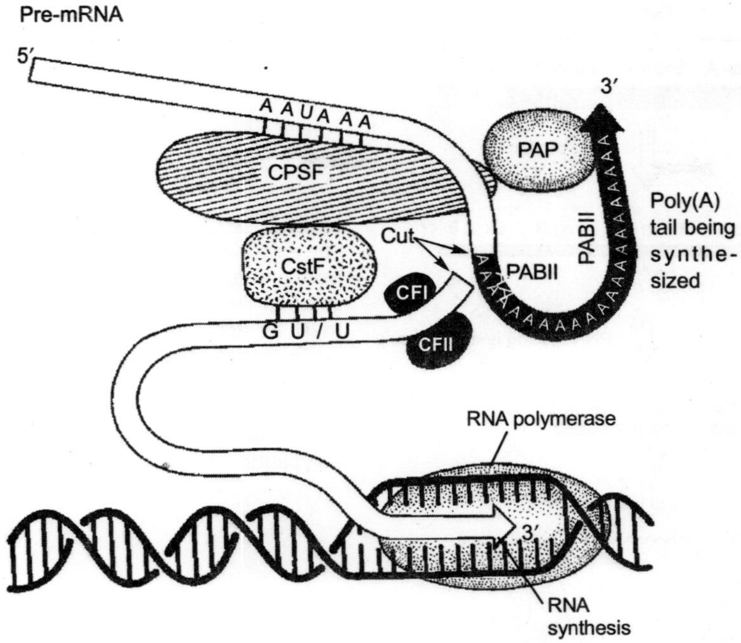

FIGURE 9.13

Schematic diagram of the 3′ end formation of mRNA and the addition of the poly(A) tail to that end. In eukaryotes, the formation of the 3′ end of an mRNA is produced by cleavage of the lengthening RNA chain. CPSF binds to the AAUAAA signal, and CstF binds to a GU-rich or U-rich sequence (Gu/U) downstream of the poly(A) site. CPSF and CstF also bind to each other, producing a loop in the RNA. CFI and CFII bind to the RNA and cleave it. Poly(A) polymerase then adds the poly(A) tail to which poly(A) binding proteins (PAP) attach.

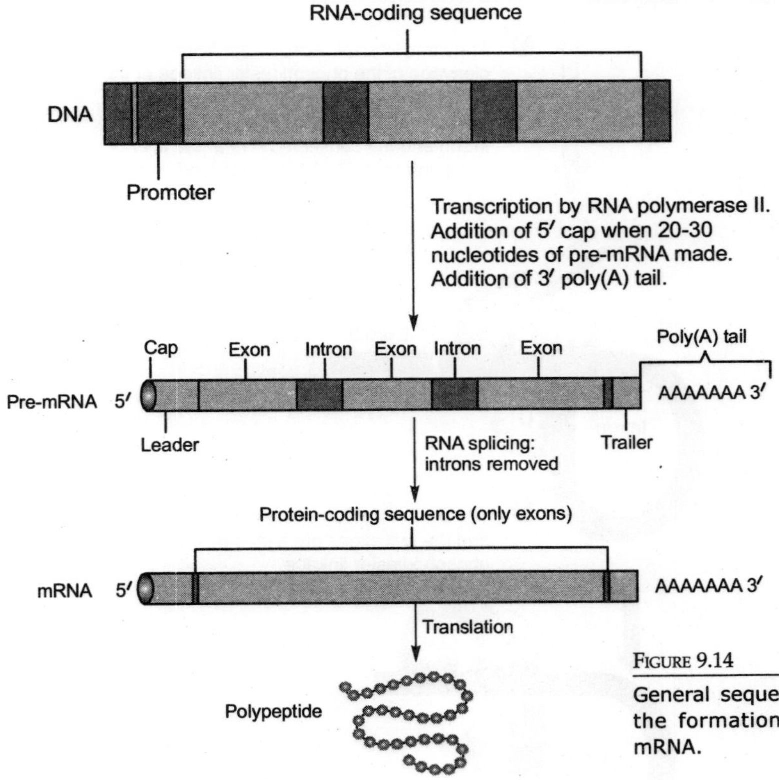

FIGURE 9.14

General sequence of steps in the formation of eukaryotic mRNA.

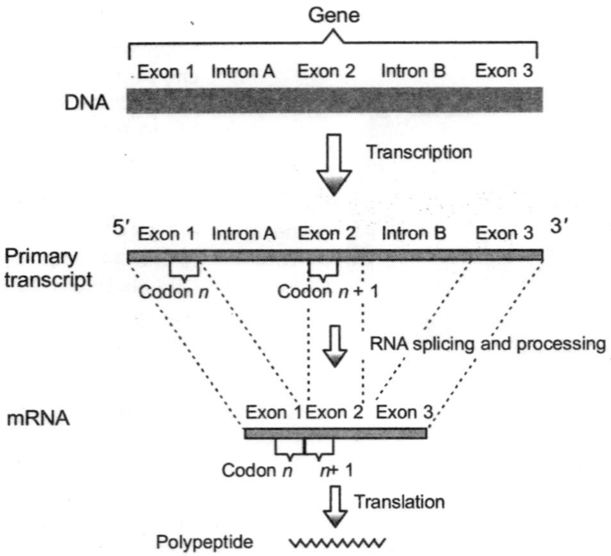

FIGURE 9.15(a)

The excision of intron sequences from primary transcripts by RNA splicing. The splicing mechanism must be accurate to the single nucleotide to assure that codons in downstream exons are translated correctly to produce the right amino acid sequence in the polypeptide product.

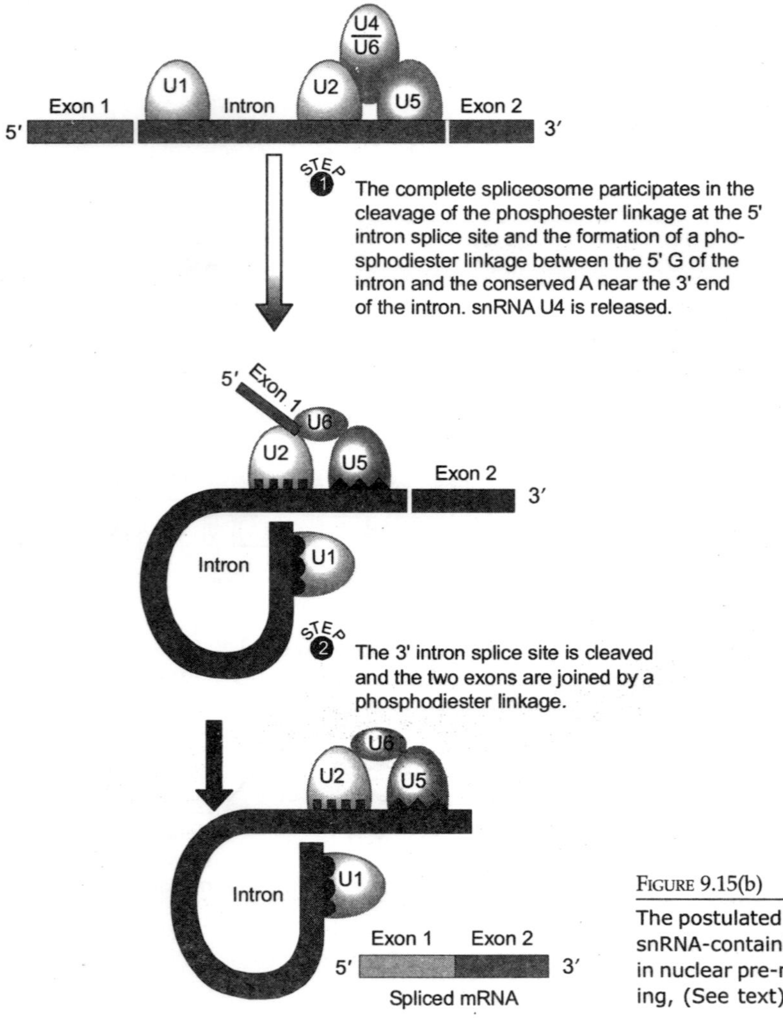

FIGURE 9.15(b)

The postulated roles of the snRNA-containing snRNPs in nuclear pre-mRNA splicing, (See text).

gene-encoded mRNA. Besides, Adenine (A) may be changed to Guanine (G) in mRNAs, or the editing of pyrimidine bases may often be noticed in tRNAs.

9.6.4.3 Transcription of rRNA

The rRNA is transcribed by RNA polymerase I to produce a single *precursor rRNA* (p30S) which contains a 5′ spacer, the 16S, 23S and 5S r RNA sequences, and a 3′ spacer. The RNA polymerase I (present in the nucleolus) transcribes the 18S, 5.8S and 28S r RNA sequence. RNA polymerase III cleaves p-30S to produce pre-16S, pre-23S and pre-5S. RNA polymerase III transcribes tRNA, 5SrRNA and other sn RNA genes. After transcription, processing takes place to produce functional ribosomal subunits that assemble with protein to form ribosome. (Fig. 9.16).

9.6.4.4 Transcription of tRNA

tRNAs are 75-90 nucleotide RNAs. Prokaryotic tRNA genes are found in one or

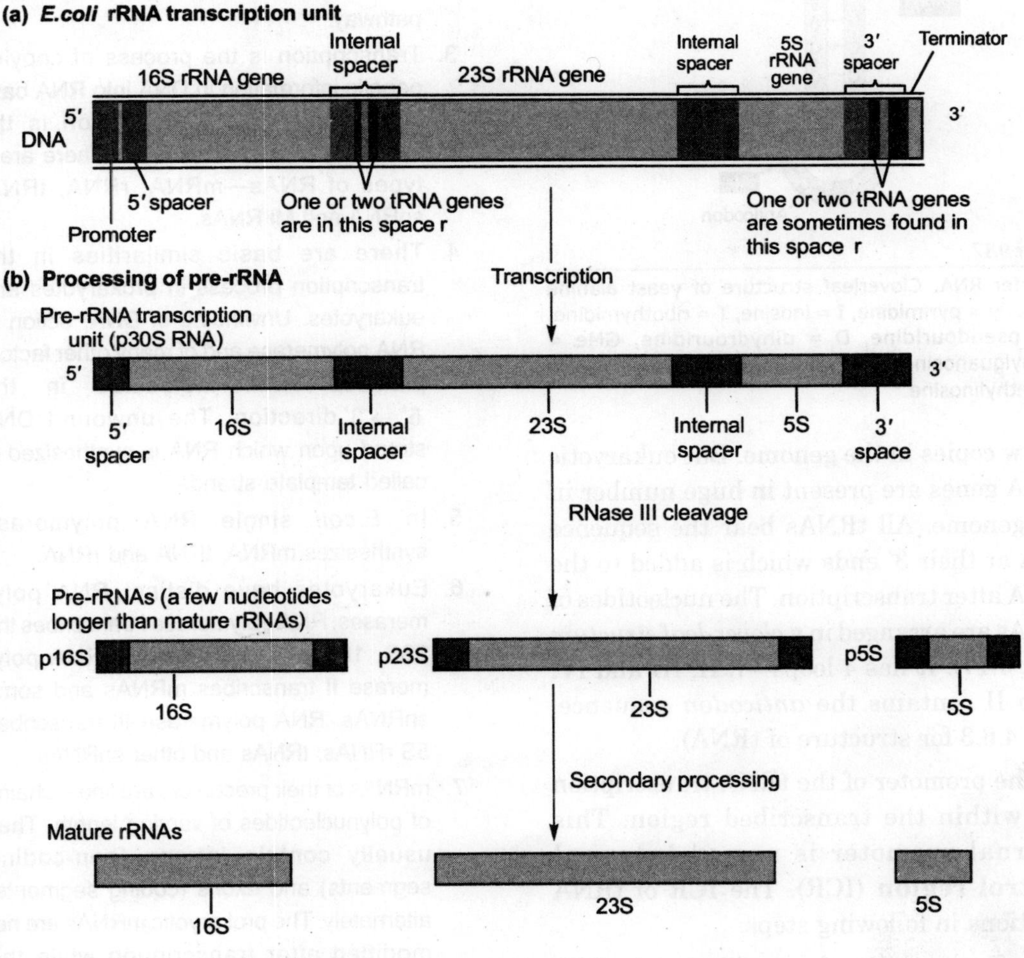

FIGURE 9.16

Production of rRNA in *E.coli.* (a) General organization of an *E.coli* rRNA transcription unit. (b) The scheme for the synthesis and processing of a precursor rRNA (p 30S) to the mature 16S, 23S, and 5S rRNAs of *E.coli* (The tRNAs have been omitted).

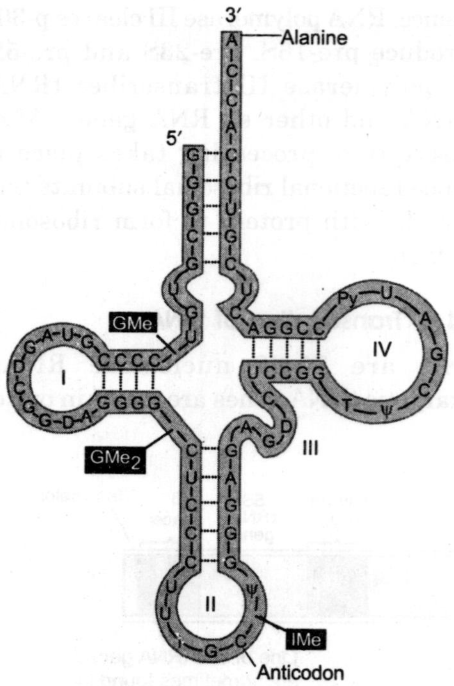

FIGURE 9.17

Transfer RNA. Cloverleaf structure of yeast alanine tRNA. Py = pyrimidine, I = inosine, T = ribothymidine, ψ = pseudouridine, D = dihydrouridine, GMe = methylguanosine, GMe_2 = dimethylguanosine, IMe = methylinosine.

in few copies in the genome. But eukaryotic tRNA genes are present in huge number in the genome. All tRNAs bear the sequence CCA at their 3′ ends which is added to the tRNA after transcription. The nucleotides of tRNAs are arranged in a *clover-leaf structure* (Fig. 9.17). It has 4 loops—I, II, III and IV. Loop II contains the *anticodon* sequence. (See 4.8.3 for structure of tRNA).

The promoter of the tRNA transcription lies within the transcribed region. This internal promoter is named **internal control region** (ICR). The ICR of tRNA functions in following steps:

1. Transcription factors bind to the ICR which binds RNA polymerase III.
2. It initiates transcription at the beginning of the gene, which is upstream from where it is bound.
3. The transcription produces pre-tRNA, which bear extra sequences at each end.
4. These sequences are removed to produce mature tRNA.

IMPORTANT NOTES:

1. Each gene is responsible for the production of each polypeptide. It is called one gene-one polypeptide concept.
2. Metabolic disorders like alkaptonuria, phenylketonuria, tyrosinosis or albinism are due to defects in gene expression or gene-function somewhere in the metabolic pathway.
3. Transcription is the process of copying genetic information in DNA into RNA base sequences, i.e., transcription is the synthesis of RNA from DNA. There are 5 types of RNAs—mRNA, rRNA, tRNA, snRNA and Mi RNAs.
4. There are basic similarities in the transcription process in prokaryotes and eukaryotes. Unwinding of DNA, action of RNA polymerase and of many other factors produce RNA molecules in the $5' \rightarrow 3'$ direction. The unwound DNA strand upon which RNA is synthesized is called template strand.
5. In *E.coli* single RNA polymerase synthesizes mRNA, tRNA and rRNA.
6. Eukaryotes have distinct RNA polymerases; RNA polymerase I transcribes the 28S, 18S and 5.8S rRNAs. RNA polymerase II transcribes mRNAs and some snRNAs. RNA polymerase III transcribes 5S rRNAs, tRNAs and other snRNAs.
7. mRNAs or their precursors are linear chains of polynucleotides of varying length. They usually contain introns (non-coding segments) and exons (coding segments) alternately. The prokaryotic mRNAs are not modified after transcription while the eukaryotic mRNAs undergo modifications by adding capping and tailing. The capping is the process of addition of a 7MG cap at the 5′ end and the tailing is the process of adding of poly(A) tail at the 3′ end.

8. Introns (the non-coding segments) are removed from the pre-mRNAs in eukaryotes. Once the introns are excised, the exons are spliced together. This is called splicing.
9. The splicing occurs in a complex called spliceosomes.
10. RNA polymerase III transcribes 5S rRNA, tRNA and some snRNA.
11. ICR is the internal promoter located within the 5S rDNA and tDNA genes.
12. All tRNAs bring amino acids to the ribosome in the translation process. So all tRNAs are almost of equal length (75-90 nucleotides); the tRNAs contain 4 loops, I, II, III and IV, of which the II loop bears the anticodon.

9.7 Translation

The process of protein synthesis in a cell is called **translation.**

Definition: The process that converts the nucleotide sequence of an mRNA into the amino acid sequence of a polypeptide (protein) is called translation.

9.7.1 Major Events in Translation

1. Proteins are formed of several subunits called **polypeptides.** Polypeptides are formed by **amino acids** linked together by peptide bonds. Amino acids are the building blocks of polypeptides or proteins.
2. The amino acid sequence of a polypeptide is determined by the **genetic code.**
3. The genetic code is the set of 3-nucleotide sequence (called **codons**) within mRNA that carries the information for specifying the amino acid sequence of a polypeptide.
4. Each of the 3-nucleotide codon is specific for one amino acid. Sometimes more than one codon may represent the same amino acid. The code is **universal** (applicable in all organism), **commaless** (without gap) and **non-overlapping** (see 9.8.1).
5. The site translation is the **ribosome.** tRNA molecules transfer specific amino acids for polypeptide synthesis from amino acid pools to the ribosomes. Specific amino acids are bound to specific tRNA, and specific anticodon of tRNA is bound to the specific complementary codon of the mRNA.
6. The initiator codon in all organisms is the **AUG** which specifics the amino acid **methionine** to start the process of translation, i.e., the protein synthesis.
7. **Peptide bonds** are formed between two adjacent amino acids, and the chain of polypeptide elongates. The ribosome has 3 adjacent sites, **A**, **P** and **E**, through which the entry, peptide bonding and exit of the amino acid occur respectively. As soon as one peptide bond is formed between two amino acids, the ribosome translocates one codon along the mRNA in preparation for the next tRNA (bound to another specific amino acid). The incoming tRNA binds to the next codon at the A site.
8. Polypeptide chain elongation goes on until any of the **terminator codons** (see 9.8) comes in the mRNA. The terminator codons are UAG, UAA and UGA. Release-factor proteins read these codons and the polypeptide is released from the ribosome.

9.7.2 Why the process of protein synthesis is called translation

In the process of protein synthesis, the codons (3-nucleotide sequence) are the "letters", the amino acids are the "words and the polypeptide is the "sentence". In the process, the **codons** (letters) recognize **amino acids** (words), which are strung together to form **polypeptides** (sentence). Thus, the genetic information stored in sequence of nucleotides in mRNAs is used to specify the sequence of amino acids in polypeptide products. Since the nucleotide sequence in the mRNA is translated into a chain of amino acids forming polypeptides, the process is called **translation**.

9.7.3 Proteins and Amino Acids

Proteins are very complex macromolecules formed of **20 different amino acids**. Proteins play many vital roles in living cells.

Proteins are composed of polypeptides. Polypeptides are long chains of 20 different amino acid sequences. The amino acids are linked together by **covalent peptide bonds**. All amino acids (except proline) contain a **free amino group** and a **free carboxyl group** (Fig. 9.18). Only the side groups (R, for radical) are varied and different in each amino acid.

The side groups are of 4 types: *hydrophobic* or *nonpolar*, *hydrophilic* or *polar*, *acidic* or *negatively charged* and *basic* or *positively charged*.

The nature of 20 different amino acids are—

1. **Hydrophobic or nonpolar**: Glycine (Gly, G), Alanine (Ala, A), Valine (Val, V), Leucine (Leu, L), Isoleucine (Ile, I), Proline (Pro, P), Phenylalanine (Phe, F) and Tryptophan (Trp, W).
2. **Hydrophilic or polar**: Methionine (Met, M), Serine (Ser, S) Threonine (Thr, T), Tyrosine (Tyr, Y), Asparagine (Asn, N) Glutamine (Gln, Q), Cysteine (Cys, C).
3. **Acidic or negatively charged**: Aspartic acid (Asp, D), Glutamic acid (Glu, E).
4. **Basic or positively charged**: Lysine (Lys, K), Arginine (Arg, R), Histidine (His, H).

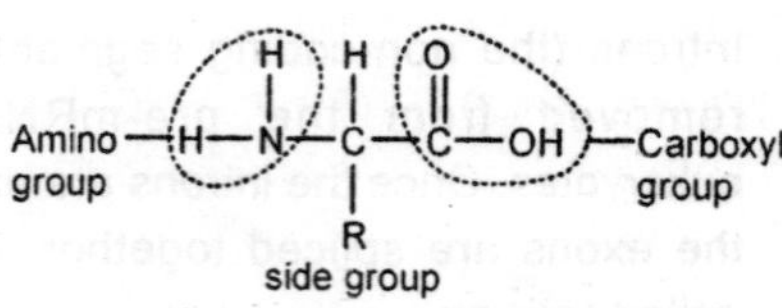

FIGURE 9.18
Structure of an amino acid.

The amino acids in the polypeptides are covalently joined by **peptide bonds**. When two amino acids are linked by a peptide bond, **a dipeptide** is formed. The formation of a peptide bond between two amino acids takes place by **connecting the amino group of one amino acid and the carboxyl group of the adjacent amino acid, and by the removal of water** (Fig. 9.19).

9.8 The Genetic Code

With the establishment of the concept of the colinearity of protein structure and gene structure, it was hypothesized that there must be a 'code' such that the sequence of amino acids in the proteins would be found in the sequence of nucleotides in the mRNA. The information written in a four-letter language (A, C, G, U) is *translated* into a 20-letter language (20 amino acids). We have learned that the codons are 'translated' into polypeptides in the process of protein (polypeptide)–synthesis. We also know that there are 4 nitrogenous bases (A, C, G and U) in the mRNA and 20 amino acids in the proteins of living cells (See 9.7.3). Thus the question arises that how do the only 4 nucleotides in the mRNA specify the 20 different amino acids in the proteins? It was observed that if there were a one-letter code, then only 4 amino acids could be encoded, and if there were a two-letter code, then only 16 (i.e. 4 × 4) amino acids could be encoded.

Both are insufficient to specify the 20 different amino acids. Thus it was revealed that, only a 3-letter code may generate 64 (i.e. 4 × 4× 4) possible codons that may sufficiently encode all the 20 amino acids. Not only that, this 3-letter code (**triplet codon**) is **universal** in its amino acid specification in all organisms and it was established that some amino acids are specified by more than one codon, called **degeneracy**.

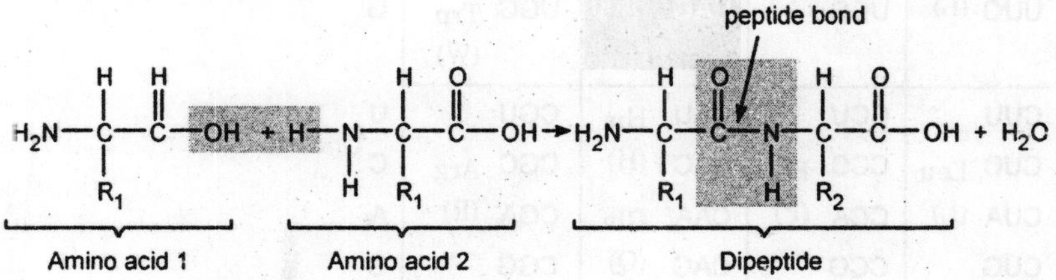

FIGURE 9.19

The formation of a peptide bond between two amino acids by the removal of water. Each peptide bond connects the amino group of one amino acid and the carboxyl group of the adjacent amino acid.

A series of experiments were performed by F. Crick, L. Barnett, S. Brenner and R. Watts-Tobin in 1960s through which it was found that the genetic code is a **triplet code** (3-letter code), called **codon** which is a set of three consecutive nucleotides in mRNA, and can specify one amino acid.

In 1961, Nirenberg and Matthaei manufactured some poly Uridylic acid (poly-U), and found that the addition of poly-U and a mixture of amino acid to a protein-synthesizing system resulted in the formation of only **Phenylalanine**. Under the triplet codon hypothesis, it was established that UUU must have coded for the phenylalanine (Phe). Soon, by similar experiments, it was revealed that AAA code for lysine, CCC for proline and GGG for glycine.

The real relationship of 64 codons to 20 amino acids was determined by M. Nirenberg and H.G. Khorana (1968–Nobel Laureates). Various experiments done by them and by many others in the following years established that out of 64 codons, 61 can specify amino acids and the other 3 cannot do that. These 3 codons (UAA, UAG and UGA) are called **non-sense codons** or **stop codons** or **terminator codons** (Fig. 9.20).

9.8.1 Characteristics of the Genetic Code

1. The genetic code is a **triplet codon**, formed by 3 successive nucleotides in the mRNA. It specifies an amino acid during translation.
2. The code is *commaless*; i.e., it is continuously read, without skipping any one. As for example, AUGAA ACAUGAAU will be read as AUG AAA CAU GAA U.... .
3. The code is **non-overlapping**. Successive groups of 3 nucleotides are used, no one is overlapped, i.e., taken twice or thrice. As for example, AUGCCCGAU will never be read as AUG GCC CGA... etc.
4. The code is **universal**, i.e., one codon specifies the same amino acid almost in all organisms.
5. There is a **degeneracy** in the code; i.e., more than one codon may specify a single amino acid (only exception is that, only the AUG can code methionine, and only the UGG codes for tryptophan). Thus the code is **degenerate** (Fig. 9.20).
6. In translation, the code AUG has a starting signal and any of UAA, UAG and UGA bears stop-signal. Codes with stop signals are called **stop codons** or **terminal codons**.
7. Francis Crick proposed that an inconsistency or **wobble occurs in anticodon**. This is called **wobble**

Second letter

First letter	U	C	A	G	Third letter
U	UUU Phe	UCU	UAU Try	UGU Cys	U
	UUC (F)	UCC Ser	UAC (Y)	UGC (C)	C
	UUA Leu	UCA (S)	UAA Stop	UGA Stop	A
	UUG (L)	UCG	UAG Stop	UGG Trp (W)	G
C	CUU	CCU	CAU His	CGU	U
	CUC Leu	CCC Pro	CAC (H)	CGC Arg	C
	CUA (L)	CCA (P)	CAA Gln	CGA (R)	A
	CUG	CCG	CAG (Q)	CGG	G
A	AUU	ACU	AAU Asn	AGU Ser	U
	AUC Ile	ACC Thr	AAC (N)	AGC (S)	C
	AUA (I)	ACA (T)	AAA Lys	AGA Arg	A
	AUG Met (M)	ACG	AAG (K)	AGG (R)	G
G	GUU	GCU	GAU Asp	GGU	U
	GUC Val	GCC Ala	GAC (D)	GGC Gly	C
	GUA (V)	GCA (A)	GAA Glu	GGA (G)	A
	GUG	GCG	GAG (E)	GGG	G

■ = Chain termination codon (stop)

FIGURE 9.20

The genetic code. Of the 64 codons, 61 specify one of the 20 amino acids. The other 3 codons are chain-terminating codons and do not specify any amino acid. AUG, one of the 61 codons that specify an amino acid, is used in the initiation of protein synthesis.

hypothesis which explains how one tRNA may recognise two codons. In codon-anticodon pairing between mRNA and tRNA, the first two bases of mRNA codon and anticodon pair properly, but the third base in the anticodon acts inconsistently (or wobbly) for which it may pair with more than one base.

IMPORTANT NOTES:

1. Proteins are composed of polypeptides.
2. Polypeptides are composed of small building blocks, the amino acids, which are of 20 different types.
3. The amino acids are linked together by peptide bonds to form long chain of polypeptides.
4. The genetic code is a triplet code. Each codon in an mRNA is composed of a set of 3 consecutive bases, and it specifies one amino acid.
5. There are 64 codons of which 61 can specify amino acids and hence are called sense codons; The rest 3 cannot specify an amino acid, and hence are called non-sense or stop or terminator codons.
6. The code is degenerate; it means some amino acids are specified by more than one codon.
7. The code is almost universal.
8. The code is commaless; that is, it is read continuously.
9. The code is non-overlapping.
10. There are specific codons for starting and terminating transtation.

9.9 The Process of Protein Synthesis or Translation

The genetic information encoded in the codons of mRNA is translated into protein on the ribosomes. The principal facts observed in the process of protein synthesis are:

1. The code in the mRNA is translated in 5′ → 3′ direction.
2. Polypeptide is produced by the arrangement of amino acids in its N-terminal to C-terminal direction.
3. tRNA molecules bind the amino acids and transport them from amino acid pool to ribsome for sequencing with the mRNA codon.
4. Each amino acid is bound to its own specific tRNA.
5. The codon in mRNA is bound to the complementary anticodon in tRNA, and this results into the correct sequence of amino acids in the polypeptide.

9.9.1 Steps in Protein Synthesis

There are 3 basic steps in protein synthesis the **initiation, elongation of polypeptides** and **termination.** Before starting of translation, two prerequisite events, the **recognition** and **activation of amino acid** take place.

- **Recognition**: Ehrenstein, Weisblum and Benzer established that (a) **the mRNA codon recognizes the tRNA anticodon** and (b) **the specificity of codon recognition lies in the tRNA molecule.**

- **Activation of amino acid and joining of it to the tRNA, or charging or aminoacylation**:

The enzyme *aminoacyl-tRNA synthetase* catalyses the attachment of a correct amino acid to the tRNA. This action is called **charging** or **aminoacylation.** There are 20 different aminoacyl tRNA synthetases for 20 different amino acids. After attachment of a specific amino acid to the tRNA, the complex is called **charged-tRNA** or the **aminoacyl-tRNA.** It is achieved by the following process (Fig. 9.21).

9.9.1.1 Initiation of Translation

After **recognition** and **activation**, the first basic steps of translation, the *initiation* starts. The process of protein synthesis is almost same in all organisms regardless of prokaryotes and eukaryotes. Initiation includes—

(a) At first, an **initiation complex** is formed between the **smaller ribosomal subunits, mRNA** and **an initiation tRNA molecule.** In *E.coli* (prokaryote), the initiation tRNA molecule carries the amino acid **methionine** with a formyl group (CHO) attached to the amino-nitrogen, and is called **formyl methionine (f-met) tRNA.**

(b) A special codon AUG in the mRNA acts as **initiator codon** that marks the **initiation site.**

(c) The initiation site is now supplied with 3 protein factors, called **initiation factors,** viz., *the* **specific initiator tRNA, GTP** and **Mg⁺⁺.**

(d) Along with the initiator codon AUG, a sequence upstream called the **ribsome binding site** (RBS) helps ribosomal subunit to bind to the mRNA. This region of mRNA is purine-rich and was discovered by Shine and Dalgarno, and is known as the **Shine-Dalgarno sequence** (This sequence is not present in eukaryotes).

(e) After binding the ribosomal subunit at Shine-Dalgarno sequence, the initiator tRNA binds to the AUG to which the 30S ribosomal subunit is bound.

(f) As the AUG specifies methionine in both prokaryotes and eukaryotes, the new protein that will be synthesized, will begin with methionine in all cases. The methionine may be removed later in necessity.

(g) The formylmethionine (f-met) is brought to the ribosome attached to a special initiator tRNA, the **tRNA-f-met,** which bears the anticodon 5′-CAU-3′ to bind the AUG of mRNA.

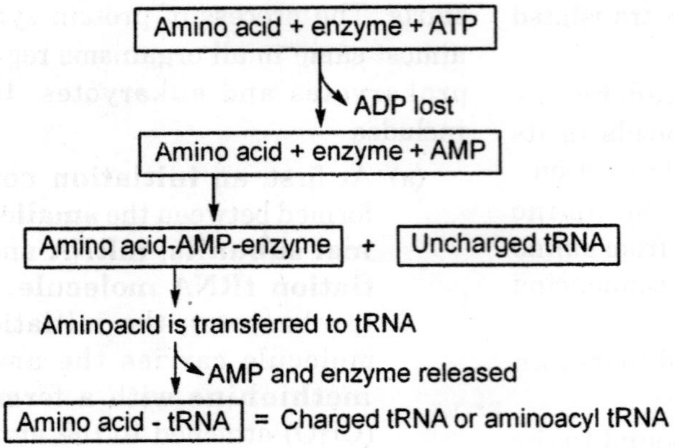

FIGURE 9.21
Formation of charged tRNA molecule

(h) When the f.met-tRNA binds to the AUG of the 30S-mRNA complex, initiator factor-3 (IF-3) is released, and a **30S initiation complex** is formed. This complex consists of 30S ribosomal subunit, mRNA, f.met-tRNA, IF-1 and IF-2.

(i) In the next step, 50S subunit binds to the 30S initiation complex which leads to GTP hydrolysis and release of IF-1 and IF-2.

(j) This results in the formation of **70S initiation complex**.

(k) There are 3 sites in the 70S ribosome complex—the amino acyl site (A), the peptidyl site (P) and the exit site (E). (Fig. 9.22).

(l) It is important to note that, in eukaryotes, instead of Shine-Dalgarno sequence, the AUG is embedded in a short sequence called **Kozak sequence**, for initiation.

(m) Several proteins, such as initiation factors (eIFs), eIF-4F, cap-binding protein (CBP), 40S subunit, Met.tRNA and GTP are involved in initiation in eukaryotes.

(n) When the 40S ribosomal subunit attaches to the AUG, the 60S also binds to it, releasing the eIFs (excluding eIF-4F). It forms an **80S initiation complex** in eukaryotes.

(o) The 80S initiation complex and the initiator Met.tRNA now bind to the mRNA in the P site of the ribosome.

(p) The poly (A) tail of the eukaryotic mRNA binds to eIF-4G and makes a loop at the 3′ end of the mRNA which also stimulates the initiation of translation in eukaryotes.

9.9.1.2 Elongation of Polypeptide Chain (Fig. 9.23)

The process of elongation of polypeptide chain is almost similar in prokaryotes and eukaryotes. It involves 3 steps:

i. Binding of **amino acyl-tRNA** to the A-site.

ii. Transfer of the growing polypeptide chain **from the tRNA in the P-site to the tRNA in the A-site**, by forming a new peptide bond, and

iii. **Translocation** of the ribosome along the mRNA **to place the next codon in the A site**. At the same time the nascent polypeptide-tRNA and the uncharged tRNA are translocated from A-and P-sites to the P-and E-sites respectively.

These 3 steps are repeated cyclically until the termination of elongation.

Step-I. Binding of aminoacyl-tRNA

(a) An aminoacyl-tRNA (charged) enters and binds to the A-site of the ribo-

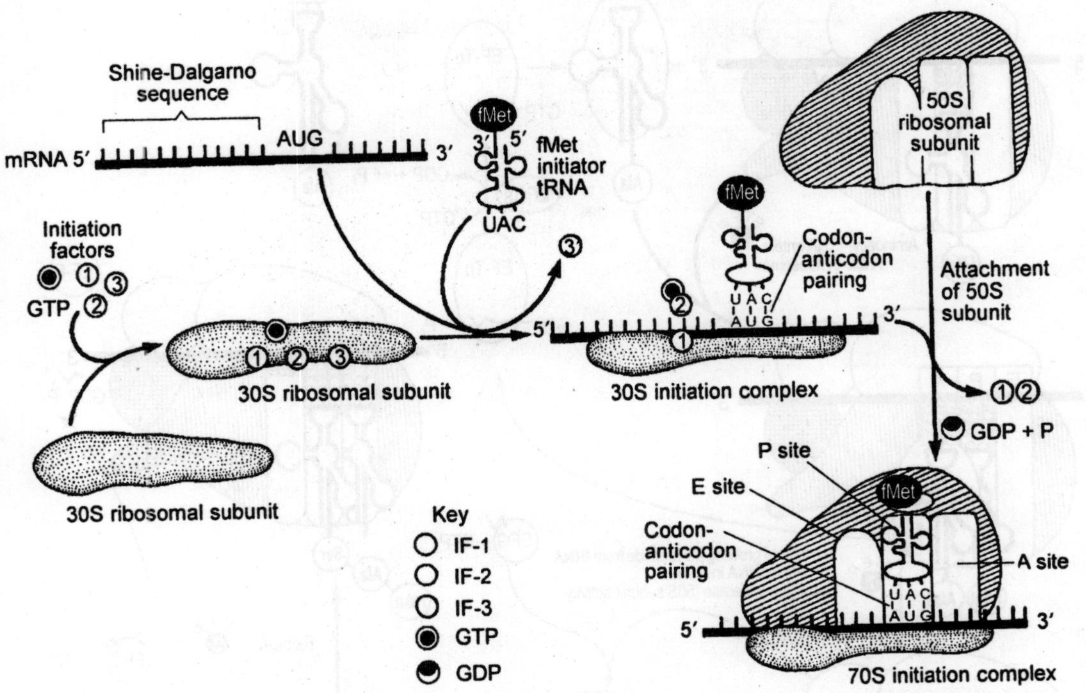

FIGURE 9.22

Initiation of protein synthesis in prokaryotes. A 30S ribosomal subunit, complexed with initiation factors and GTP, binds to mRNA and fMet-tRNA to form a 30S initiation complex. Next, the 50S ribosomal subunit binds, forming a 70S initiation complex. During this event, the initiation factors are released and GTP is hydrolyzed.

some, with specific mRNA codon. The three nucleotides in the anticodon of this tRNA pair with the nucleotides of the mRNA codon at the A-site.

(b) In this binding, the **elongation factor Tu** with a molecule of **GTP** (EF-Tu- GTP) is required. After the formation of a peptide bond, the GTP is cleaved.

(c) After the cleavage of GTP, the EF-Tu-GDP is released from the ribosome. This EF-Tu-GDP is converted to the active EF-Tu-GTP by elongation factor Ts (EF-Ts).

Step-II. Transfer of growing polypeptide chain from P-site to A-site

(a) A peptide bond is formed between the amino group of the aminoacyl-tRNA in the A-site and the carboxyl group of the growing polypeptide chain attached to the tRNA at P-site.

(b) The formation of peptide bond detaches the growing chain from the tRNA at P-site and covalently joins the chain to the tRNA at the A-site. The enzyme **peptidyl transferase** catalyses the reaction. The GTP is hydrolyzed at the time of peptide bond formation.

Step-III. Translocation of the ribosome to place the next codon of mRNA in the A-site

(a) The peptidyl-tRNA is **translocated** from the A-site to the P-site.

(b) The uncharged tRNA is **translocated** from P-site to E-site. Both the translocation is due to the three-nucleotide movement of ribosome toward the 3′ of the mRNA.

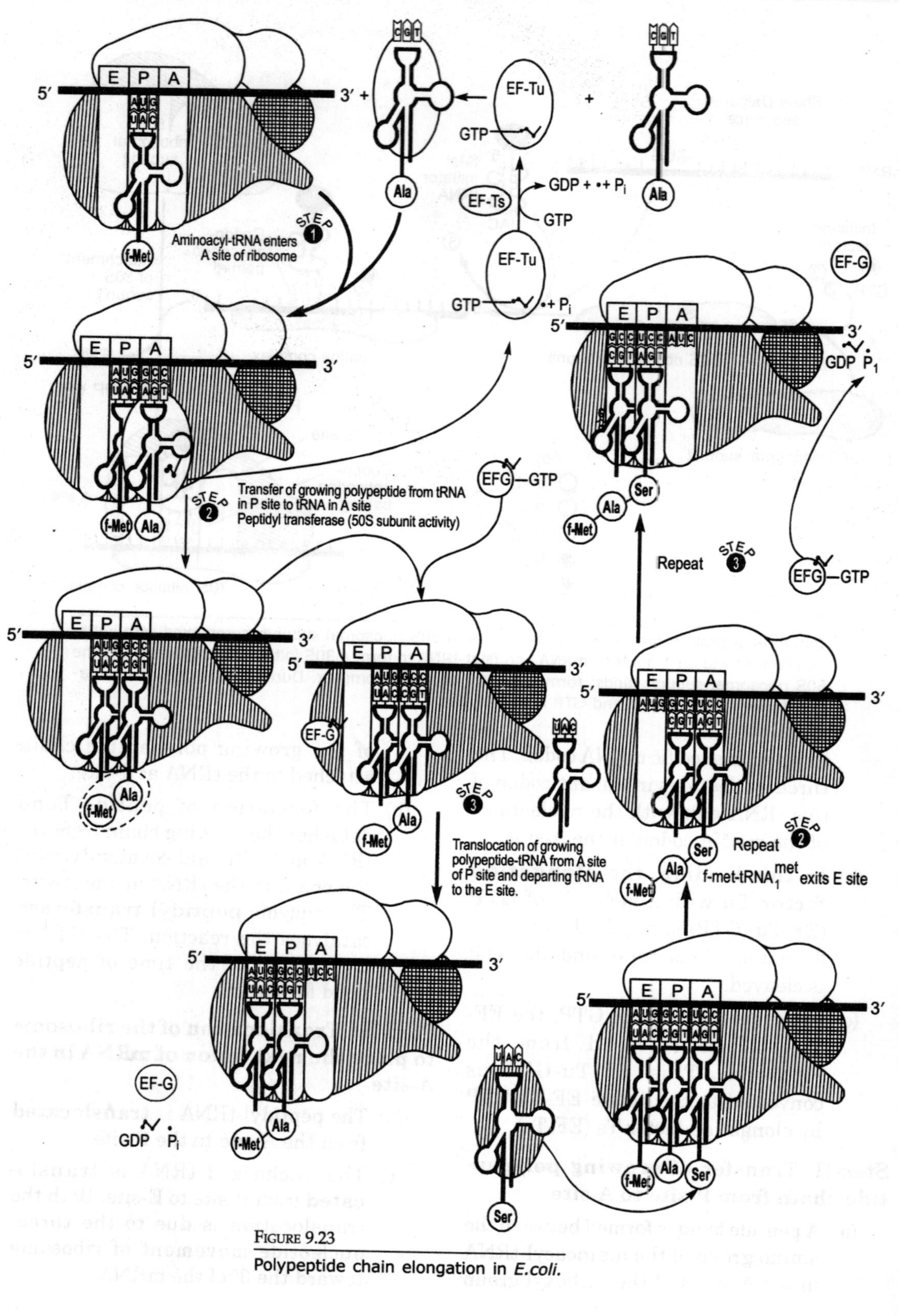

FIGURE 9.23
Polypeptide chain elongation in *E.coli.*

(c) GTP provides necessary energy for the movement of ribosome, and elongation factor G (EF-G) is required for the chain elongation.

(d) As the peptidyl-tRNA has translocated from A-site to P-site, the former remains empty or blank and becomes ready to accept another charged tRNA.

(e) The chain elongation is a rapid process. In *E.coli*, every one amino acid is added to the polypeptide chain in about 0.05 second.

9.9.1.3 Termination of Translation (Fig. 9.24)

(a) Termination of polypeptide chain occurs only when any of the 3 *terminator codons* or *stop codons* or *nonsense codons* (UAA, UAG or UGA) enters the A-site of the ribosome. The stop codons are recognized by *release factors* (RFs).

(b) In *E.coli*, there are two release factors—RF-1 and RF-2. The former recognizes UAA and UAG while the latter recognizes UAA and UGA.

(c) In eukaryotes, there is a single release factor (eRF) that can recognize all the three stop codons.

(d) Termination is completed by the release of the mRNA molecule from the ribosome. Then the ribosome splits into its subunits.

The whole process of protein synthesis is then repeated for each type of protein needed by the cell.

Important Notes:

1. Genetic information is carried by the triplet codon in the mRNA.
2. The codon is translated into amino acid sequences to synthesize polypeptides.
3. The process of polypeptide synthesis is conducted in the ribosome.
4. Various types of RNA (rRNA, mRNA, tRNA) and various types of protein factors (IF, Tu, RF etc.) along with GTP are required in the process of translation.
5. The whole process of translation is carried out in 3 steps: initiation, elongation of polypeptide chain and termination.
6. The process is governed by the specific cations of the genetic code.
7. All the codons except UAA, UAG and UGA are responsible for elongation; The UAA, UAG and UGA are responsible for termination. The AUG is responsible for initiation.

9.10 Regulation of Gene Expression

It is now known that all the genes of a nucleus are never expressed at the same time, hence an individual shows different phenotypes at different times. Only a small fraction (less than 10%) are expressed in any one tissue at a certain time. Certain *Key genes* that regulate specific patterns of morphogenesis in *Drosophila melanogaster*, have been discovered. Researches have unfolded that the development of this fly is under the control of a *cascade of regulatory genes*. These genes in turn, add new pattern of information to the original blue-print. There are some "house-keeping genes" that encode rRNAs, ribosomal proteins and tRNAs which are expressed in all cells. But many other genes are expressed for only a short period in certain cells at a certain stage of development. Thus, the expression of genes is regulated, and the regulation is sequential. The regulation of sequential pathways of gene-expression is the primary cause for the diversity of cell types in higher organisms.

9.10.1 Genetic Control of Development in *Drosophila*

Morphology is the genetic reflection in an organism. Morphogenesis is the result of regulated pathways of gene expression. The process and timings of gene expression and the pathways of gene-product action in *Drosophila* are very complex.

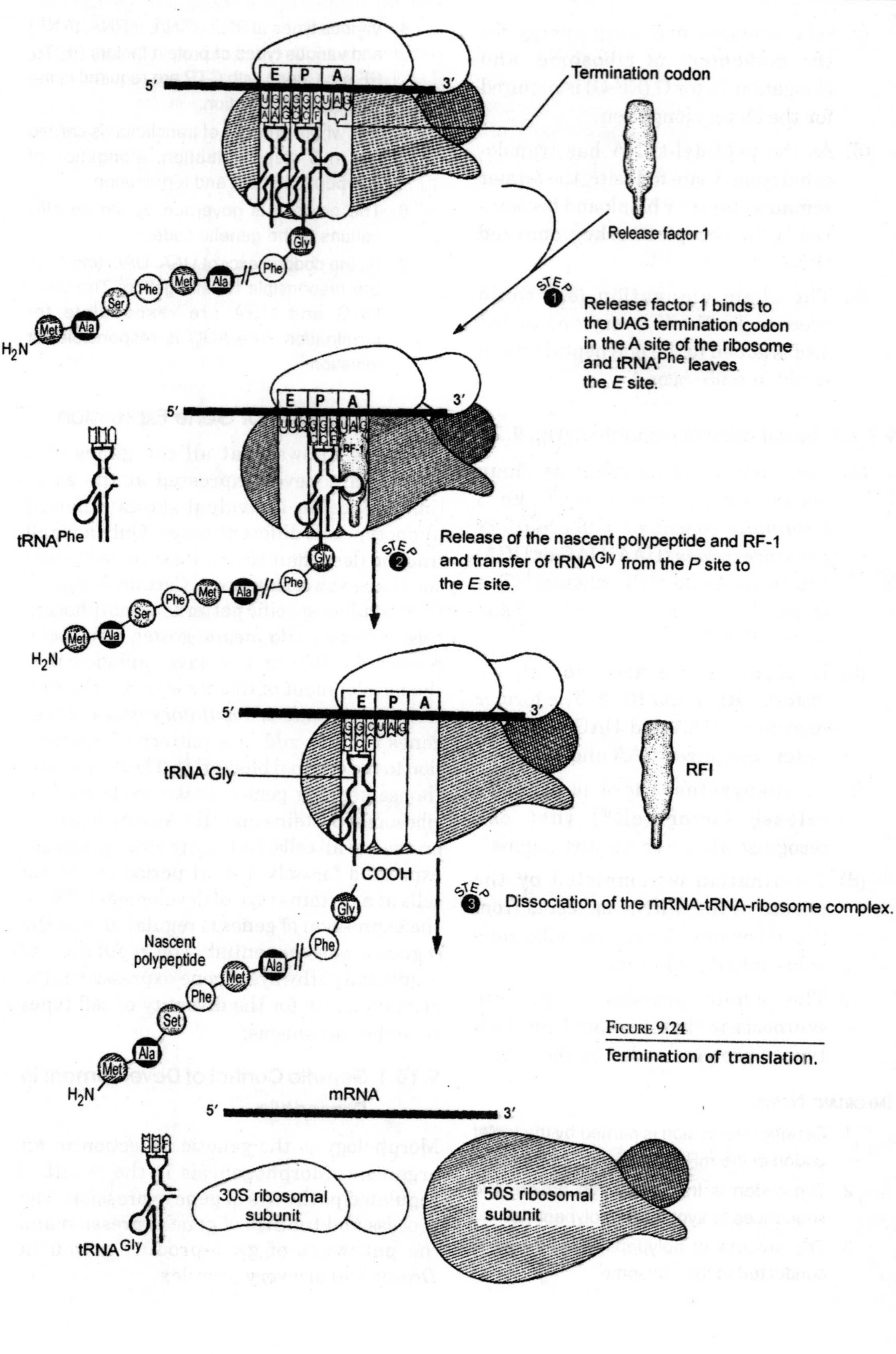

FIGURE 9.24
Termination of translation.

Development in *Drosophila* is regulated by a cascade of regulatory genes, resulting in different patterns of gene expression on various parts of its zygote, blastula, gastrula and so on, until the total morphology of the adult is achieved. Differences in the patterns of gene-expression in the different tissues occur at different times. The different cell lineages that produce different tissues and organs gradually diverge according to the gene expression of the cascade of regulatory genes.

The developmental process in *Drosophila* is shown in the Fig. 9.25a and 9.25b. The genetic programmes are predetermined for every stage. Metamorphosis from larva to pupa, and from pupa to adult requires changes in the programme of gene expression. The changes are triggered by a hormone called *ecdysone*, which stimulates the transcription of different sets of genes in different tissues.

The cells of blastoderm of *Drosophila* differentiate into two types of cells(1) Cells that give rise to larval tissue and (2) Cells that will form various tissues and organs of the adult.

9.10.1.1 Imaginal Discs

Certain undifferentiated cells form some structures called *imaginal discs*. The number of such imaginal discs are 17. Each of the imaginal discs gives rise to a specific organ of the adult insect.

The discs are—

1. 1 pair of labial discs
2. 1 pair of antenna discs
3. 1 pair of eye discs
4. 1 pair of wing discs
5. 1 pair of haltare discs
6. 3 pairs of leg discs
7. Single genital disc.

9.10.1.2 Determination

All the cells of discs have definite destiny, and they irreversibly determine a specific pattern of **differentiation**. This is called **determination**. The imaginal disc determination occurs in the blastoderm stage, and it can be worked out by chalking out **fate maps**. The fate maps are the diagrams that relate the fates of different cell-lineages.

9.10.1.3 Homeotic Mutants

The differentiation by determination of the committed cells of discs is under genetic control. The achievement of specific pattern is predestined, as this is regulated by genes.

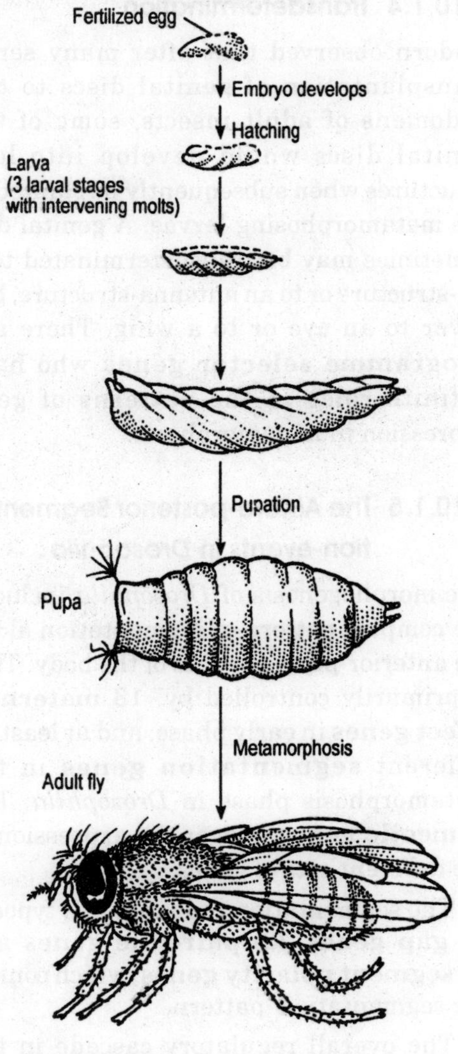

FIGURE 9.25(a)

Stages of development in *Drosophila*.

Some mutations, called *homeotic mutations* may cause altered states of determination in specific imaginal discs. They result in altered structures, called *homeotic mutants*. A large number of homeotic mutants have been discovered in *Drosophila*. There are two clusters of homeotic genes on the chromosome 3, viz., **Antennapedia complex (ANT-C)** and **Bithorax complex (BX-C)**. Fig. 9.26 shows the body segments of *Drosophila* primarily affected by the ANT-C and BX-C.

9.10.1.4 Transdetermination

Hadorn observed that after many serial transplantation of genital discs to the abdomens of adult insects, some of the genital discs would develop into leg-structures when subsequently implanted in the metamorphosing larvae. A genital disc sometimes may be transdeterminated to a leg-structure or to an antenna-structure, but never to an eye or to a wing. There are **programme selector genes** who have definite roles in the patterns of gene expression in metamorphosis.

9.10.1.5 The Antero-posterior Segmentation-events in *Drosophila*

The morphogenesis of *Drosophila* includes the complex pattern of segmentation along the anterior-posterior axis of the body. This is primarily controlled by, **18 maternal-effect genes** in early phase, and at least 20 different **segmentation genes** in the metamorphosis phase in *Drosophila*. The **homeotic genes** trigger the expression of the different sets of genes.

The segmentation genes are of 3 types—(1) **gap genes**, (2) **pair-rule genes** and (3) **segment polarity genes**, which control the segmentation pattern.

The overall regulatory cascade in the control of development in *Drosophila* can be summarized as below (Fig. 9.27). The arrows

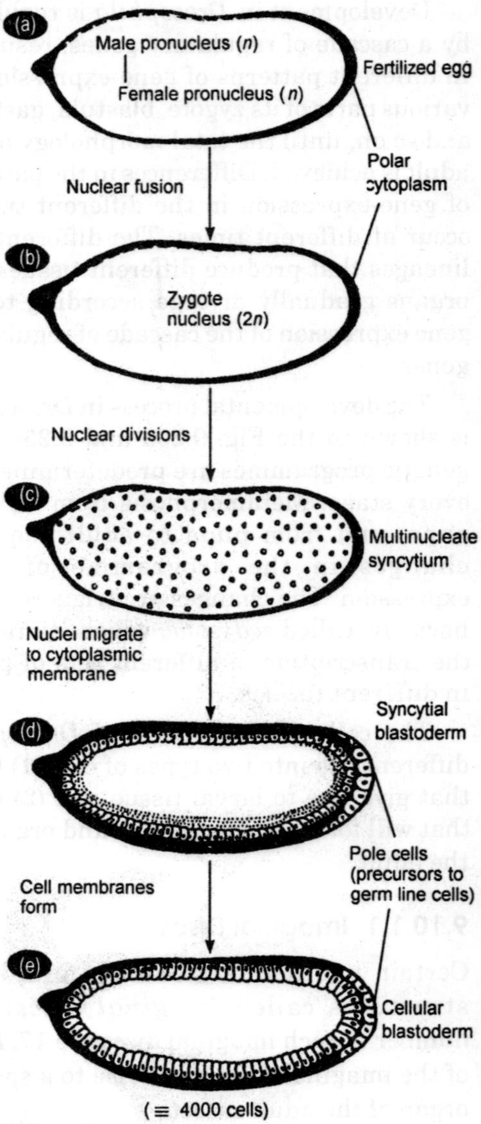

FIGURE 9.25(b)

Embryonic development in *Drosophila.* (a) The fertilized egg initially contains two haploid (n) nuclei, one from the egg and one from the sperm. (b) The two haploid nuclei then fuse to form the diploid nucleus ($2n$) of the zygote. (c) The zygotic nucleus divides very rapidly to produce a single cell (syncytium) with many nuclei. (d) After about 9 nuclear divisions have produced about 512 nuclei, the nuclei migrate to the periphery of the syncytium and continue to divide (about 4 more times each). A few nuclei migrate into the polar cytoplasm and form pole cells, which are the progenitors of all the germ line cells of the fly. (e) Cell membranes then form around the nuclei to produce the cellular blastoderm, composed of about 4000 cells (After V.e. Foe and B. M. Alberts, *J. Cell Sci.* 61: 31-70, 1983).

indicate the pathway of gene regulation that affect the expression of one or more genes.

9.10.1.6 The "Homeobox" Domain and Transcription

Usually the homeotic genes, gap genes and segmentation genes have a highly conserved region of 180 bp length. This conserved structure is called "**homeobox**". In all cases the homeobox lies within the coding region of the gene and specifies a 60-amino acid segment, called **homeodomain**. These act as transcription-regulators. The homeobox sequences are essential for DNA-binding activity.

9.10.1.7 Hormonal Regulation of Gene Expression

Steroid hormones are known as the major regulators of transcription in the eukaryotes. In some cases, they are known to bind with receptor proteins which then activate the transcription.

Much of the regulation of gene-expression in eukaryotes occurs during transcription. Some occurs at the level of RNA-processing. Further, the transcription in eukaryotes are regulated by the **enhancers** and **silencers** that control gene expression from a distance.

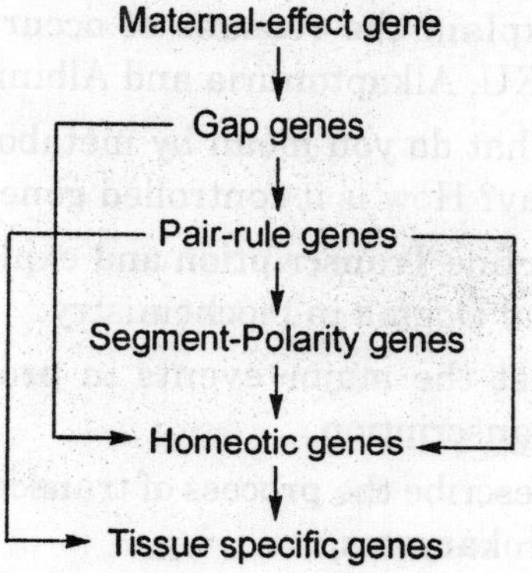

FIGURE 9.27

Regulatory cascade of the genes in the development of *Drosophila*.

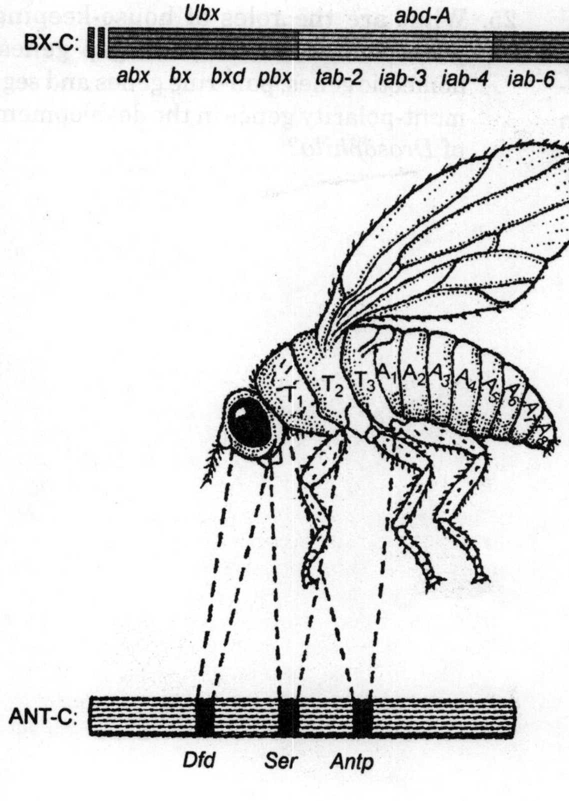

FIGURE 9.26

Diagram showing the body segments of *Drosophila* primarily affected by homeotic mutations in the two major clusters of homeotic genes on chromosome 3. The *Bihorax* complex (BX-C) contains genes that control the development of thorax segments 2 and 3 and the segments of the abdomen (A1-A8). The *Antennapedia* complex (ANT-C) contains three genes (*Dfd, Ser,* and *Antp*) essential for development of the head and thorax segments. The dotted lines indicate the zones of influence of the various genes of BX-C and ANT-C. Each of the homeotic mutants shown converts a particular body segment to a phenotype more like an adjacent segment.

EXERCISE

1. How do the metabolism and genes remain related?
2. What do you understand by one gene-one polypeptide concept?
3. What are "inborn errors"?
4. Explain the reasons of occurrence of PKU, Alkaptonuria and Albinism.
5. What do you mean by metabolic pathway? How is it controlled genetically?
6. Define Transcription and explain Central Dogma in biochemistry.
7. List the major events in prokaryotic transcription.
8. Describe the process of transcription in prokaryotes.
9. What is RNA polymerase? What role does it play in transcription?
10. Describe transcription in eukaryotes.
11. What is the role of eukaryotic RNA polymerase? How many type of RNA polymerases occur in eukaryotes?
12. What are RNA processing and RNA-maturation? Describe the process in an eukaryote.
13. What is spliceosome? What is RNA splicing? Describe the process of splicing.
14. What is translation?
15. List the major events in translation.
16. What do you know about the Genetic Code?
17. List and explain the properties of genetic code.
18. Describe the process of protein synthesis.
19. What is 'charging'? How a tRNA is charged?
20. Describe the 'initiation' of translation.
21. How does a polypeptide elongate?
22. How does a polypeptide formation terminate?
23. Describe the method of genetic control of development of *Drosophila*.
24. What are 'imaginal discs' and determination in *Drosophila*?
25. What are the roles of house-keeping genes, homeobox domain, gap genes, homeotic genes, pair-rule genes and segment-polarity genes in the development of *Drosophila*?

CHAPTER 10

GENETIC ANALYSIS OF SEX DEVELOPMENT (SEX-DETERMINATION) IN *DROSOPHILA* AND MAMMALS

Sex-determination in *Drosophila*, Dosage Compensation in *Drosophila*, Sex-determination in mammals, Dosage compensation in mammals.

Development of an organism results from differential gene activity of its genome. There is a constant amount of DNA in the genome from the beginning to the end of development of an organism. Both genes and environmental factors affect the phenotype of an individual.

In *Drosophila*, the sex is determined by the ratio of the number of X-chromosomes to the number of sets of autosomes (See chapter 7.4). The different ratios of the two types of chromosomes result in different levels of development through the proteins encoded by genes. These proteins influence a master regulatory gene for sex-determination that controls a *cascade of regulated alternative RNA splicing events* that leads to differentiation into female-specific or male-specific cells.

In mammals, the sex of the individual is determined by the presence or absence of the Y chromosome. If the Y chromosome is present, the **SRY** gene of it produces a factor that regulates certain genes, which are necessary to form a **testis**. In the absence of Y (i.e., in the absence of SRY gene), the gonad develops into an **ovary** as is found in an XX female.

In mammals, a dosage-compensation mechanism equates expression of X-linked genes in males and females (See chapter 7.2.1). It involves the *inactivation* of one X chromosome in the female in its early development. The process of inactivation is complex involving the transcription of **XIST** gene on the chromosome to be inactivated.

The mechanism of dosage compensation in *Drosophila* relates to the molecular steps for sex-determination. The absence of **Sxl** gene and of the protein made by it in males, helps the formation of a key protein which associates with other proteins to form a complex. This complex binds to various points on the X chromosome and triggers chromatin remodeling events in it. The chromatin remodeling mechanism increases the transcriptional level of *male X-linked genes two-fold* to match the gene expression of X-linked genes in the female.

10.1 Development of Sex in *Droso-phila*

In chapter 7.4 we have discussed that the sex in *Drosophila* is determined by X : A ratio (Ratio of the number of X chromosomes to the number of sets of autosomes). It is now evident that there is a **regulatory cascade of genes** that determines the sex of an individual in *Drosophila*. This is called *regulatory cascade model.* The model explains that—(Fig. 10.1).

1. The X : A ratio is read at the beginning of the sex-development. In female *Drosophila* (XX), the ratio is 1 (2X : 2A), and in males (XY), the ratio is 0.5 (1X : 2A).
2. Getting this information, the sex determining genes make the choice of one pathway—either the female-ward developmental pathway or the male-ward developmental pathway. This is called choice of alternative pathways which starts from the master regulatory gene **Sxl** (or, sex lethal gene).
3. The **Sxl** is essential in the development of **female sex**, but it is not necessary for the development of male sex. In fact, Sxl mutants are lethal in males, and so it must be inactive in males.
4. Alternative splicing of the Sxl pre-mRNA starts to function in two different pathways. If the ratio of X : A chromosomes of the embryo is 1, the gene switches toward femaleness, and if the ratio is 0.5, the gene switches toward maleness.

   [There are numerator and denominator genes in the X and A chromosomes (autosomes) respectively, that produce respective proteins. In females (XX) the excess of numerator proteins produces numerator-numerator dimers that activate Sxl, and choice of female alternative pathway is selected. In males (XY) the Sxl remains inactive due to want of numerator dimers.]
5. In early embryogenesis, translation of certain Sxl-mRNA in females produces the **SXL early protein** after proper splicing. In males, Sxl proteins are not produced.
6. In late embryogenesis in females, the **SXL early protein** binds to the Sxl pre-mRNA and causes regulated splicing. There are 8 exons in the late pre-mRNA from which the exons E1 and E3 are excised. Thus, a mature mRNA is produced with exons L1, 2, 4, 5, 6, 7 and 8. Translation of this mRNA produces the **SXL-late protein**.
7. In males, the **absence of SXL early protein** results in a mature mRNA that includes exon 3. This exon bears a stop codon and thus, **no SXL-late protein** is produced in males.
8. The regulation cascade model (Fig. 10.1) shows that in female embryo (X : A = 1), the SXL late protein regulates splicing of **tra** (transformer) gene, which subsequently produces **TRA protein**.
9. In males, due to absence of SXL late protein, a default splicing occurs, that is, **no TRA protein** is produced.
10. The TRA protein splices the pre-mRNA of **dsx** (double sex) **gene**, and thus in females, **DSX-F protein** is produced. This protein is a transcription factor that **represses male-specific gene expression** in all cells. This results into the differentiation of **female-specific cells** in the body.
11. In males, due to absence of TRA protein default splicing of dsx pre-mRNA occurs. As a result **DSX-M protein** is produced. **It represses the female specific gene expression** in all cells. This results into the differentiation of *male-specific cells* in the body.

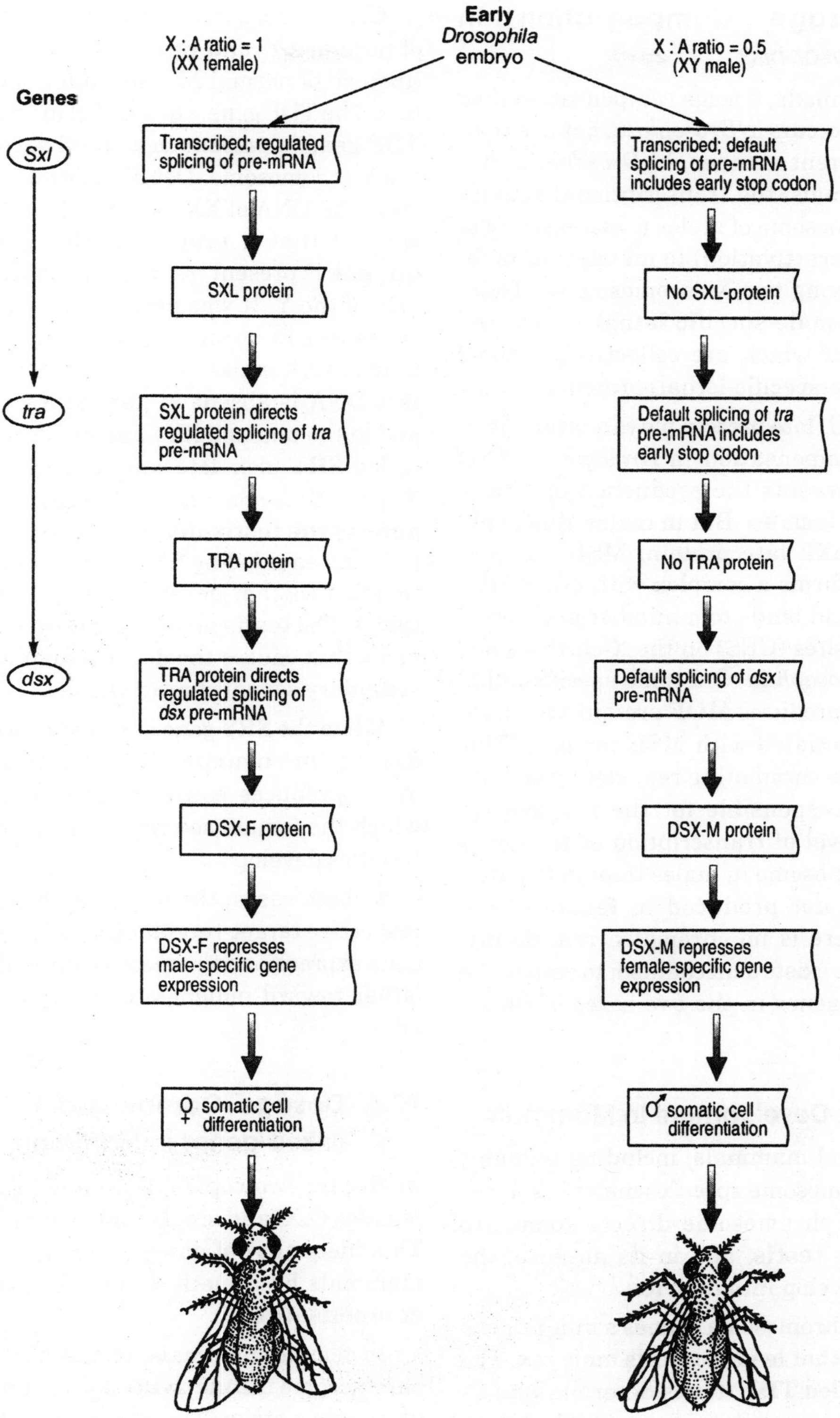

FIGURE 10.1

Development of sex in *Drosophila* (Regulation cascade Model).

10.2 Dosage Compensation in *Drosophila*

As in mammals, dosage compensation (See 7.2.1) also occurs in *Drosophila*, but in a complete different manner. In *Drosophila,* unlike mammals, the transcriptional activity in X-chromosome of males is increased two-folds (hyperactivation) to match that in females (having two X-chromosomes). There are some male-specific lethal genes, the products of which are collectively named MSL (male-specific-lethal) proteins.

The SXL late protein plays a vital role in dosage compensation in *Drosophila*. This protein prevents the production of MSL-2 protein in females. But in males, due to absence of SXL late protein, MSL-2 is produced. It forms a complex with other MSL proteins and binds to a number of chromatin entry sites (CES) on the X-chromosome of male *Drosophila*. One male-specific-lethal gene **mof** produces **MOF** protein, which remains associated with MSL complex. This protein has chromating remodeling activity which is responsible for the two-fold increased level of transcription of the genes on X chromosome in males than in females. MSL-2 is not produced in females, and hence, there is no chromatin remodeling. Thus there exists a mechanism to equate the dosage of genes in the two sexes of *Drosophila*.

10.3 Sex-Development in Mammals

In placental mammals, including humans, the Y chromosome specifies maleness. Presence of Y chromosome directs gonads to develop as **testis**, and in its absence, the gonads develop into **ovaries**.

The Y chromosome carries a unique gene (or genes) that leads towards male sex. The gene is called **TDF** (testis-determining-factor gene) because it produces **testis-determining factor** that causes the gonads to differentiate into **tests**, not ovaries.

Certain rare males having XX (instead of normal XY) and rare females having XY (instead of normal XX) are called *sex-reversals*. The TDF gene was studied in them. The TDF gene is present in a small segment of the Y chromosome. Detail molecular analysis of the DNA of XX males and XY females showed that a **male-specific gene sequence** is present near the end of the small arm of the Y. It was noticed that this DNA is present in normal XY males and in the unusual XX males; it was also noticed that this DNA is absent in normal XX females and in the unusual XY females. This gene is called **SRY** (**s**ex determining **r**egion of the **Y**) gene. *It is now established that the* **SRY gene is the testis-determining gene**. The proteins encoded by SRY are the transcription factors that develops a gonad into the **testis**. The **testis** produces testosterone hormone that causes the development of male-secondary sexual characteristics.

When the SRY gene is absent, the gonads develop into **ovaries** instead of testis. The ovary produces estrogen hormone due to which the female-secondary sexual characteristics develop.

In both cases, the hormones bind to the respective target-receptor cells and regulate gene expression in respective direction, i.e., either toward maleness or toward femaleness.

10.4 Dosage Compensation for X-linked genes in Mammals

There are two copies of X-linked genes in females (XX) and single copy in males (XY). This inequality of doses is compensated in mammals by a mechanism called **dosage compensation**.

In female mammals, one of the two X-chromosome is **inactivated** in somatic cells at an early stage. The choice of X-chromosome i.e., which one is to be inactivated, is apparently random.

The steps involved in the inactivation process of X-chromosome are:

1. Inactivation begins at a particular site of the chromosome. This site is called *X-inactivation centre* (XIC); then the inactivation spreads in opposite direction of the chromosome until it reaches the ends.
2. While all genes become inactive in that X chromosome, one gene remains active. This active gene is called **XIST**, which is located at the XIC region.
3. In human females, the XIST gene encodes an RNA that remains restricted to the nucleus and particularly localized *to inactivated X-chromosome only*.
4. It is known from the experiments on female mice that, one X-chromosome (that chromosome whose Xist gene is active) becomes coated with respective RNA, and the other X remains uncoated.
5. Thus, most of the genes on the coated chromosome become repressed, and the chromosome becomes **inactive**.
6. Thus, in mammalian dosage compensation mechanism, the active X chromosome represses its **Xist** gene.

IMPORTANT NOTES:

1. In *Drosophila*, a ratio of (X : A = 1) results in a *female* and a ratio of (X : 2A = 0.5) results in a *male*.
2. The ratio results in different amounts of proteins encoded by numerator genes on the X-chromosomes versus a denominator on an autosome.
3. In females, the transcription factors resulted from excess numerator proteins activate a master regulatory gene, Sxl, for sex determination.
4. In males, no activation of Sxl occurs.
5. This regulatory event creates a cascade of regulated alternative RNA splicing events, which lead to differentiation of female-specific or male-specific cells.
6. In mammals, Y-chromosome determines sex; its presence determines male sex and its absence causes development of female sex.
7. The SRY gene on the Y-chromosome produces a transcription factor that regulates certain genes to produce protein factors that direct the gonad to form testis. Thus SRY is equivalent to TDF (testis determining factor).
8. In XX individuals, the SRY is absent, testis is not formed, and the gonad develops into ovary.

EXERCISE

1. Explain the term TDF and discuss its role in sex determination in mammals. What is SRY gene and how does it relate to TDF?
2. How sex is determined in *Drosophila* in molecular level?
3. What is dosage compensation? Why it is needed? Discuss the mechanism in *Drosophila*.
4. Discuss the dosage compensation mechanism in mammals.
5. How sex is determined in humans in molecular level?
6. Discuss the role of Y-chromosome in mammalian sex-development.
7. What is sex-reversal? Why are they important in study of sex determination in mammals?
8. Write notes on: SRY and TDF, XIC, inactivation of X-chromosome in mammals, hyperactivation, SXL-protein.

CHAPTER 11

MUTATION

Definition, Kinds of mutation, Somatic and germinal mutation, Back mutation, Suppressor mutation, Gross mutation, Spontaneous mutation, Induced mutation, Point mutation, Molecular basis of mutation. Types of point mutation, Mechanism involved in mutation, Ionization and Transition, Depurination and Deamination, Radiation-Mutation, Chemical Mutagens and Mutation, Method of Detection of Sex-linked mutation (ClB method), Detection of mutation involving attached-X chromosome, Significance of mutation.

In genetics, **mutation** means change, but not a simple change. It is a genetic change which, when transmitted to offspring gives rise to heritable variations. **Mutation is a heritable change in DNA.** More accurately, it is a change in the DNA at a particular locus in an organism. It may occur within or between genes and is the source of all new genetic variations.

11.1 Definition

Mutation is any detectable and heritable change in the genetic material not caused by genetic recombination.

The term 'mutation' was first used by **de Vries**, when he described the phenotypic changes in the flower colour of evening primrose (*Oenothera lamarckiana*); those changes are now known to be due to ploidy and chromosomal aberration (see chapter 8). First report of 'true mutation' was that of Morgan in 1910, when some white-eyed *Drosophila* were observed among the culture of red-eyed *Drosophila*. The red-eye was the wild character (normal character) and th[e] white eye was a **mutant** (result of mutation[)] character.

11.2 Kinds of Mutation

Various nomenclature are encountered f[or] mutation according to their nature of occur[]rence:

1. **Somatic Mutation**: When th[e] **change** occurs in a somatic cell (n[on] reproductive cell) such that, [it] produces daughter cells like itself b[ut] not the whole organism (that mea[ns] it does not affect the germ cells), th[e] change is called **somatic mutatio[n].** As the change is confined only to th[e] somatic cells, the role of somat[ic] mutation in evolution is insignifican[t]. In vegetables, somatic mutati[on] produces high quality of fruits, a[nd] in humans, it may causes sever[e] fatal diseases.
2. **Germinal Mutation**: Heritab[le] change occurring in germ cell is call[ed]

germinal mutation. It may be *dominant or recessive*. Dominant germinal mutations are expressed immediately in the progeny. But recessive germinal mutations remain obscured in diploids. If the mutation occurs in a **gamete**, only a single individual of the progeny may bear the mutant gene, but if it occurs in the **primordial germ cells**, several gametes will receive the mutant gene and it will be spread in the progeny.

3. **Chromosomal Mutation** (See chapter 8).
4. **Back or Reverse Mutation**: Mutations are often reversible. As for example, we may consider the case of human eye colour. Both the alleles for brown and blue eye colours are considered wild; but in a population of maximum brown-eyed people, appearance of a blue-eyed person is considered a mutation in the allele. If, subsequently, this change (i.e. from brown to blue) is again mutated to the original (i.e., brown), it is called a **back or reverse mutation**.
5. **Suppressor Mutation**: When a 'second' mutation "suppresses" the effects of original mutation, it is called **suppressor mutation**. It may occur in the same gene of the original mutation or in different genes, or even in genes in different chromosomes.
6. **Gross Mutation**: When changes involve rearrangement of genes by movement, transfer or translocation of genes, they result in **gross mutation or multiple mutation**.
7. **Spontaneous Mutation**: Mutations occurring without a known cause are called **spontaneous mutations**. They may result from an inherent metabolic error, or may be caused by unknown agents present in the environment.
8. **Induced Mutation**: Mutations caused by certain induction or exposure of organisms to physical and chemical agents (mutagens) are called **induced mutation**. The mutagenic agents are called **mutagens**, and they bring change in DNA.
9. **Point Mutation**: The heritable change occurring in a single or few nucleotides or nucleotide pairs of DNA (i.e., in a particular point of DNA) is called *point mutation*. In viruses, this change occurs in a single or few nucleotides of RNA.

11.3 Molecular Basis of Mutation

The structures of the bases in DNA (or RNA) are not static. Hydrogen atoms can move from one position to another position of a base (purine or pyrimidine). Such chemical fluctuations are called **tautomeric shifts**. These shifts change the configuration of bases and can alter their pairing potentials. The usual pairing for common and stable forms of bases are that the adenine (A) always pairs with thymine (T), and guanine (G) always pairs with cytosine (C) (See chapter 4).

There are two forms of bases: the **more stable forms and less stable forms**. The more stable form of T and G are **keto forms** and that of A and C are **amino forms**. They may sometimes undergo tautomeric shifts and may form the stable **enol** and **imino** forms respectively (Fig. 11.1). The **keto form** of T and G and the **amino form** of C and A are common, while the **enol form** of T and G and the **imino form** of C and A are rare or uncommon. The **tautomeric shifts** may create the rare imino or enol states (called tautomers), which lead to A : C and G : T base pairs (Fig. 11.2). These are rare base pairing. The net effect of this type of pairing is the detachment of this "mismatched" condition in the subsequent repli-

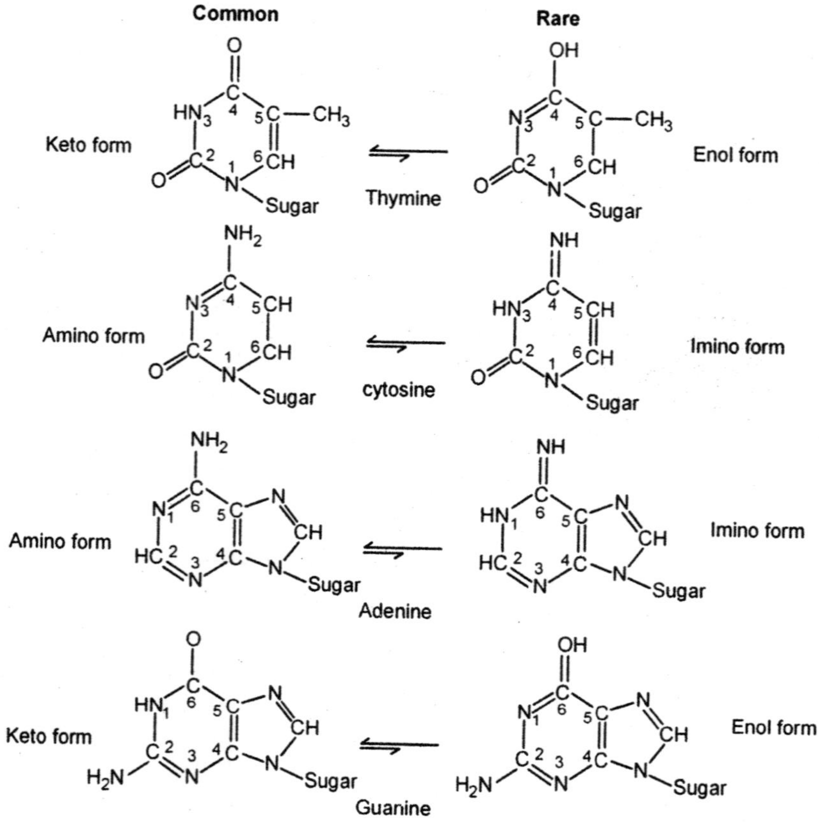

FIGURE 11.1

Tautomeric forms of the four common bases in DNA. The shifts of hydrogen atoms between the number 3 and number 4 positions of the pyrimidines and between the number 1 and number 6 positions of the purines change the base-pairing potential of the bases.

cation of the DNA. This alters A : T to G : C or G : C to A : T base pair substitution (Fig. 11.3). This results in *mutation* usually called "**point mutation**".

11.3.1 Types of Point Mutations

There are two main types of point mutation—(a) **Base-pair substitution Mutation** and (b) **Base-pair Insertion or Deletion Mutation.**

(a) Base-pair Substitution Mutation: It involves the change in the DNA in such a way that one base-pair is *replaced* (substitution) by another. This replacement of base results in two kinds of mutation: **Transition Mutation and Transversion Mutation (Fig. 11.4)**

- **Transition Mutation**: This type of mutation results from the change of one purine-pyrimidine base pair to the other purine-pyrimidine base pair i.e. purine is replaced by purine and pyrimidine is replaced by pyrimidine. The mutation may be from A : T to G : C, G : C to A : T, T : A to C : G, and C : G to T : A.
- **Transversion Mutation**: It is a mutation from a purine-pyrimidine base pair to a pyrimidine-purine base pair (i.e., a purine is replaced by a pyrimidine, and a pyrimidine is

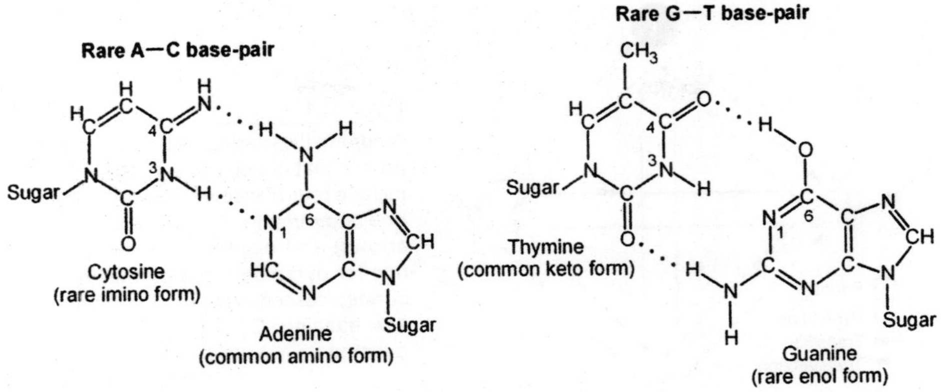

FIGURE 11.2

Examples of mismatched AC and GT base-pairs that can form when one of the purines or pyrimidines exists in a rare tautomeric form. Similar mismatched base-pairs can form when thymine exists in its rare enol form or adenine exists in its rare imino form.

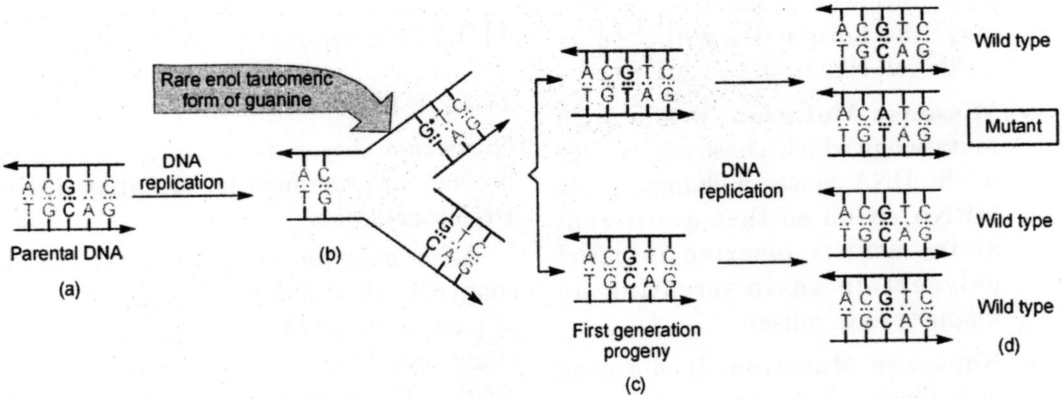

FIGURE 11.3

Mutation via tautomeric shifts in the bases of DNA. A guanine (a) undergoes a tautomeric shift to its rare enol form (G*) at the time of replication (b). In its enol form, it pairs with thymine (b). During the next replication (c to d), the guanine shifts back to its more stable keto form. The thymine incorporated opposite the enol form of guanine (b) directs the incorporation of adenine during the subsequent replication (c to d). The net result is a GC to AT mutation. If a guanine undergoes a tautomeric shift from the common keto form to the rare enol form at the time of incorporation it will be incorporated opposite thymine in the template strand and cause an AT to GC mutation.

replaced by a purine). The change may be of 8 types; viz., A : T to T : A, T : A to A : T, G : C to C : G, C : G to G : C, A : T to C : G, C : G to A : T, G : C to T : A and C : G to A : T.

An example of transversion mutation can be cited from human hemoglobin (Hb). Analysis of **normal human Hb** and **sickle cell Hb** shows that they differ by only a single amino acid. Normal Hb has **glutamic acid** at a certain point while sickle cell Hb has **valine** at the same point. The genetic code for glutamic acid is GAA or GAG, while that of valine is GUU or GUC or GUA or GUG. If the middle 'A' of

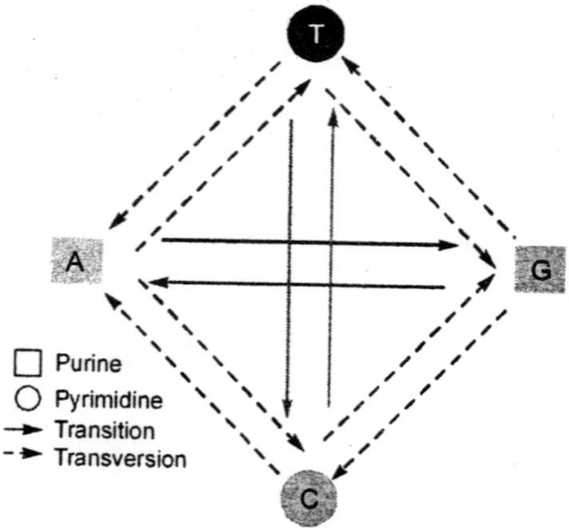

FIGURE 11.4

Diagram illustrating the base substitutions that are possible in DNA. These include four transitions (purine for purine or pyrimidine for pyrimidine; solid arrows) and eight transversions (purine for pyrimidine and pyrimidine for purine; dashed arrows). T = thymine, A = adenine, C = cytosine, and G = guanine.

GAA or GAG is replaced by 'U' and forms GUA or GUG, the **transversion mutation** will occur (a change from purine to pyrimidine). In this way, the normal Hb will produce sickle cell Hb.

- **Missense Mutation**: It is a point mutation in which a base pair change in the DNA causes a change in the mRNA codon so that a different amino acid is inserted into the polypeptide chain (creating an inappropriate sense).
- **Nonsense Mutation**: It is a gene mutation in which a base-pair change in the DNA changes an mRNA codon to a stop codon or nonsense codon, (UAG, UAA or UGA).

(b) **Base-pair Insertion or Deletion Mutation**: When one or more base pairs are added to or deleted from the DNA, **reading frame** of the resultant mRNA will change. It is called **frame shift mutation**.

- **Frame shift Mutation: Addition** or **deletion** of one base pair shifts the mRNA's reading frame by one base, so that incorrect amino acids are added to the polypeptide chain following the mutation. It usually results in a nonfunctional protein or a shortened protein (by bringing stop codon abruptly) or a longer protein (by avoiding stop codon) etc.

11.3.2 Mechanism Involved in Mutation

11.3.2.1 Tautomerization

Nitrogenous bases of rare states (tautomers) provide a mechanism for mutation during DNA replication.

As for example, if adenine (A) in an old chain of DNA is in the rare state at the time of pairing with the complementary new chain, cytosine (C) can pair with it. The result of this pairing forms a DNA molecule that contains an 'exceptional' or 'forbidden' base pair. This 'pair' is not stable, and at the next replication, 'A' is expected to return to its normal state and to pair naturally with thymine (T). 'C' introduced in the complementary strand (due to the tautomeric shift in adenine of the old chain), will now pair with guanine (G). Thus, two kinds of DNA molecules will be produced—one, identical to the original DNA and the other, that has undergone a base-pair substitution of G : C for A : T. This is a case of **transition** (Fig. 11.5).

When such mutation occurs the two kinds of DNA molecules will eventually be separated into different cells forming different cell-linage or clones.

In this example, the cell-line containing the *transition* has a G : C pair in place of A : T pair. This can result in the coding for a different amino acid and brings a detectable mutation. Such change during DNA replication is called **copy-error mutation**. The steps of the *transition* following *tautomerism* are shown in the Fig. 11.5. It is important to note that any base of rare state can become involved in a transition.

11.3.2.2 Transition due to Ionization

Like tautomerization, **ionization** of a base at the time of DNA replication may also result in **transition** mutation. Ionization involves the loss of hydrogen atom from the number 1 nitrogen of a base.

In ionized state, thymine (T) can pair with guanine (G) in its common state. Similarly, ionized state of guanine (G) can pair with thymine (T) in its common state. These unstable base pairs result in transition. The formation of 'forbidden' T : G and G : T base pairs following ionization has been shown in Fig. 11.6.

11.3.2.3 Depurination and Deamination

Some spontaneous chemical changes in the base also may bring mutation. As for

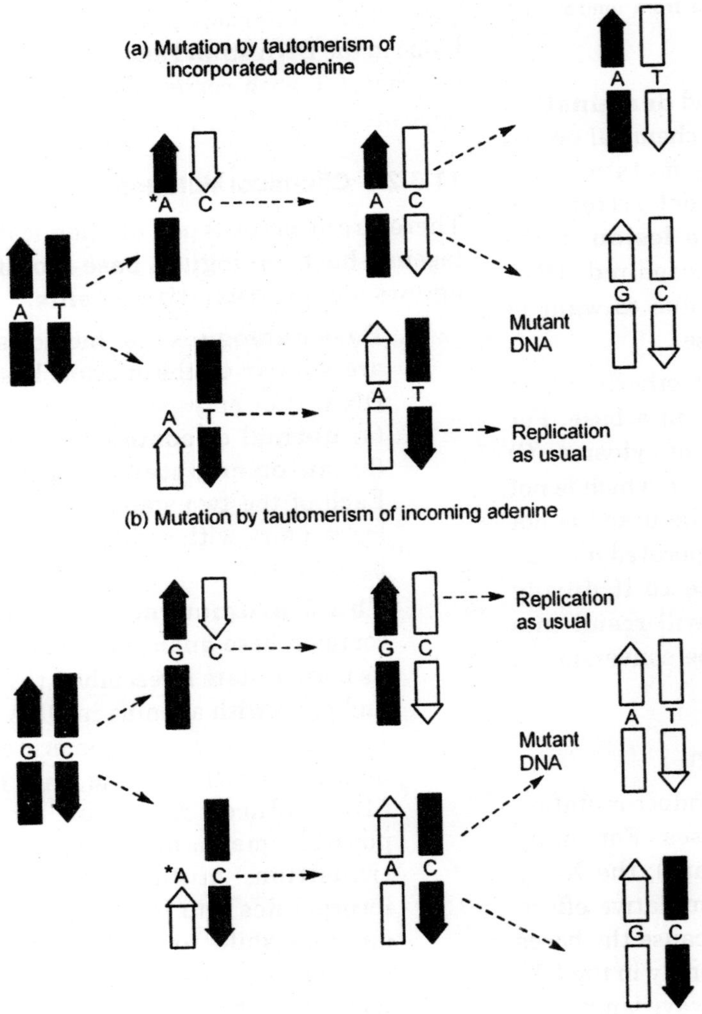

FIGURE 11.5
Consequences of tautomerization of an adenine residue (*) at the time of DNA duplication.

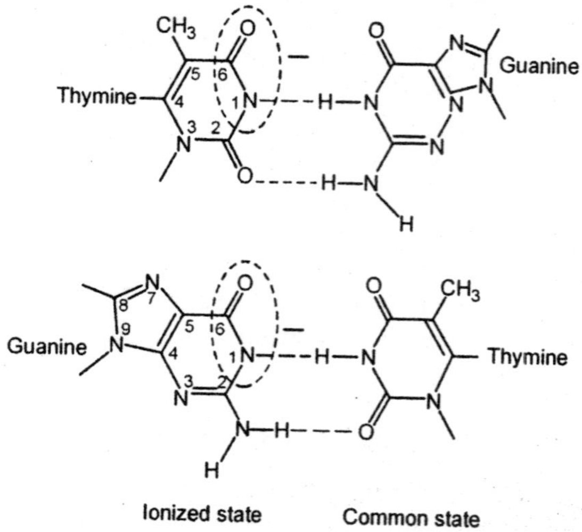

FIGURE 11.6

"Forbidden base pairs" resulting from ionization of the number 1 nitrogen.

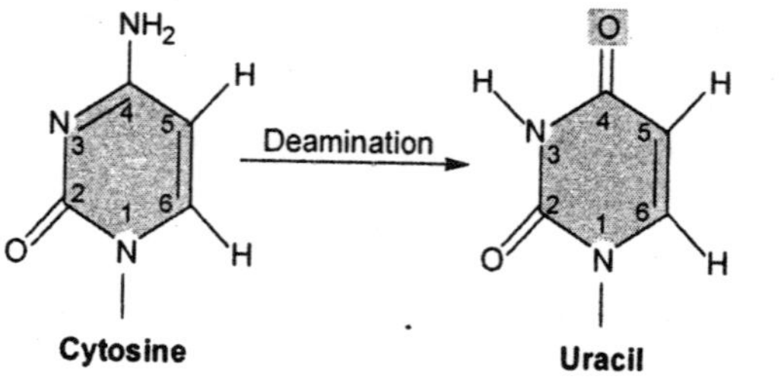

FIGURE 11.7

Deamination of Cytosine.

example, **depurination** and **deamination** of bases are most common chemical events that produce spontaneous mutations. In **depurination**, a purine (A or G) is removed from the DNA. It creates a **lesion** in the DNA. If the lesion is not repaired, DNA replication is interrupted due to want of specific complementary base.

In **deamination,** on the otherhand, an amino group is removed from a base. For example, the deamination of cytosine (C) produces uracil (U) (Fig. 11.7), which is not a normal base in DNA. If the uracil is not replaced, an 'A' will be incorporated into the new DNA strand opposite to it during replication. Ultimately it will result in a *transition mutation,* i.e., the conversion of the C : G to T : A base pair.

11.3.2.4 Radiation Mutation

Both X-ray and UV-ray can induce mutation causing **ionization** of bases. For many organisms (including humans), the X-ray ionizing radiation has a cumulative effect. UV-ray causes mutation because the bases in the DNA absorb light strongly in the UV-range (254-260 nm). At this wave length UV light induces point mutation by causing photochemical changes in the DNA. In this reaction, abnormal chemical bonds are formed between adjacent pyrimidines on the opposite strands. The bonds between adjacent thymines form **thymine dimers** ($\widehat{TT}$). This abnormal pairing produces a bulge in the DNA strand and inhibits normal pairing of T with corresponding A on the opposite strand (Fig. 11.8).

11.3.2.5 Chemical Mutagens and Mutation

There are 3 main types of chemical mutagens—**base analogues, base-modifying agents** and **intercalating agents**.

1. **Base analogues** are chemicals that are similar to the normal bases of DNA. They may exist in two states—(a) **normal** or **common state** and (b) **tautomeric state** or **tautomer**. Each of the two states of base analogue pairs with a different normal base in DNA.

   The 5-bromouracil (5BU) is a common base-analogue mutagen. In its normal state it resembles thymine and pairs with adenine in DNA (Fig. 11.9). In its rare or tautomeric state, 5BU pairs with guanine (Fig. 11.9). 5BU induces mutation by two possible mechanism—It induces transition mutation when it incorporates into DNA in one state, and then shifts to its alternate state during the next round of DNA replication (Fig. 11.10).

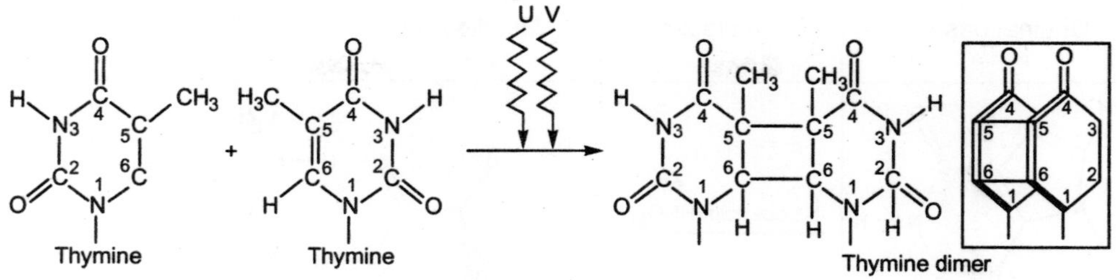

FIGURE 11.8

Production of thymine dimers by ultraviolet light irradiation.

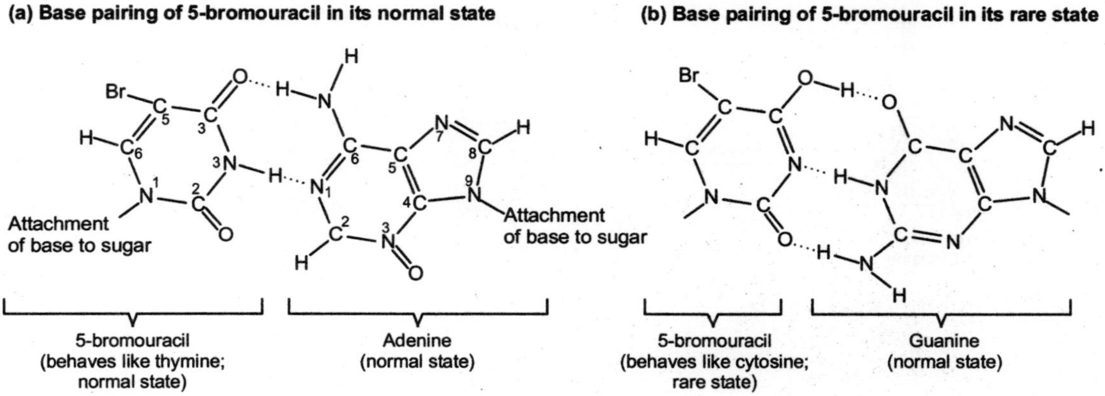

FIGURE 11.9

Mutagenic effects of the base analog 5-bromouracil (5BU). (a) In its normal state, 5 BU pairs with adenine. (b) In its rare state, 5 BU pairs with guanine.

- **Mutagenic Action of 5BU**

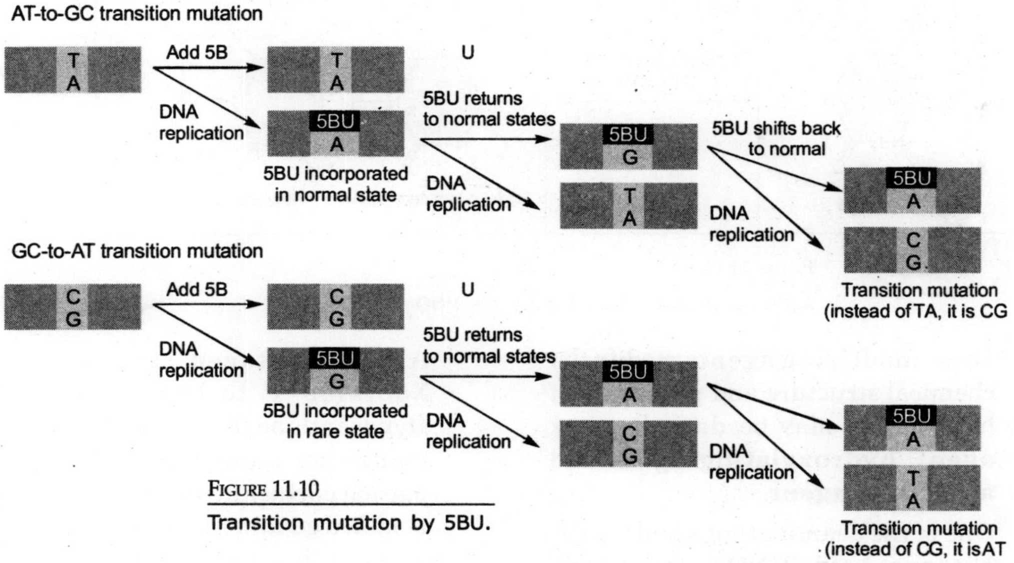

FIGURE 11.10

Transition mutation by 5BU.

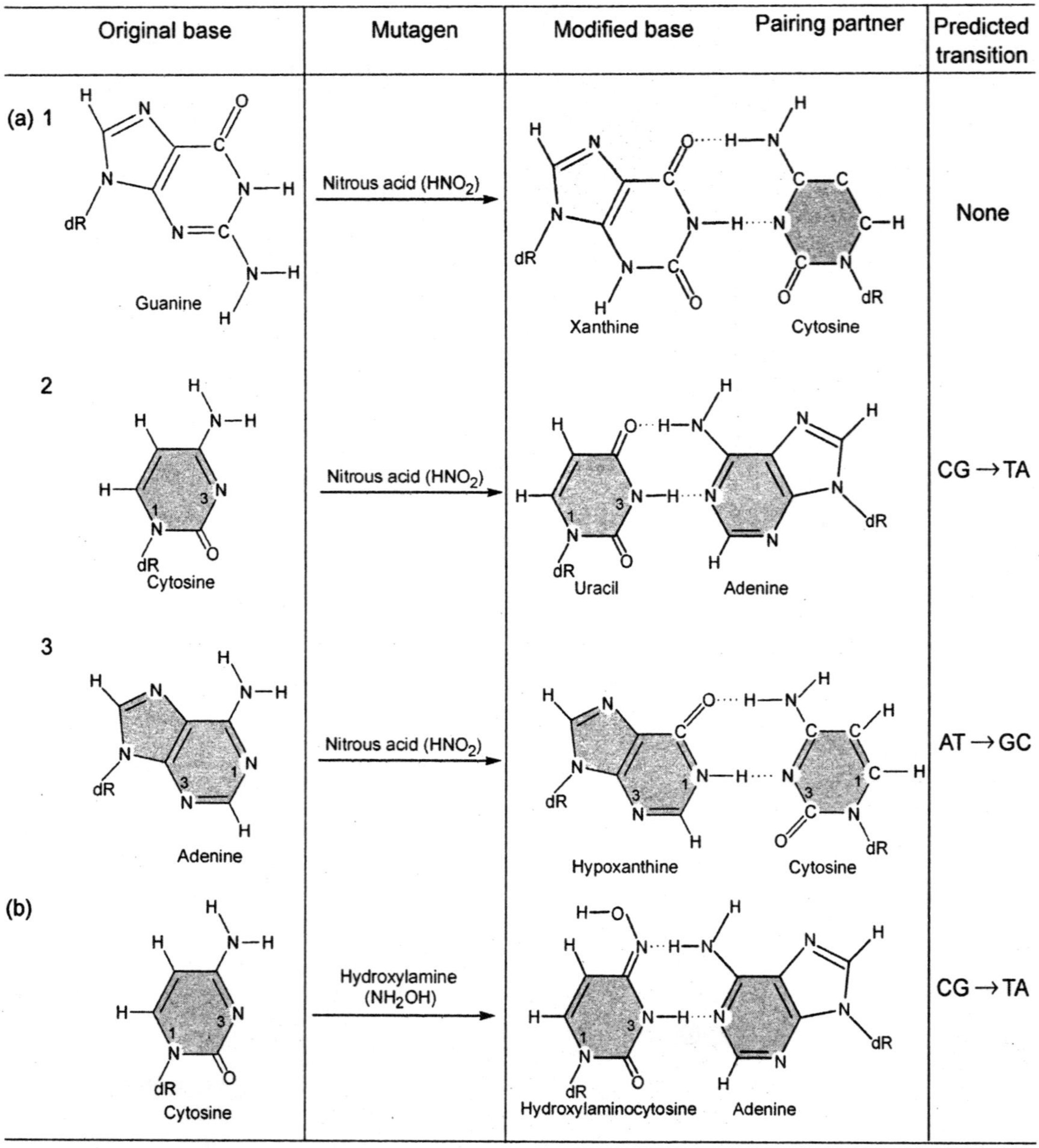

FIGURE 11.11

Action of 2 base-modifying agents HNO_2 and NH_2OH.

2. **Base-modifying agents** modify the chemical structure and properties of bases. They may be **deaminating agent, hydroxylating agent** and **alkylating agent**.

One of the deaminating agents is the **Nitrous Acid** (HNO_2). It removes amino groups (—NH_2) from G, C and A. This mutagen modifies G to Xanthine, C to Uracil and A to Hypoxanthine (Fig. 11.11) and brings transition mutation by changing pairing partner in case of C and A (the modified guanine, xanthine, however, pairs with cytosine and transition does not occur).

Another mutagenic (hydroxylating) agent **hydroxylamine** (NH_2OH) modifies C to hydroxylaminocytosine and brings transition mutation by changing the pairing partner (Fig. 11.11).

The alkylating agent, **Methyle Methane Sulfonate** (MMS) modifies G into O^6-methylguanine

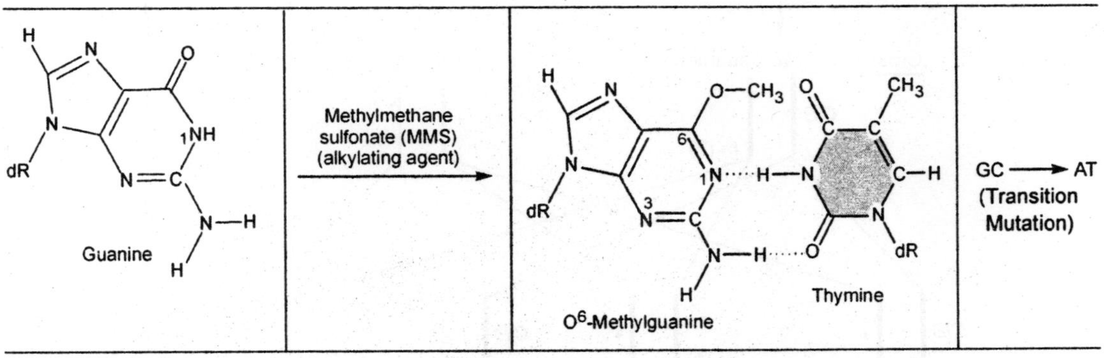

FIGURE 11.11(a)

Action of MMS on Guanine, showing Transition Mutation.

and brings transition mutation by changing the pairing partner [Fig. 11.11(a)].

3. **Proflavin acridine and ethidium bromide** are *intercalating agents.* They insert themselves between the adjacent bases and bring changes during replication of DNA.

11.4 Method of Detection of Sex-Linked Mutation (ClB method)

H.J. Muller (1927) was the first person to demonstrate that X-ray can cause mutation. He showed that X-ray treatment significantly increased the frequency of **sex-linked recessive lethal mutation** in *Drosophila melanogaster.*

He developed a unique technique to identify lethal mutations in the X-chromosome of *Drosophila*. This technique is known as **CLB method**, in which he used female *Drosophila* heterozygous for a normal X chromosome and a special X chromosome (called ClB chromosome), designed for the experiment.

- **The ClB Chromosome**

This special X chromosome has 3 essential components: (1) the 'C' for **cross-over suppressor**. (It is a long inversion of chromosome that is thought to prevent crossing over between genetic markers on the ClB chromosome and the alleles of the normal X chromosome. Actually it does not prevent crossing over but produces unviable gametes). This inversion (C) is required in the experiment to assure that the markers in the ClB chromosome remain together during meiosis. (2) The 'l' refers to a recessive gene for **lethal**. (3) The 'B' refers to the presence of partially dominant mutation that produces **bar eye**. Its presence helps to identify the females heterozygous for the ClB chromosome. Both the recessive lethal (l) and the bar eye mutation are present within the inverted segment of the ClB chromosome.

- **Muller's experiment (ClB method)**

Follow (Fig. 11.12). Muller mated heterozygous ClB females with males, which were treated with X-ray beforehand, so that

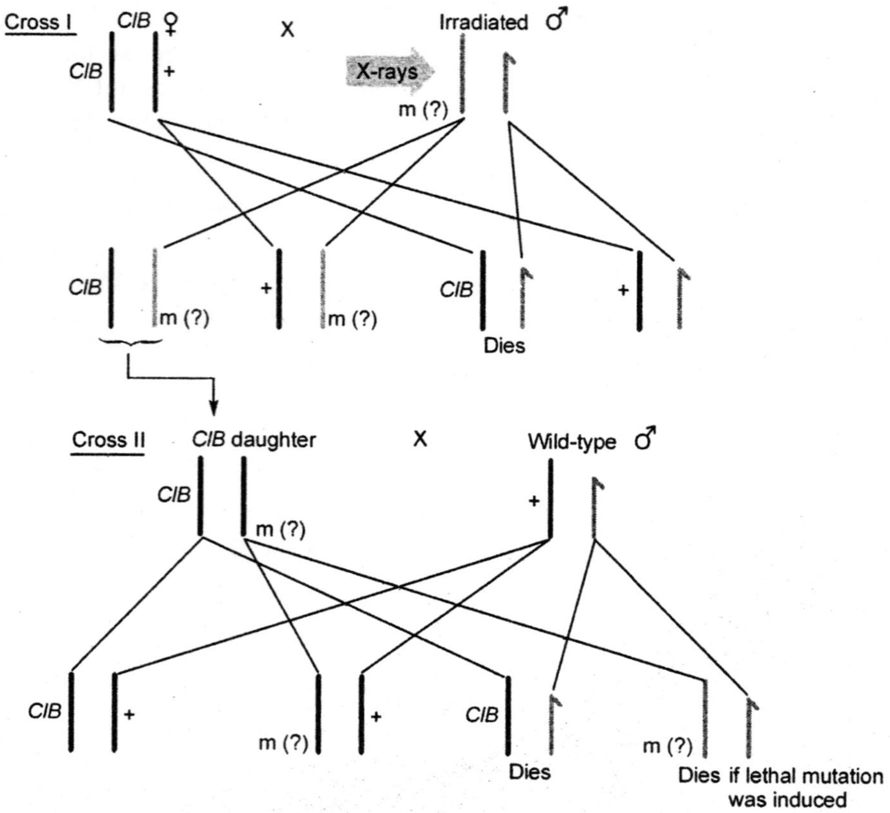

FIGURE 11.12

Muller's *CIB* method for detecting sex-linked recessive lethal mutations in *Drosophila*. Females heterozygous for the *CIB* X chromosome are mated with irradiated males (Cross I). *CIB* daughters, which will always carry the *CIB* chromosome of the male parent, are then mated with wild type males (Cross II). This second mating will produce only female progeny if an X-linked recessive lethal has occurred on that particular X chromosome. It will produce both female and male progeny (in a ratio of 2 : 1) if no X-linked lethal was induced on the irradiated X chromosome that the *CIB* daughter received. Scoring for the presence or absence of sex-linked recessive lethal mutations thus involves simply scoring for the presence or absence of male progeny from matings of Cross II.

radiation can cause mutation in some of their sperms (Cross I). The resulting zygotes of this mating were of 4 types; *one type* were *bar eyed* females that carried the ClB chromosome of the female parent and the *irradiated* X chromosome of the male parent. These are ClB-daughters. *The second types* were normal females with one normal X and one irradiated X chromosome received from the parents.

Another type, the ClB males failed to survive because they contained one X chromosome with the recessive lethal which expressed itself in hemizygous condition. Lastly, the *4th type*, the male zygotes (with normal X chromosome) survived to develop into a male.

The ClB daughters (carrying ClB chromosome and irradiated X chromosome) were then crossed (Cross II) with wild type males. This second cross would produce *only female* progeny if an X-linked recessive lethal has occurred on that particular X-chromosome. It would produce *both female and male* progeny (in a ratio of 2 : 1) if no X-linked lethal was induced on irradiated X-chromosome that the ClB daughter received from first cross. The presence or absence of male progeny after the second cross unambiguously proves the lethal mutation due to X-ray radiation; if males are produced, no mutation occurred, if males are not produced, mutation definitely occurred.

By using this method, Muller was able to demonstrate an increase in mutation rate after X-ray treatment.

11.4.1 Detection of Mutation involving attached X-chromosome

Another method for detection of mutation in *Drosophila* involves the use of **attached-X chromosomes**. In this case, sex-linked mutations are phenotypically visible.

Attached-X chromosomes undergo compulsory nondisjunction because the two X chromosomes are joined to a single centromere. They are denoted as $\widehat{XX}$.

If females with two attached–X chromosomes plus a Y chromosome ($\widehat{XX}Y$) are mated to normal males (XY), any mutation (due to mutagen) that occurs on the X chromosome of the male will be visible in the viable male progeny.

In this matings, the male progeny receive their X chromosome from their **male parent** rather than from the mother. If the male parent is treated with mutagenic agent, and if a mutation with visible effect on phenotype has been induced in its X chromosome, it should be expressed in all viable male progeny (Fig. 11.13).

11.5 Significance of Mutation

Most mutations are detrimental to the organisms. They are kept in low level in the population through natural selection. Mutation is an important source of evolutionary process since changes in heritable traits result from random mutation. In a broad sense, it includes all the heritable changes that alter the phenotype of an individual. Mutation can produce drastic changes in organism or in population, or it can remain silent. Most of the mutant genes are recessive and are expressed phenotypi-

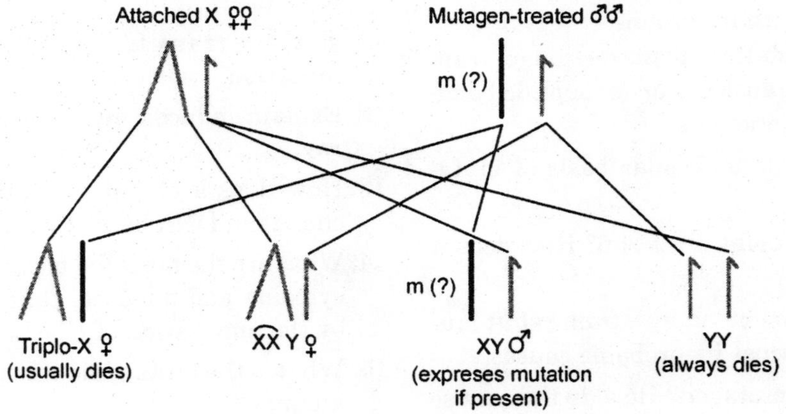

FIGURE 11.13

Attached-X method for detecting sex-linked visible mutations. The mutagen-treated males are crossed with $\widehat{XX}Y$ females (the two X chromosomes are attached to a single centromere). In such a mating the viable male progeny receive their X chromosomes from their male parents. If a mutation with a visible effect on phenotype has been included in the X chromosome of a mutagenized male, it should be expressed in all viable male progeny.

cally only in homozygous condition. Thus, mutations lead to produce variations in off-spring. And the variations are essential in evolutionary process.

Important Notes:

1. Mutation brings change in the sequence of base pairs in a DNA molecule or in the structure of a base.
2. The change may include a single base pair substitution, insertion, deletion, rearrangement and duplication in bases of the DNA, or deletion of a whole section of a chromosome.
3. Mutation may occur spontaneously or may be induced experimentally using mutagens.
4. Mutations in the sequence in genes are called gene mutation.
5. Mutations may be of various types like reverse, suppressor, silent, nonsense, missense etc.
6. Mutation-causing agents are called mutagens. They may be physical elements like X-ray, UV ray etc. or chemical agents like HNO_2, MMS etc.
7. Radiation may cause genetic damage and bring lethality through mutation.
8. Mutation producing chemical agents are HNO_2, NH_2OH, MMS etc.
9. CIB method was introduced by Muller to detect X-linked lethal mutation in *Drosophila*.
10. Other technique for detection of mutation in *Drosophila* is **attached-X chromosome** method.

EXERCISE

1. Choice the right answer—
   The cause of mutation is—(a) genetic recombination (b) heritable changes in genetic information (c) faulty transcription of the genetic code.
2. Which of the following terms are not associated with mutation?
   (a) frame shift (b) transversion (c) transition (d) translocation (e) peptide bondage (f) genetic code.
3. Describe the molecular basis of mutation.
4. What is a point mutation? How does it occur?
5. What do you mean by a frame shift mutation. Discuss its probable causes.
6. What is a mutagen? How do they cause mutation? Give example.
7. Write notes on radiation mutation.
8. How can a sex-linked mutation in *Drosophila* be detected?
9. What is a ClB technique? What is its use?
10. What is an attached-X technique? What is its use?
11. Explain transition and transversion mutation.
12. Explain induced mutation by using X-ray.
13. How does a chemical mutation bring change in DNA base? Cite example.
14. What are the modified bases of guanine, cytosine and adenine, that are formed by the application of HNO_2 on them?
15. What is the mutagenic effect of MMS on guanine?

CHAPTER 12

TRANSPOSABLE ELEMENTS

(Mobile Gene, Transposable Gene or Jumping Gene)

Transposable element, Types, Prokaryotic transposable elements, IS element, Composite transposons, Tn3 elements, Medical significance, Eukaryotic transposable elements, Ty elements, Transposable elements in Maize (Ac and Ds elements), Retrotransposons in *Drosophila*, P-element in *Drosophila*.

Conventionally it is known that genes are linearly localized on chromosomes, and they occupy a fixed position (locus); the genes on the chromosome do not move or change their places; this behaviour of genes allow geneticists to construct chromosome maps.

On the contrary to this knowledge, in 1940s and 1950s Barbara McClintock cited examples that genes can move and change their position. She did a series of experiments on corn (maize) and discovered existence of "Controlling elements"; she discovered Ds and Ac gene-loci in corn, and "established beyond a doubt that both the AC and Ds loci were capable not only of changing their positions on the genetic map, but also of inserting into loci to cause unstable mutations." The phenomenon of change of position of Ds was called "transposition" by her. But, unfortunately, her discovery was not recognized by the then geneticists. Many years later, in 1970s her discovery on 'transposition' was recognized, and it was only in 1983 (35 years after the publication of the first evidence for transposition), McClintock was awarded Nobel Prize for this outstanding work.

Today, it is an established and well known fact that certain genes are *mobile*, i.e., they can move and change position. These **mobile** genes are variously termed as: **mobile elements transposons, transposable elements, transposable genes, jumping genes, insertion sequence, IS elements, cassettes** etc.

12.1 Main Features of Transposable Elements

1. Transposable elements are normal components that are present everywhere in the genomes of prokaryotes and eukaryotes.

2. There are two main classes of transposable elements one class, present both in prokaryotes and eukaryotes, encodes proteins that move the DNA directly to a new position or replicate the DNA to produce a new element that integrates elsewhere in the genome. The other class, present only in eukaryotes, encodes a reverse transcriptase, and the resultant DNA-copies integrate at new sites in the genome.
3. In prokaryotes, transposable elements are moveable at any position on the same chromosome, or to a plasmid or phage chromosomes.
4. In eukaryotes, they can move to anywhere in any chromosome.
5. The movement of transposable elements is known as *nonhomologous recombination*, because they have no homologous sequence in the new chromosome, where they are inserted.
6. Transposons can produce mutations by inserting into genes. They are important in the evolution of the genomes through the chromosome rearrangement.
7. Various types of transposable elements have been discovered, which are structurally and functionally diversed.

12.2 Types of Transposable Elements

Based on the mechanism of transposition, the transposable elements are of three types (Table 12.1):

1. **Cut-and-paste transposons**: These are the elements where transposition involves excision of an element from its position in a chromosome and insertion of it into another position.
2. **Replicative transposons**: Transposition involves replication of the transposable elements' DNA.
3. **Retrotransposons**: When the transposition involves the insertion of copies of an element that were synthesized from the elements' RNA, it is called retrotransposon.

12.3 Prokaryotic Transposons

There are three main types of transposable elements in bacteria—**IS elements** (insertion sequence elements), **Composite Transposons** and **Tn 3 elements**.

TABLE 12.1: Types of Transposable Elements

Category	Example	Host organism
Cut -and- paste transposons	IS elements (e.g. IS 50)	Bacteria
	Composite transposons (e.g. Tn 5)	Bacteria
	Ac/Ds elements	Maize
	P elements	*Drosophila*
Replicative transposons	Tn 3 elements	Bacteria
Retrotransposons:		
(a) LTR	Ty 1	Yeast
	Gypsy	*Drosophila*
(b) Retroposons	F,G and I elements	*Drosophila*
	Telomere-specific retroposons	*Drosophila*
	LINEs (e.g. L1)	Humans
	SINEs (e.g. Alu)	Humans.

12.3.1 IS Elements

These are the simplest transposons in bacteria. They belong to *cut-and-paste* category of transposons, and are found in bacterial chromosomes and plasmids. IS elements are the *insertion sequence*, which can insert at many different sites in the chromosome and plasmids. It was first discovered in *E.coli*. There are many kinds of it, of which the smallest one is the IS1. It is a 768 nucleotide-pair long element. Each type of IS is an element that has a short identical sequences at its ends; the terminal sequences are in inverted orientation with respect to each other, and are called **inverted terminal repeats.** These repeats are 9-40 nucleotide-pair long. When IS elements insert into a chromosome they form a **duplication** of part of the DNA sequence where they were inserted. One copy of the duplication is located on each side of the element. These short, directly repeated sequences are called **target site duplication.** They are thought to arise from staggered cleavage of the double stranded DNA molecule (Fig. 12.1). The transposition on IS element requires an enzyme called *transposase*.

12.3.2 Composite Transposons

When two IS elements insert near each other, the **composite transposons** are produced. The sequence between the two inserted IS is transposed by the joint action of the flanking elements. The composite transposon carries a gene for **antibiotic resistance**; as for example the **Tn5** carries the Kanamycin resistance gene, the **Tn9** carries chloramphenicol resistance gene, and the **Tn10** carries tetracycline resistance gene.

In Tn9, the flanking IS elements are in direct orientation with each other but in Tn5 and Tn10, the orientation is inverted.

12.3.3 Tn3 Elements

These transposons have inverted terminal repeats (each 38-40 np long). They produce target-site-duplications of 5np after insertion.

The Tn3 elements are larger (5000 or more np long) than the IS element. The transposition of Tn3 occurs in two steps. First, the enzyme *transposase* mediates the fusion of two molecules (one from donor plasmid and the other from recipient plasmid) and produces a *cointegrate*. In the second step, another enzyme, *resolvase* mediates a site-specific recombination of the two elements, so that a recombination event occurs between the two elements. The donor and the recipient plasmids are then separated, each with a copy of Tn3 (Fig. 12.2).

12.3.4 Medical Significance of Bacterial Transposons

The bacterial transposable elements are responsible for the '*drug and antibiotic-*

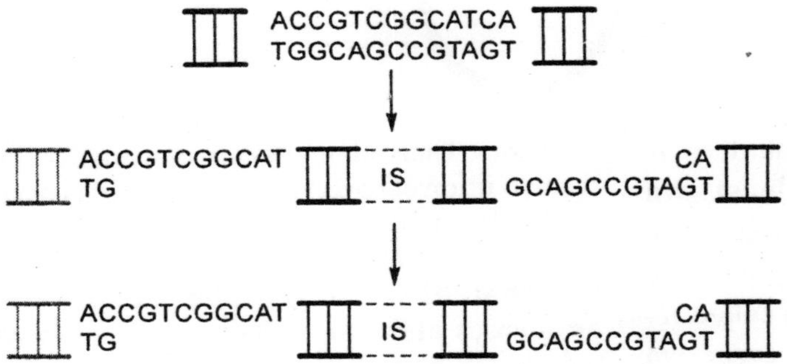

FIGURE 12.1

Production of target site duplications by the insertion of an IS element. The element causes breaks in the target DNA at the sites indicated by the arrows. After insertion, DNA synthesis fills in the gaps left by the staggered breaks, producing direct repeats on each side of the element (After S. N. Cohen and J. A. Shapiro, *Sci. Amer.* 242: 40-49, 1980).

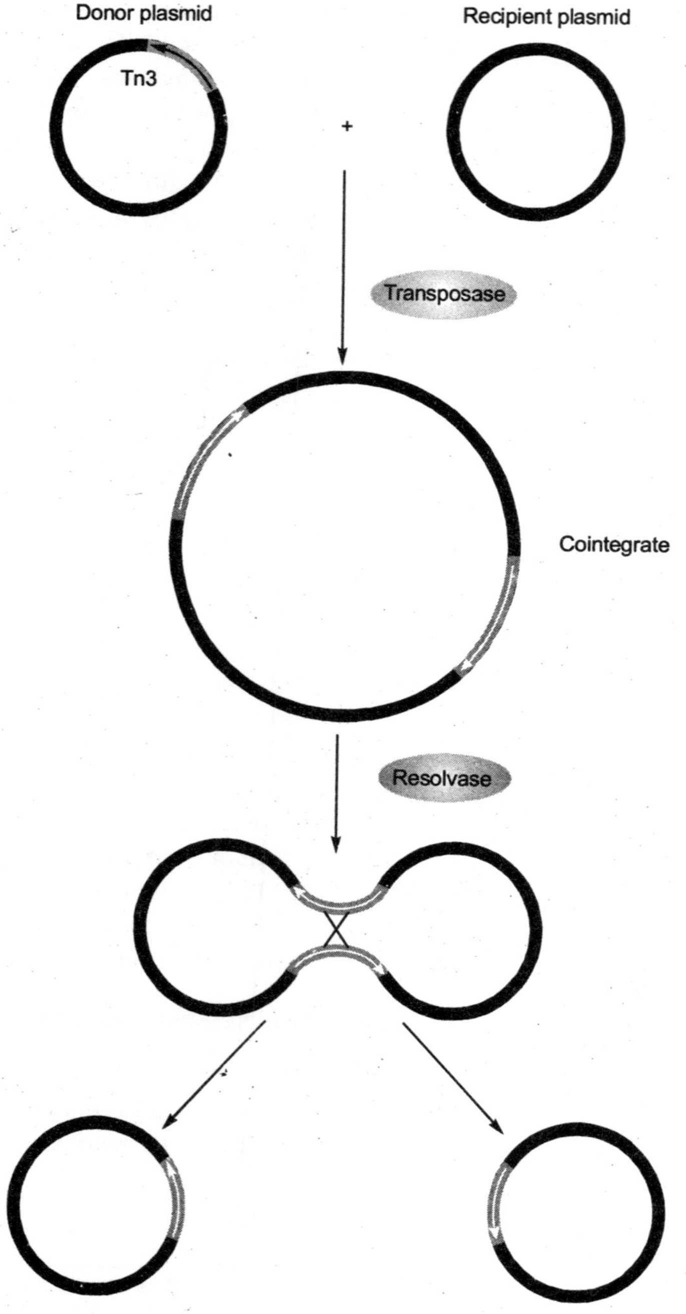

FIGURE 12.2

Transposition of Tn3. The donor plasmid carries a single copy of Tn3 and the recipient plasmid lacks Tn3 altogether. A transposase produced by the *tnpA* gene of Tn3 catalyzes the formation of a cointegrate in which the two plasmids are fused. During this process, Tn3 is also replicated so there is a copy of the element at each junction in the cointegrate. A resolvase produced by the *tnpR* gene resolves the cointegrate by mediating a recombination event between the two Tn3 elements. Donor and recipient plasmids are thereby separated, each with a copy of.Tn3.

resistance in bacteria. All conjugative **R plasmids** (See 14.6) have at least two segments; one segment for carrying genes for DNA transfer, and the other segment for carrying genes for resistance, are called **R-determinants**. Many R plasmids carry two or more R-determinants, each flanked by IS elements. These elements are responsible for the evolution of bacterial plasmids that carry multiple antibiotic and drug resistance factors. This phenomenon alarms with the fact that "if we do not restrict use, the antibiotics and drugs that are so effective today may have little, if any, utility in the future".

12.4 Eukaryotic Transposable Elements

The original discovery of transposable elements was done by Barbara McClintock in maize in 1940s and 1950s. Besides maize, these elements are also recorded in yeast, *Drosophila* and in other eukaryotes.

12.4.1 Yeast-Ty-elements

The transposable element of yeast (*Saccharomyces cerevisiae*) is called **Ty**; they are about 5900 np long and includes two directly repeated terminal sequences called long terminal repeats (LTR) or deltas (δ). The Ty elements encode a single mRNA that begins at the promoter in the delta at the 5′ end. The mRNA transcript contains two open reading frames designated by two genes **Ty A** and **Ty B**. These genes encode two different proteins needed for transcription (Fig. 12.3).

The genetic structure of Ty resembles that of eukaryotic retroviruses. The transposition of Ty elements involves an RNA-intermediate. After its formation, a product of the Ty B gene uses the RNA to make a double stranded DNA by reverse transcription process. The newly synthesized DNA is now inserted somewhere in the genome. For its similarity to the function of retrovirus, the Ty is also called **retrotransposons**.

12.4.2 Transposable Element in Maize (Ac and Ds)

B. McClintock discovered (1948) two factors Ds and the Ac (Dissociation and activator respectively) in chromosome 9 in maize and provided an explanation for the genetic instability of that chromosome.

The Ac/Ds family of maize comprises numerous elements scattered throughout the genome. The Ac is functionally autonomous element. It consists of 4563 np bounded by an 8 np direct repeat. This direct repeat is created at the time of insertion of the element on a chromosome. Other repeats are found within the element itself. At the ends of the Ac element, 11 np sequence is repeated in the opposite orientation. These inverted terminal repeats (ITR) play important role in transposition.

In Ds element, on the other hand, structural diversities are noticed. Some are non-autonomous and are derived from Ac by the deletion of internal sequences; others are aberrant Ds element having the characteristic 11 ITR and some subterminal sequences of Ac, but with a different DNA in the middle. Another Ds element is called 'Double Ds element', in which one Ds element has inserted in the middle of another Ds element, but in inverted orientation (Fig. 12.4).

The protein *transposase* is involved in the transposition of the Ac/Ds family.

12.4.3 Retrotransposons in *Drosophila*

It is now known that the transposable elements are present in many organisms, and in *Drosophila*, almost 15% of its DNA is mobile.

The **retrotransposons** of *Drosophila* are retrovirus-like element. They have 5000-15000 np length, and they resemble the Ty elements of yeast. The either end of the

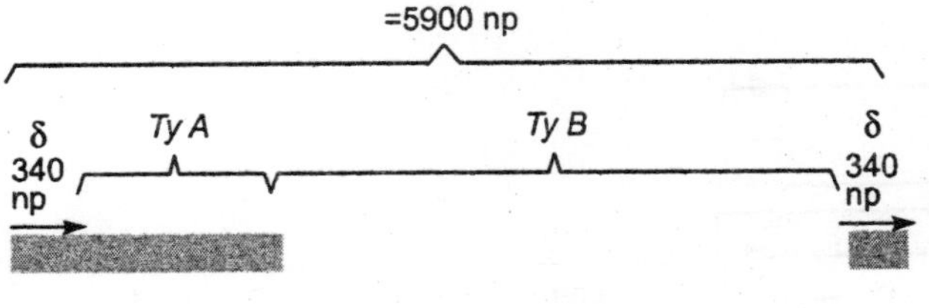

FIGURE 12.3

Genetic organization of yeast *Ty* elements. The long terminal repeat (LTR) sequences are denoted by the Greek letter δ. *Ty* elements apparently contain two genes, *TyA* and *TyB*, as shown. Sequence lengths are given in nucleotide-pairs (np).

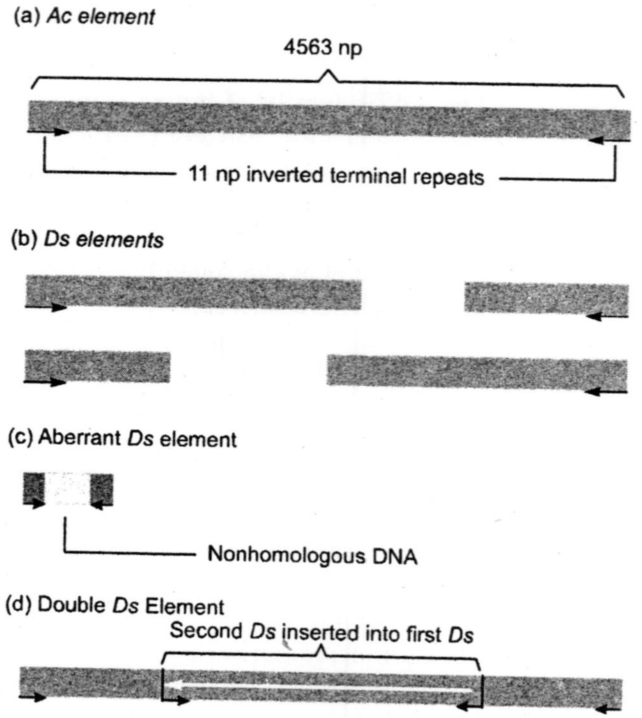

FIGURE 12.4

Structural organization of autonomous and nonautonomous members of the *Ac/Ds* family of transposable elements in maize. The terminal inverted repeats are indicated by the short arrows under each element. (a) *Ac*, the autonomous element. The lengths of DNA sequences are given in nucleotide-pairs (np). (b) Nonautonomous *Ds* elements derived from *Ac* by the deletion of internal sequences (gaps). (c) An aberrant *Ds* element with terminal and subterminal sequences from *Ac* flanking non-*Ac* DNA. (d) A double *Ds* element, in which one *Ds* element has inserted in the middle of another *Ds* element, but in inverted orientation. (After N. V. Federoff, *Cell* 56: 181-191, 1989.)

retrotransposon is demarcated by LTR (Long terminal repeat) sequence, and these are bounded by short repeat sequences, oriented in opposite directions. When a retrotransposon inserts into a chromosome, it produces a target site duplication, with one copy on each side of the transposon. They are responsible for many mutations in *Drosophila*.

12.4.4 P-element Family in *Drosophila*

The P-element family includes a large number of transposable elements in *Drosophila*. The largest of them is 2907 np long, its terminal inverted repeats being 31 np long. They carry a gene that encodes the transposase protein, and when this protein attaches to the element, it can move the element autonomously to another position in the genome (Fig. 12.5).

The P-elements have been studied for **hybrid dysgenesis** in *Drosophila*. Crosses between certain strains of *Drosophila* produce hybrids with some aberrant traits including mutation, chromosome breakage and sterility. These traits together are called **dysgenesis**. The transposable P-elements are responsible for hybrid dysgenesis. There are two strains of *Drosophila*, the **P-cytotype** and **M-cytotype**. The P-cytotype represses P-element movement, and the M-cytotype permits the movement.

When P-elements are combined with the M-cytotype by mating, they are induced to

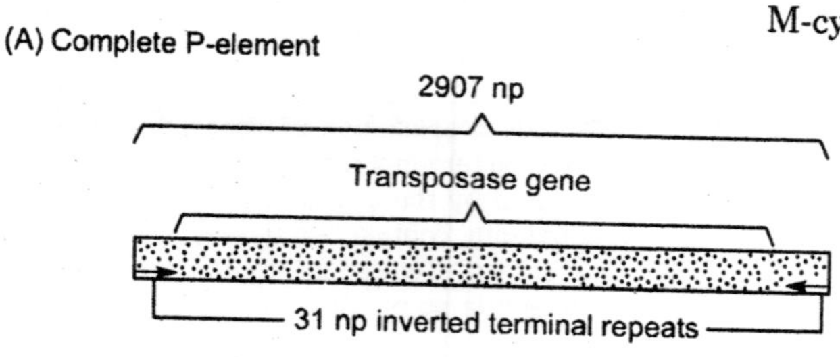

FIGURE 12.5

Structure of P-element in *Drosophila*.

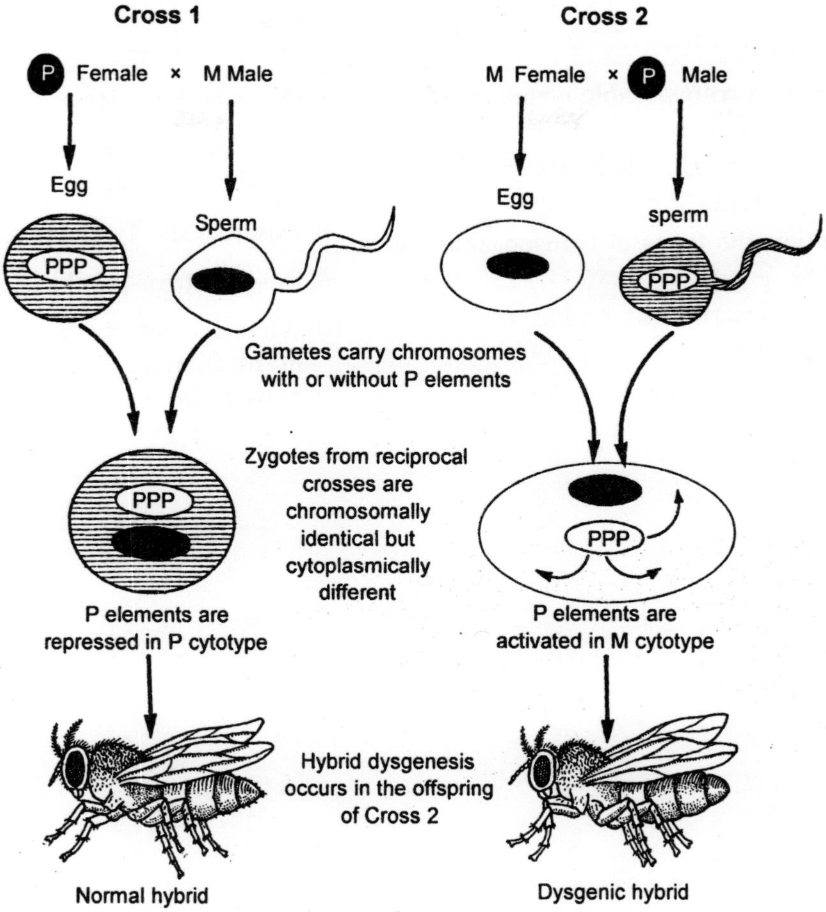

FIGURE 12.6
Reciprocal crosses between P and M strains in *Drosophila*.

transpose. The cross of the two strains of *Drosophila* in the diagram (Fig. 12.6) illustrates how hybrid dysgenesis develops.

IMPORTANT NOTES:

1. Transposable elements are mobile parts of DNA that can insert in any position of the genome and can cause mutation.
2. There are various names for transposable elements, viz., transposons, mobile genes, jumping genes, satellites, insertion sequence, mobile elements etc.
3. The first study on transposon was done in maize by Barbara McClintock in 1948. The mechanism of transposition in plant is similar to that of bacterial IS elements.
4. The bacterial transposons are medically important due to antibiotic-and drug-resistance incidence.
5. In eukaryotes, transposons like Ty, transpose, through an RNA intermediate (resembling retrovirus), and are called retrotransposons incidence.
6. The *Drosophila* P-element is a kind of transposon that is responsible for hybrid dysgenesis.
7. Transposable elements are used in genetic analysis.
8. Transposable elements are also useful in the genetic transformation of higher organisms.

EXERCISE

1. What is a transposable element? What are its synonyms?
2. What are the main features of transposable elements?
3. Mention the types of transposable elements.
4. Briefly describe the prokaryotic transposable elements.
5. Describe IS element.
6. Describe Tn3 element.
7. What is the medical significance of transposable elements?
8. Describe the Ty element.
9. Describe transposable element in maize.
10. What are retrotransposons and P-element family? Describe their role in *Drosophila*.

CHAPTER 13

CANCER GENETICS

Cancer is a genetic disease, Types, Cell-cycle and cancer, Genes related to cancer, Proto-oncogene, Oncogene and v-oncogene, Retrovirus and cancer, DNA tumor-virus, Tumor suppressor genes, Retinoblastoma tumor-suppressor gene, Breast-cancer, Mutator genes, Colorectal cancer, Telomere, Telomerase and cancer, Chemicals, Radiations and cancer.

It is often said that mutations are deleterious and detrimental; it is absolutely true in case of cancer; **mutations** in genes responsible for regulation of cell growth and cell division are the cause of **cancer**. When critical genes in cell-cycle regulation are mutated, cancer develops.

Certain terms are often used in cancer genetics. Let us understand them:

1. **Hyperplasia**: It is the uncontrolled growth of cells.
2. **Anaplasia**: The structural abnormality of cell is called anaplasia.
3. **Tumor** (*or*, **neoplasm**): Unregulated proliferation of cells pile up on top of each other and form tumors. The tumors are of two types: **benign** and **malignant**. When the tumor-cells do not invade surrounding tissue, the tumor is called **benign** but when the cells detach from a tumor and invade the surrounding tissues, the tumor is called **malignant**, i.e., cancerous.
4. **Metastasis**: When a malignant tumor spreads to other sites of the body, and forms a secondary tumor, the process is called **metastasis** ('change' of 'state').
5. **Carcinogens**: Certain agents like radiation, certain chemicals, and certain type of viruses can permanently transform normal cells into cancerous cells. These agents are called carcinogens.
6. **Oncogenesis**: The initiation of tumors in an organism is called **oncogenesis** (onkos = mass, bulk).

13.1 Cancer is a Genetic Disease

It is now established that cancer is a genetic disease. It develops due to genetic malfunctions. The causes of the malfunctions are uncertain; but whatever may be the remote reason, the result is a **mutation** in critical genes that are responsible for cell division.

Due to mutation, *the cells lose control over regulation in cell-division* and they divide continually forming a tumor. When cells form the tumor, they detatch, spread and invade the surrounding tissue and form secondary tumors (metastasis) in other places of body; this is called cancer. Thus, oncogenesis and mutation of certain genes are correlated.

13.2 Types of Cancer

There are many different kinds of cancers developing in different tissues of the body. Cancers may be called 'groups of diseases'. The most common cancers are prostate-, breast-, lung-, colon-, lymphoma-, bladder-, ovarian and blood-cancers (leukemia). Some cancers are *hereditary* (familial) and others are not.

Epithelial cells and tissues of many organs are the most prevalent source of cancer. Some forms of cancers can be kept arrested after early detection while other forms have no remedy till now.

13.3 Cell-Cycle and Cancer

Cell-cycle in somatic cells includes four stages—Gap-1 (G_1), Synthesis (S), Gap-2 (G_2) and Mitosis (M). [See chapter 3]. The cell-cycle is continued under the control of many genes in a system of **checks** and **balances**.

- **Checkpoints**: There are certain checkpoints in a cell-cycle that control the continuity of the cycle. These points determine whether the cell division should proceed or should be stopped. The major checkpoints are—

 (a) **G_1-to-S Checkpoint** (in mammals) or **START** (in yeast): It determines whether the cell will proceed from G_1 to S phase or not. If there is an irregularity in the genes concerned, it will either be repaired or destroyed by the action of the checkpoint.

 (b) **G_2-to-M Checkpoint**: It is present at the junction between G_2 and M. It checks whether all the DNA of the dividing cell has replicated.

 (c) **M Checkpoint**: It occurs during the mitotic (M) phase. It checks whether the chromosomes have attached properly to the mitotic spindle.

- **Proteins associated to checkpoints**

Cyclins and **Cyclin-dependent kinases** (Cdks) are two important protein factors for the action of checkpoints (Fig. 13.1).

The balanced action of checkpoints allows or stops continuation of cell-cycle in normal cells receiving stimulatory and inhibitory signals respectively from outside. A **neoplastic or cancerous cell** cannot respond to the outside signals and loses the control over cell-division and multiplies without check or constraints. Mutation of the genes which are responsible for stimulation or inhibition or for signal transduction can cause uncontrolled cell-division and result in development of cancer.

13.4 Genes Related to Cancer

Usually three types of genes mutate to produce cancers. These are **proto-oncogenes, tumor-suppressor genes** and **mutator genes**.

13.4.1 Proto-oncogene, Oncogene and V-oncogene

Mutation of proto-oncogene results into **oncogene**, that is, mutant proto-oncogenes are called oncogenes. Some viruses, cause transformation of cells to produce tumors. These are called **tumor viruses**. They may contain either DNA or RNA. The genome of **RNA-tumor-virus** possesses **viral oncogenes**, whose action stimulates uncontrolled cell proliferation, and these are the altered forms of normal host-cell genes. An RNA-tumor virus (retrovirus), like **Rous**

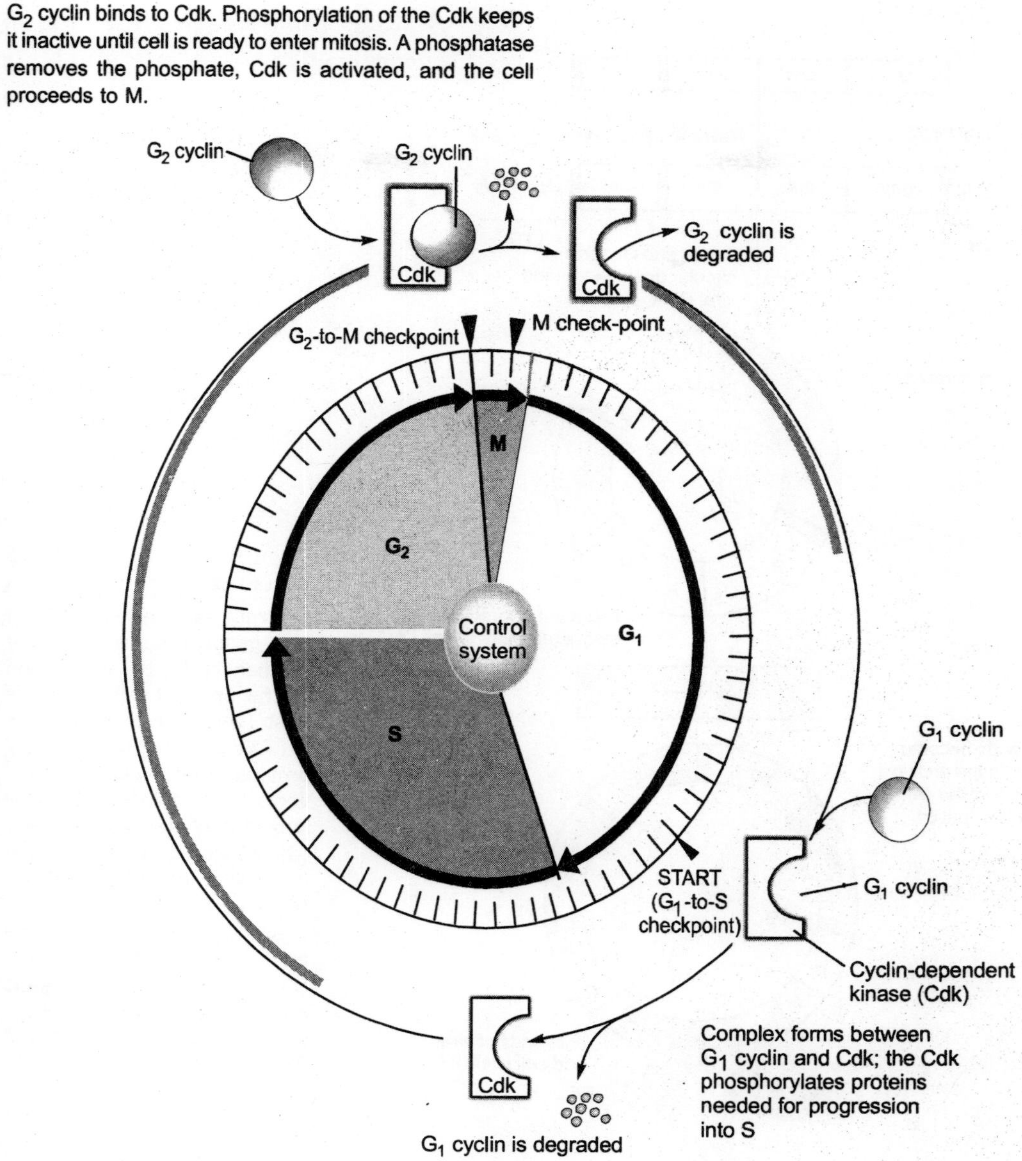

FIGURE 13.1

Checkpoints and cell cycle.

Sarcoma Virus (RSV) causes tumor in chicken called **sarcoma**. The RSV may undergo the following changes during its life-cycle (Fig. 13.2).

1. An RSV at first infects a host-cell.
2. The RNA genome is released from the virus.
3. Through the action of *reverse transcriptase* enzyme, a proviral DNA is produced.
4. The RSV proviral DNA dissolves their ends and ligates together to form circular-double stranded viral DNA.
5. Now, viral DNA is integrated to the cellular DNA by forming staggered nicks in both type of DNA.

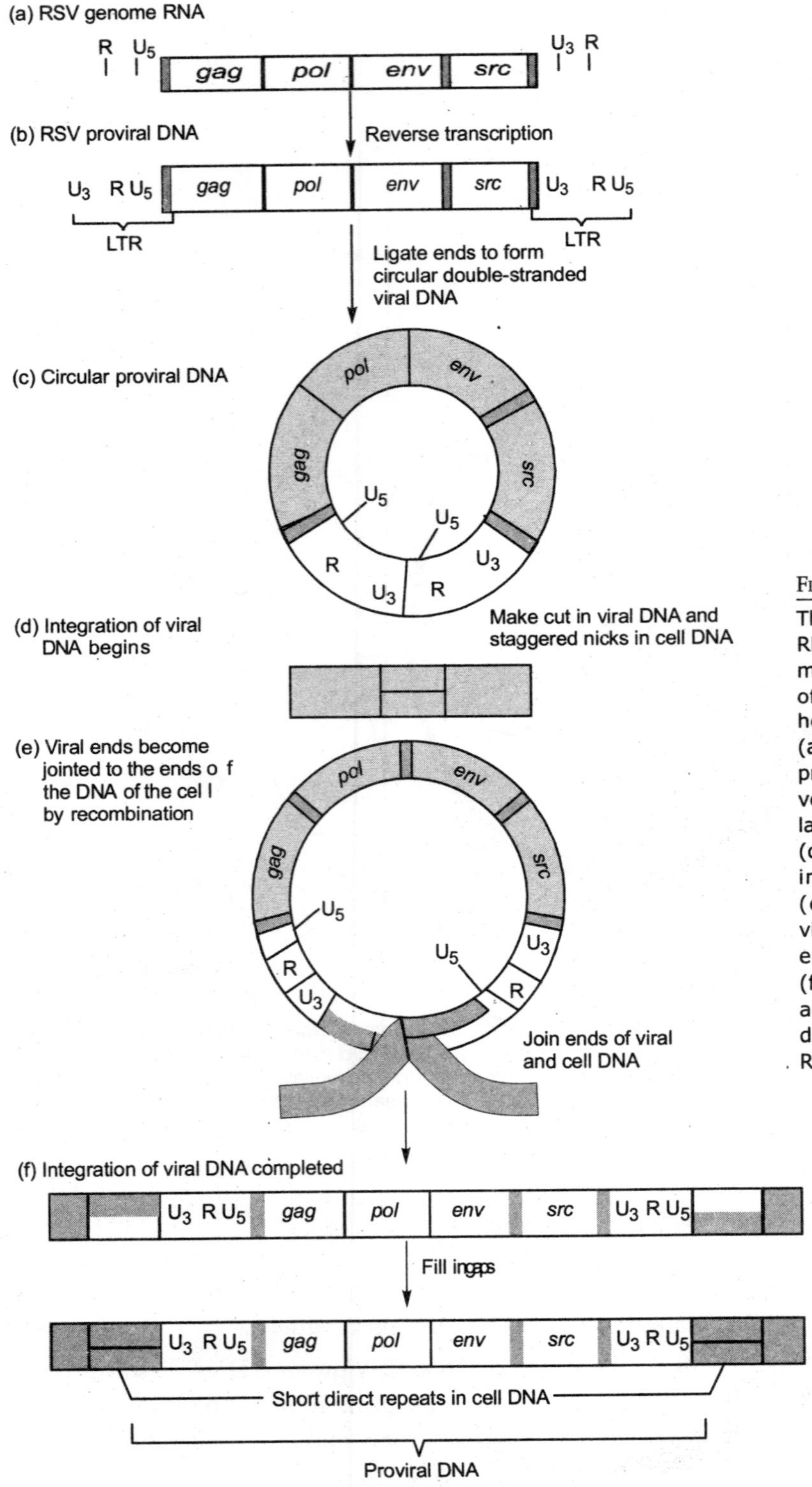

FIGURE 13.2

The rous sarcoma virus (RSV) RNA genome and a suggested mechanism for the integration of the proviral DNA into the host (chicken) chromosome. (a) RSV genome RNA. (b) RSV proviral DNA produced by reverse transcriptase. (c) Circularization of the proviral DNA. (d) Staggered nicks are made in viral and cellular DNAs. (e) By recombination, the viral ends become joined to the ends of the DNA of the cell. (f) The single-stranded gaps are filled in, and a complete, double-stranded, integrated RSV provirus results.

6. The viral-DNA-ends become joined to the ends of the DNA of the cell by recombination.
7. Then, by using host transcriptional machineries, viral genes are transcribed.
8. The single stranded gaps are filled in, and a complete double stranded, integrated **provirus** is produced.
9. Progeny RNA genomes are produced by transcription of the entire, integrated viral DNA and packaged into new viral particles.
10. They exit the host cell and infect other cells.

13.4.1.1 Viral Oncogene (V-onc)

The viral oncogenes (V-onc) are responsible for various types of cancers. The phage-viruses may pick up cellular genes (including oncogene) and can transduce the genes (including oncogene) to other cells. These are called **transducing retroviruses**. Some transducing retroviruses and their viral oncogenes are shown in (Table 13.1).

13.4.1.2 Cellular Proto-oncogenes

Normal animal cells contain genes with DNA sequences that are very much related to the viral oncogenes (Bishop, Varmus and others 1970s). These genes are called proto-oncogenes. A variety of human-tumor-cells also contain oncogenes (Weinberg, Wigler 1980s). These are also similar to the viral oncogenes, and are closely related to the human proto-oncogenes. In fact, **human oncogenes are mutant forms of normal proto-oncogenes**. The oncogenes present in a virus is called **V-onc** and that in a cell is called **C-onc**, or cellular oncogene. The V-onc-carrier retrovirus is a transducing virus, and when it infects a normal cell of a host, the recombined oncogene of the host cell and V-onc together transform the normal cell into a cancer cell.

13.4.1.3 Change of Proto-oncogenes

A proto-oncogene may be changed into an oncogene by three methods. The methods are **point mutations**, **deletion** and **gene-amplification**.

13.4.1.4 Retrovirus Inducing Cancer

The retrovirus that carries **V-onc** may be of transducing type and can cause cancer if the V-onc is expressed. After retrovirus infection, induction of tumor can take place due to the activity of V-onc of its genome. Retroviruses that carry oncogene are called transducing retroviruses; the cellular proto-oncogenes have similar DNA sequences like that of viral oncogenes. The proto-oncogenes are called C-onc, which may induce tumors.

13.4.1.5 DNA Tumor Virus

DNA tumor viruses can transform normal cells to cancer cells. Some of the DNA tumor viruses are papoviruses, hepatitis-B viruses,

TABLE 13.1

Retrovirus	V-onc.	V-onc. origin	Cancer
1. Rous Sarcoma virus (RSV)	src	chicken	Sarcoma
2. Abelson Murine Leukemia virus	abl	mouse	Pre-B cell leukemia
3. Avian Erythroblastosis virus	erb A	chicken	Erythroblastosis and sarcoma
4. Avian Myeloblastosis virus (AMV)	myb	chicken	Myeloblastosis
5. MC 29	myc	chicken	Sarcoma, Carcinoma and Myelocytoma

herpes viruses, adenoviruses and pox-viruses. In the host cells, they produce a viral protein that activates DNA replication in the host cell. Now, by using host-proteins, the viral genome (DNA) replicates, transcribes and produces a large number of progeny viruses; this causes lysis and death of the host-cell. The released viruses infect new cells. But, occasionally, the viral DNA is not replicated and becomes integrated into the host-cell-genome. If the viral protein that activates the replication of host-cell DNA is now synthesized, it will transform the host cell into a cancer cell, by uncontrolled proliferation of the cell in question.

13.4.2 Tumor Suppressor Genes

Normal cells contain such genes which are able to suppress the uncontrolled cell proliferation that produces tumors. These genes are called **tumor suppressor genes.**

Some tumor suppressor genes and the types of cancer caused by their mutation are as follows (Table 13.2).

13.4.2.1 Retinoblastoma Tumor Suppressor Gene (RB): Knudson's Two-hit Mutation Model

The retionoblastoma (eye tumor) is a cancer caused by two mutational events. It was proposed by Knudson in 1971 that the **sporadic** and **hereditary** retinoblastoma are the results of two independent mutation, (Fig. 13.3), but they occur in different way in the two types.

The two-hit model explains that in **hereditary** (familial) retinoblastoma, one mutation is inherited, and the person is predisposed to cancer (RB). When the second mutation (Fig. 13.3) occurs later in somatic cells, cancer develops. But in **sporadic** (nonhereditary) retinoblastoma, both mutations occur in somatic cells.

13.4.2.2 Breast Cancer (caused by Tumor Suppressor gene)

The most common cancer in human is the breast cancer. The average age of attack is 55. Approximately 5% of breast cancers are hereditary.

Two **tumor-suppressor genes**, the BRCA 1 and BRCA 2 play roles in breast cancers. It is believed that hereditary (familial) breast cancer results from mutation of the tumor suppressor genes, BRCA 1 and BRCA 2.

The BRCA 1 gene, located at the chromosome $17q^{21}$, can be mutated and cause cancer of breast and also the cancer of ovary.

The chromosome location of BRCA 2 is $13q^{12}$-q^{13}, and is generally associated with breast cancer alone.

When the allele of a tumor-suppressor gene are inactivated or lost due to mutation,

TABLE 13.2

Gene	Cancer	Hereditary Syndrome
APC	Colon carcinoma	Familial adenomatous polyposis (FAP)
BRCA 1	Breast cancer	Breast and ovarian cancer
BRCA 2	Breast cancer	Breast cancer
DCC	Colon carcinoma	Colorectal cancer
RB	Retinoblastoma	Retinoblastoma
VHL	Kidney carcinoma, pancreatic tumors	von Hippel-Lindau syndrome

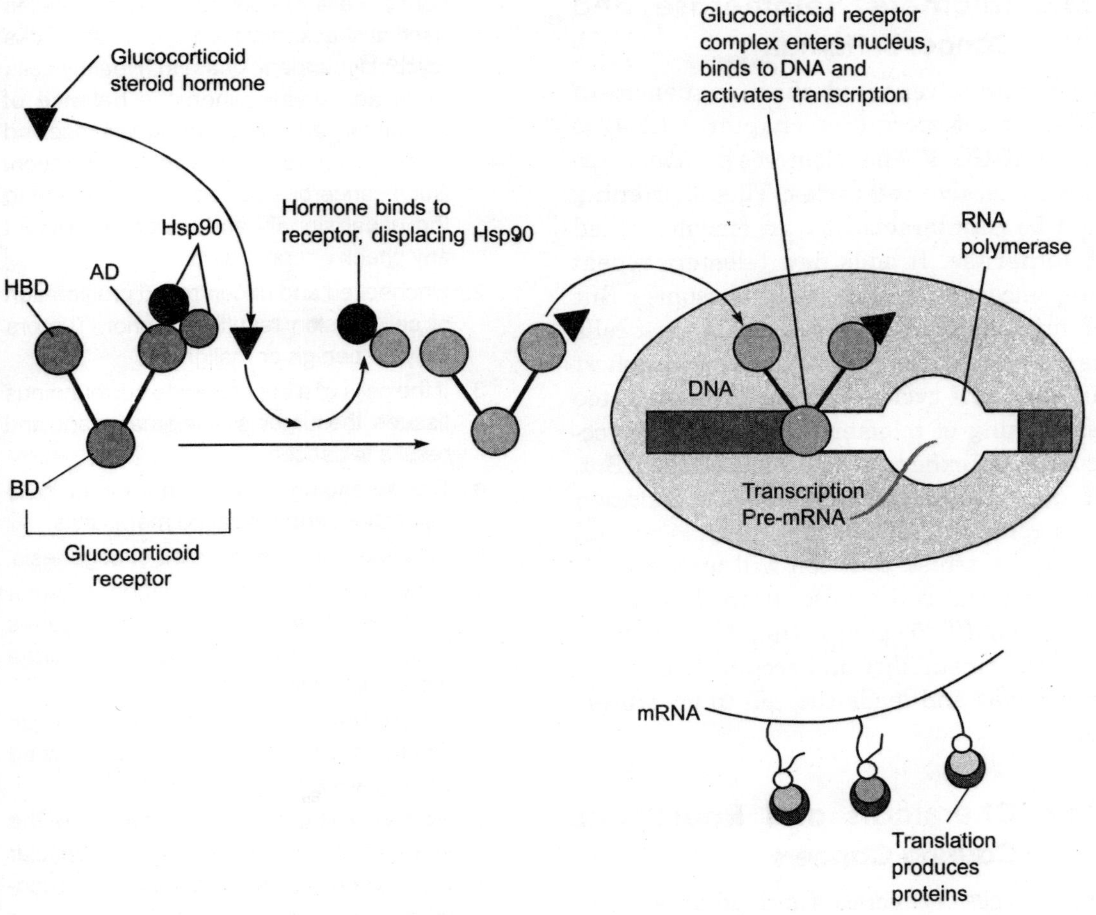

FIGURE 13.3

Model for the action of the steroid hormone glucocorticoid in mammalian cells. For the receptor, AD = activation domain, BD = DNA-binding domain, and HBD = hormone-binding domain.

the inhibitory or suppressing activity of it is lost, and uncontrolled cell division and cell proliferation starts, leading to formation of cancer.

13.4.3 Mutator Genes

A mutator gene can be defined as "a gene that, when mutant, increases the spontaneous mutation frequencies of other genes." Mutations of mutator genes hamper their normal functions of DNA replication and DNA-repair.

13.4.3.1 Colorectal Cancer (caused by Mutator Gene)

The human **hereditary nonpolyposis colon cancer** (HNPCC) is an autosomal dominant genetic disease causing colorectal cancer.

When any of the genes h MSH2, h MLH1 and h PMS2 in humans is mutated results in HNPCC. The h MSH2 and h MLH1 are responsible for 90% HNPCC. Mutation of these genes make the DNA replication faulty, and mutation rates are significantly higher.

13.5 Telomere, Telomerase and Cancer in Human

In humans, repeated short sequences of DNA in telomere (See chapter 3.1.2.4) is 5′-TTAGGG-3′. The telomeres get *short* during successive cell-cycles. This shortening can be counteracted by an enzyme called **telomerase**. It adds new telomere repeat sequences to the ends of chromosomes. But in human, telomerase activity in most cells is not significant, and the telomeres got short in each cell cycle. Eventually, due to too shortening of telomeres, DNA damage occurs, and further cell division is blocked. But, if, one of these cells undergoes a mutation in a gene (viz., TP53) that controls normal cell-cycle arrest, that cell will undergo cell division and cell-proliferation. Instead of mutation if TP53 gene is lost, it brings chromosomal instability and accumulates other mutations and leads the cell to be cancerous.

13.6 Chemicals and Radiations Causing Cancers

Many types of chemicals and radiations may act as **carcinogens**. They bring changes in the genome of the cell. Some act directly, others act indirectly. Examples of chemical carcinogens are alkylating agents, which acts directly. The polycyclic aromatic hydrocarbons, azo dyes and natural metabolites and nitrosamines are indirectly acting carcinogens.

The melanoma skin cancer may be induced by the exposure of the skin to U-V ray. Cancer may also be induced by ionizing radiation from X-ray. The radiation carcinogens act directly to cause cancer.

Important Notes:

1. Cancers are genetic diseases, Check-points in cell cycle determine whether a cell has DNA damage or defective cell-cycle machinery and permits only the normal cells to continue. The molecules used at checkpoints are cyclins and Cdks (cyclin dependent kinases). Healthy cells grow and divide when the balance of stimulatory and inhibitory signals received from outside favour the cell proliferation. But a cancerous cell does not respond to the usual signals and multiplies without any check or constraints.
2. Unchecked and uncontrolled proliferation of cell division results in tumor. Tumors may be benign or malignant.
3. If the cells of a tumor invade surroundings tissues, the tumor is called malignant, and results in cancer.
4. The spreading of malignant tumor cells throughout body is called metastasis.
5. Initiation of tumors is called oncogenesis.
6. If the genes like proto-oncogenes, tumor suppressor genes and mutator genes undergo mutation, then they can change a normal cell to a cancer-cell.
7. Certain DNA-viruses and RNA viruses can cause cancers. The RNA tumor-causing viruses are retroviruses.
8. Normal cells contain genes similar to the viral oncogenes. These are called cellular proto-oncogenes. When a proto-oncogene is mutated, it produces a cellular oncogene, which induces a tumor.
9. Retroviruses (RNA viruses) replicate via a DNA-intermediate. When a retrovirus infects a cell, the RNA is released, and through action of reverse transcriptase, a proviral DNA is synthesized. This DNA integrates with the DNA of the host-cell, and using the transcriptional machineries of the host, is transcribed. Progeny viruses are produced that infect other cells.
10. After retrovirus infection, cancer can be induced as a result of the activity of the V-onc (viral oncogene). Proto-oncogenes are present in normal cells which are similar to V-onc. Mutated proto-oncogenes are called C-oncs, which have potency to induce tumors.
11. Mutated proto-oncogenes, tumor-suppressor genes and mutator genes can cause cancer.

12. Proto-oncognes may change to oncogenes by point-mutation, deletion or gene amplification.
13. Interaction of tumor-suppressor genes is involved in the development of various human cancers including breast, colon and lung cancer.
14. The tumor-suppressor genes RB (Retinoblastoma), BRCA 1, BRCA 2, APC, VHL etc. can cause cancer through mutation.
15. Knudson proposed that retinoblastoma develops through two mutational events. This is called 'two-hit mutation model' of cancer. RB may be of two types, sporadic and familial. In familial RB, one mutation is inherited, and the second mutation occurs later in somatic cells developing cancer. But in sporadic RB, both mutations occur in somatic cells, and cancer develops later in life.
16. Certain chemicals like alkylating agents and radiations like X-ray and U-V ray may act as carcinogens.

EXERCISE

1. What is cancer? How does it develop?
2. Define malignant and benign tumor.
3. Define metastasis.
4. Discuss the relation between cell-cycle and cancer.
5. What are checkpoints? What is their role?
6. Which type of genes are related to cancer? What is proto-oncogene, oncogene and V-oncogene?
7. How does a retrovirus induce cancer?
8. Explain the nature and function of tumor-suppressor genes. Mention examples.
9. What is RB? How does it develop?
10. What is Knudson's two-hit model of cancer?
11. How does the breast-cancer develop?
12. What is mutator gene? What is its role? Explain its role in a typical cancer.
13. How is the telomere of human chromosome related to cancer?
14. Give examples of radiations and chemicals that act as carcinogens.

CHAPTER 14

THE BACTERIAL AND VIRAL GENETICS

Transformation, Transduction, Conjugation, F-factor, Merozygote, Recombination in merozygote, Mapping the bacterial chromosome, Plasmids and Episomes, Recombination in phage-virus, Lysogeny and Prophage, Reversal of lysogeny.

Bacteria contains 3 types of genetic materials: **a single main chromosome**, **episomes** and **plasmids**.

The bacterial **chromosome** contains a single circular DNA. A few thousand genes are present in the DNA. The **plasmids** are smaller in size, and may contain from 3 to several hundred genes. The **episome** may be present or absent in a bacterial cell. It may be inserted in a chromosome or remain independent. The F-factor of *E.coli* is an episome.

Bacteria reproduces by simple fission with an equal distribution of their genetic material to their daughter cells. Recombination of genetic materials occurs by 3 different processes: **transformation, transduction** and **conjugation**.

- **Transformation**: It involves the uptake of DNA molecule from one bacterium (**donor**) by another bacterium (**recipient**).
- **Transduction**: It is a process of transporting or carriage of bacterial genes from the **donor** to the **recipient** by a bacteriophage.
- **Conjugation**: It is the process during which DNA from donor or 'male' cell is transferred to a recipient or 'female' cell through the "conjugation tube" formed between them.

14.1 Transformation

The process of transformation in bacteria (*Diplococcus pneumoniae*) was first discovered by Griffith in 1928, that led to the discovery that the 'transforming material' is the DNA, by Avery, Macleod and McCarty in 1944 (see 4.2.1). The process of transformation involves (Fig. 14.1)—

1. Double stranded DNA from donor cell is bound by a receptor on the surface of a competent cell.
2. After entrance into the cell, one strand of DNA is degraded, and the other strand is coated by a single-strand DNA-binding protein.
3. The single strand of transforming DNA invades the chromosome of the recipient cell. There, it pairs with the complementary strand of DNA and replaces the equivalent strand.

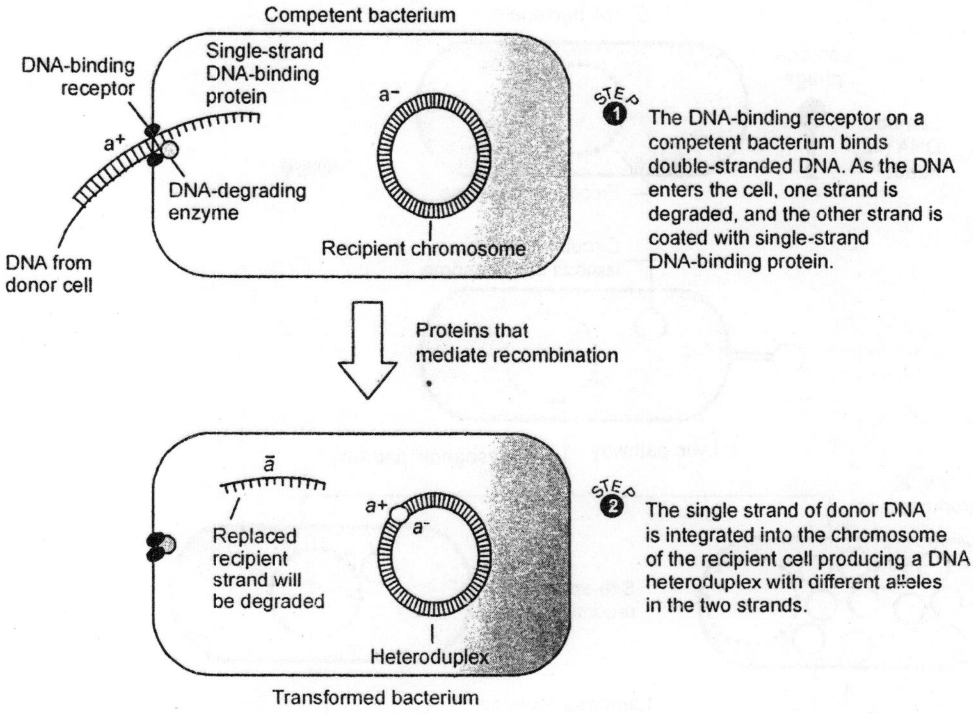

FIGURE 14.1

The mechanism of transformation in *Streptococcus pneumoniae*.

4. The replaced equivalent strand is then degraded.
5. If the donor and recipient cells carry different alleles (a^+ and a^-) of a gene, the resultant recombinant double helix will have one allele (a^+) in one strand and the other allele (a^-) in the second strand.
6. This type of DNA double helix is called **heteroduplex**.

14.2 Bacteriophage

Bacteriophages are the carrier of genetic materials during the process of transduction in bacteria. The phages are of two types, the **virulent** phage and **temperate** phage. The virulent type multiplies and *lyse* the host cell after infection. On the other hand the temperate type may enter the *lytic cycle* or the **lysogenic pathway**. In the lysogenic pathway, the chromosomes of the phage are integrated with that of the host, and they are replicated like the chromosome of the host (Fig. 14.2).

14.2.1 Lysogeny and Prophage

The **temperate phages** do not always cause lysis of the host-bacteria. Instead, the DNA of the virus becomes integrated into the bacterial chromosome. Once the viral DNA becomes a part of the host-chromosome, it replicates along with the bacterial DNA. In this way, as host cell reproduces, more and more bacteria carry the virus in its quiescent form. A bacterium carrying viral DNA in its genome is referred to as *lysogenic* and the viral DNA itself is called a **prophage** (Fig. 14.2).

14.2.2 Reversal of Lysogeny

A prophage is harmless to its host. But after several generations, it returns to a

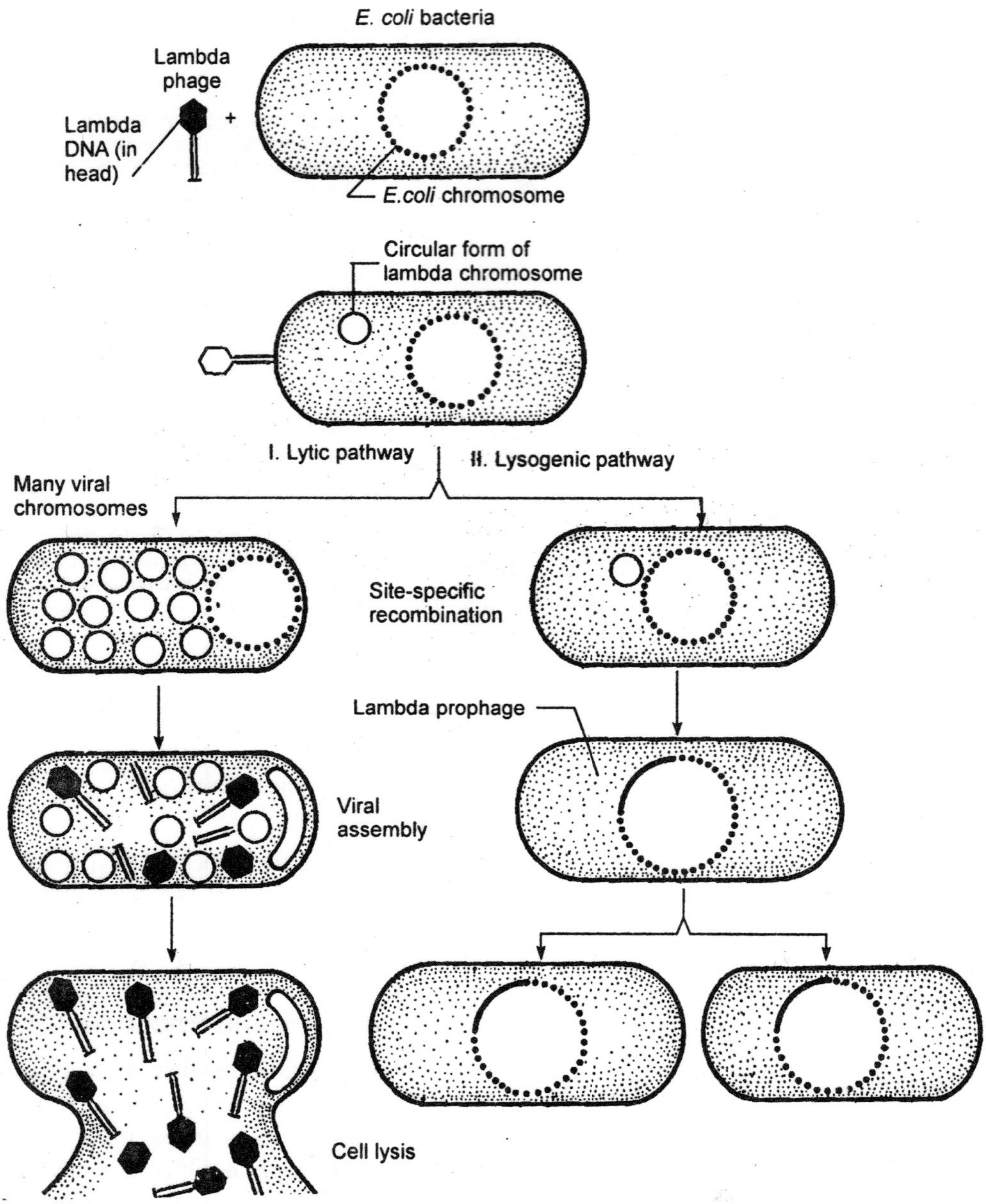

FIGURE 14.2

Schematic diagram illustrating the two alternative life-styles of temperate bacteriophage the *E.coli* phage lambda (λ). After the injection of the linear lambda chromosome into an *E.coli* cell, the chromosome is converted to its circular form. It then either (I) commences replicating autonomously and lyses the host cell, releasing 100 or more progeny viruses (the ***lytic pathway***), or (II) integrates via a site-specific recombination event into the *E.coli* chromosome (the ***lysogenic pathway***). In the integrated *prophage* state, the lambda chromosome replicates as part of the host chromosome and is transmitted to both progeny cells at each cell division.

vegetative or **infective** state. Then the phage reverts to the **lytic cycle**, reproducing itself and causing the *lysis* of the host cell. This spontaneous change from prophage to virulent state, can be induced by U-V rays also.

14.3 Transduction

The process of transduction in bacteria was discovered by **Zinder** and **Lederberg** in 1952. Transduction may be of two types: **generalized** and **specialized** or **restricted**.

14.3.1 Generalized Transduction

In generalized transduction—

1. Some virulent and certain temperate bacteriophages are involved.
2. Transducing particles are produced during the lytic cycles.
3. The T2, T4 and T6 bacteriophages, for example, degrade the host DNA and reutilize the mononucleotides produced for the synthesis of phage DNA.
4. The host DNA is not present in the progeny particles.
5. After a transducing phage injects a fragment of host DNA into a recipient cell, that DNA may be integrated into the host chromosome or remain free.
6. The genes located on the tranduced chromosomes may be expressed, even if they are not integrated.

14.3.2 Specialized or Restricted Transduction

In specialized or restricted transduction—

1. Bacteria (*E.coli*) lysogenized by λ phage indicates that these prophages are *restricted* to specific loci on the bacterial chromosome, at which genetic exchange may occur.
2. The λ chromosome is a double stranded DNA molecule, with a single stranded projection of 12 nucleotides at each 5′ end. These projections are complementary to each other, and thus become circularized inside the *E.coli* cell. It may follow 2 alternative paths—(a) vegetative reproduction of λ can lead to the proliferation of λ particles and **lysis** of the host, or, (b) The genes responsible for vegetative growth may be **repressed** by a particular protein, and **lysogenic path** is followed leading to integration and insertion of λ in the host chromosome (Fig. 14.3).
3. After insertion into the host chromosome, pairing and recombination occur (Fig. 14.4).
4. A specific phage gene (*int*) is responsible for recombination phenomenon which leads to the integration of the **prophage** chromosome (Fig. 14.4).
5. In continued function of the λ repressor induction, a number of λ gene products are manufactured, one of which (*xis*) is responsible for the *excision* of the prophage from the host chromosome, leading to follow the **lytic pathway**.
6. Limitation of lysogeny to a 'micro-colony' is called **abortive lysogenization**.

14.4 Conjugation

During conjugation in bacteria, the DNA from donor or male cell is transferred to a recipient or female cell through the "conjugation tube" formed between them. The process of conjugation in bacteria was discovered by **Lederberg** and **Tatum** in 1946.

14.4.1 F-factor (fertility factor or sex factor)

A bacterium can only act as a donor of genetic material if it contains a **plasmid** called an **F-factor** (fertility or sex factor).

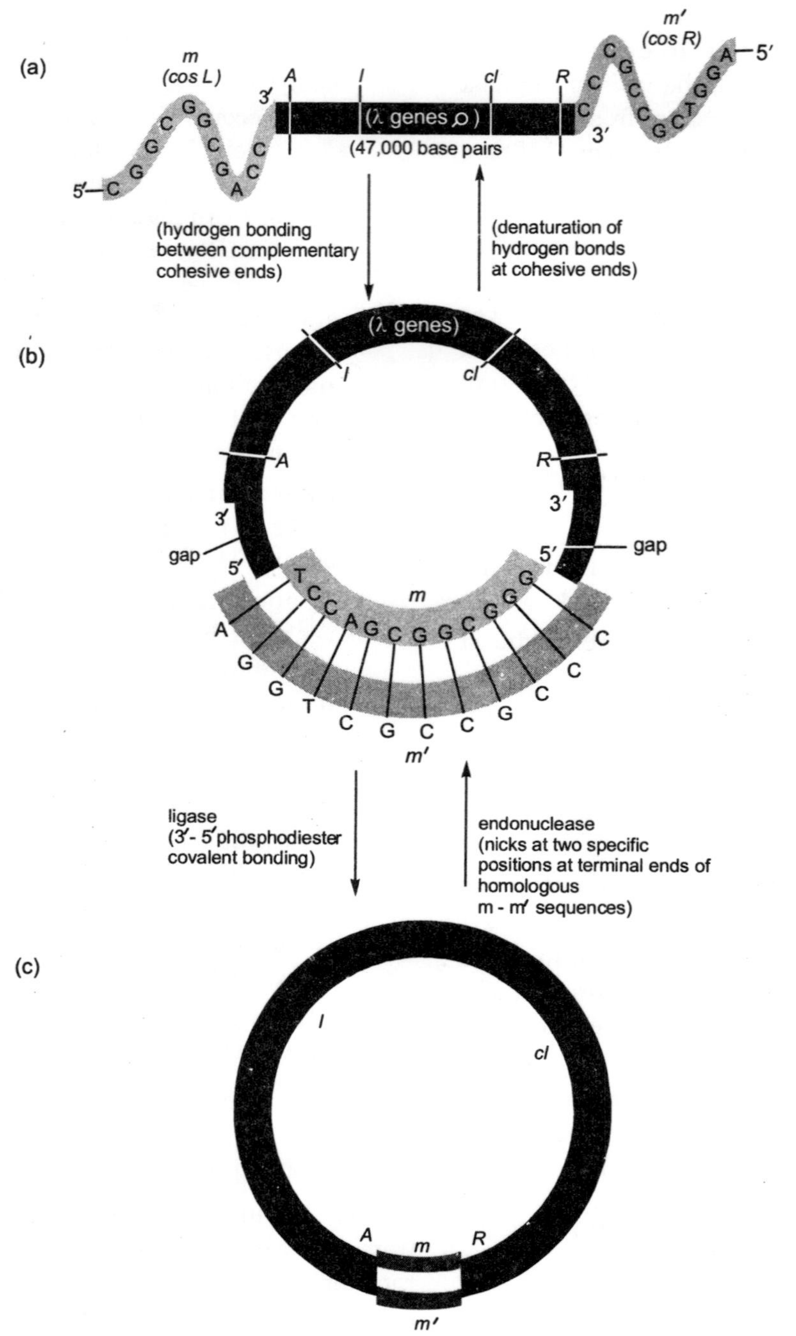

FIGURE 14.3

Diagram of the process by which the λ chromosome becomes circular or linear through pairing or nonpairing of the bases at its cohesive ends. (a) Linear structure of λ showing its single-stranded ends and their nucleotide base structure. (b) Pairing of the ends leads to a double-helical DNA structure and circularization of the λ chromosome. This process takes place immediately after entry into the host cell. (c) The phosphoester gaps remaining between nucleotides because of circularization are chemically connected (covalent bond) by the ligase joining enzyme. It is as a circular molecule that λ replicates vegetatively within *E.coli*.

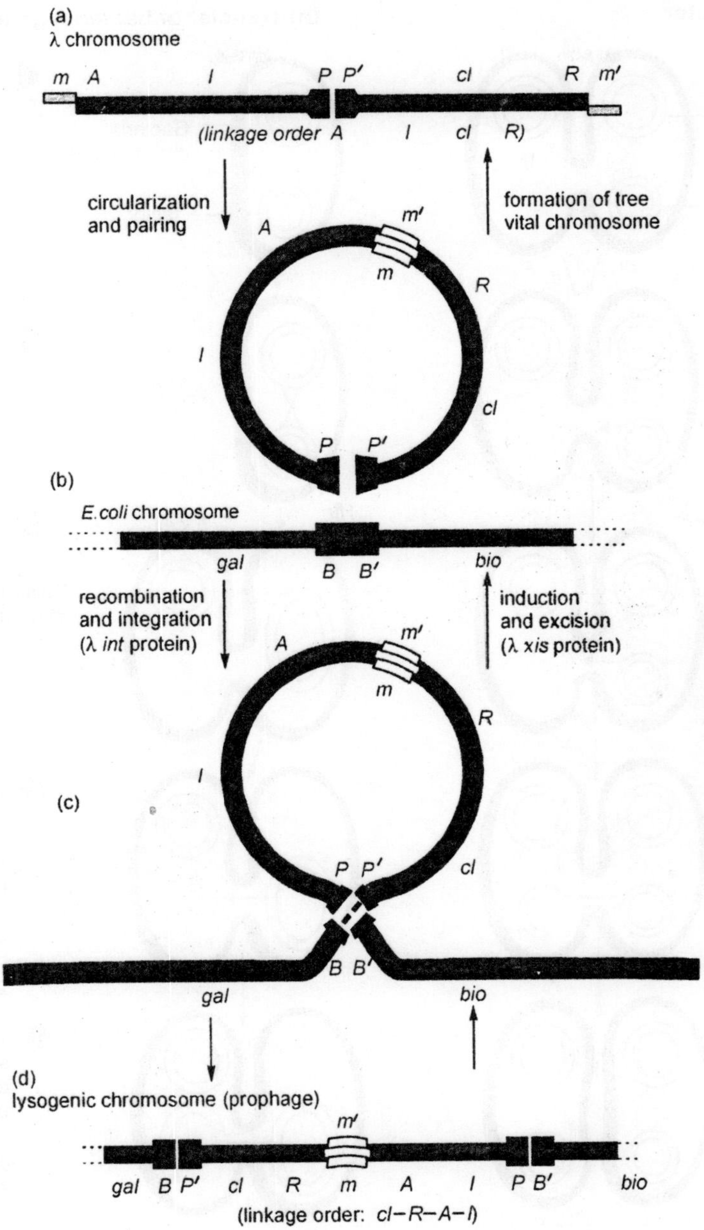

FIGURE 14.4 Lysogenization pathway for λ, leading from (a) the infective linear λ chromosome, to (b) circularization and pairing, to (c) recombination, to (d) integration.

The donor (male) cell possesses F^+ and the recipient (female) lacks it and is called F^-. F-factor has two regions—*origin* (O) and *F-pili*. When F^+ and F^- cells are mixed together, they show **conjugation**.

14.4.2 Process of Conjugation

During conjugation, genetic materials are transferred from donor to recipient. Beginning at the *origin* a single strand of DNA is transferred to the F. Once the F-factor-single-stranded-DNA enters the F^- recipient cell, the complementary strand is synthesized. When the complete F factor has been transferred, the F^- cell becomes an F^+ cell. In this $F^+ \times F^-$ cross, *no bacterial chromosome is transferred*. Only the F factor (plasmid) is transferred (Fig. 14.5).

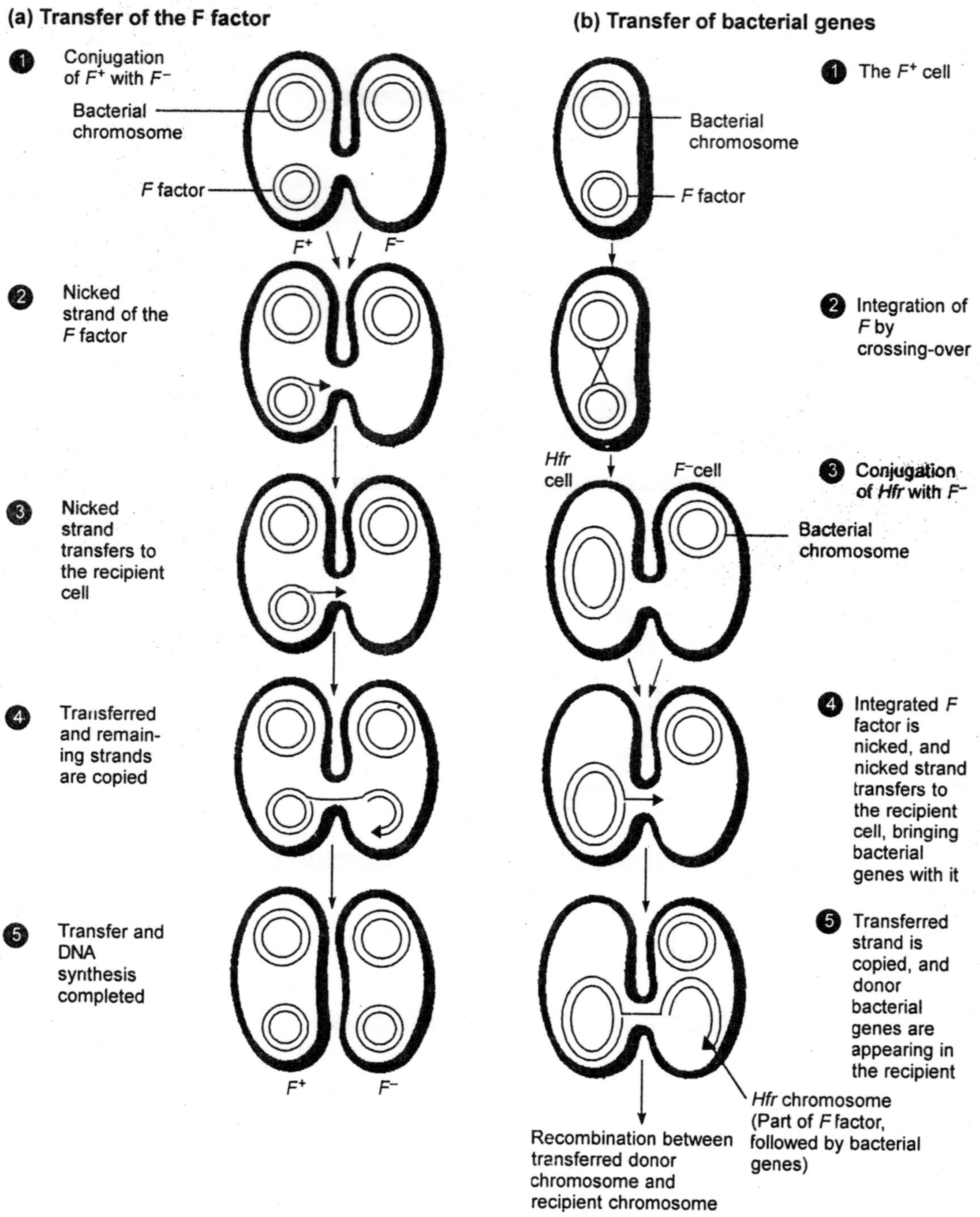

FIGURE 14.5

Transfer of genetic material during conjugation in *E.coli.* (a) Transfer of the F factor from donor to recipient cell during $F^+ \times F^-$ matings. (b) Production of Hfr strain by integration of F factor and transfer of bacterial genes from donor to recipient cell during Hfr $\times F^-$ matings.

14.4.3 High Frequency Bacteria (Hfr bacteria)

When the F factor is not free in the bacterial cell, but is joined to the bacterial chromosome, the bacterium is called **high frequency** (Hfr) bacterium.

14.4.4 The Merozygote

When an Hfr donor (F^+) and a recipient (F^-) conjugates, the F factor is transferred to the recipient. Now, the F^- cell possesses two copies of all the genes that have been donated by the Hfr parent. If the recipient, at the end of conjugation (Fig. 14.6), were to have two complete sets of genetic material, it would be like an 'eukaryotic zygote'; but it is referred to as '**merozygote**' or '**partial zygote**'. The cell has all its own genes and, in addition, some 'floating' genes from another source.

14.4.4.1 Recombination in Merozygote

Bacterial chromosome will only retain one copy of each of its genes. In fact, it may or may not retain particular donated genes. Only the genes that are recombined into the chromosome of the recipient, are retained. That is, if the donated gene is to be retained, the donor DNA has to be incorporated into the recipient DNA at the expense of the original gene or genes. The old DNA is replaced by the new (Fig. 14.7).

14.5 Mapping the Bacterial Chromosome

The diagrams showing conjugation and recombination (Fig. 14.6 and 14.7) also show that the bacterial chromosomes (and their genes) enter recipient cell in a line. Assuming the whole process takes a finite time (many minutes), the temporal order of entry of different genes should correspond to

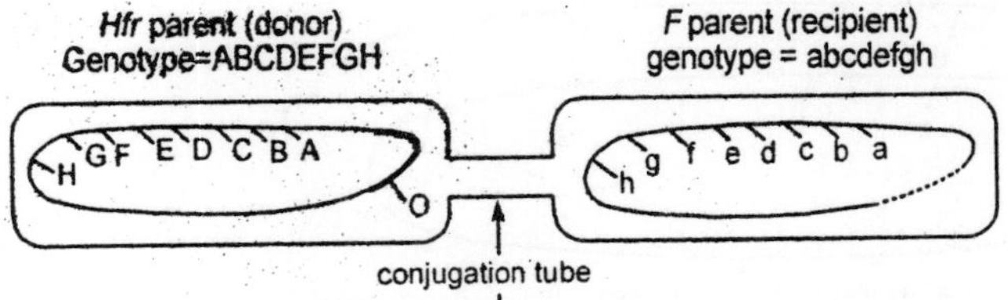

(i) The *Hfr* and *F* cells join by a conjugation tube

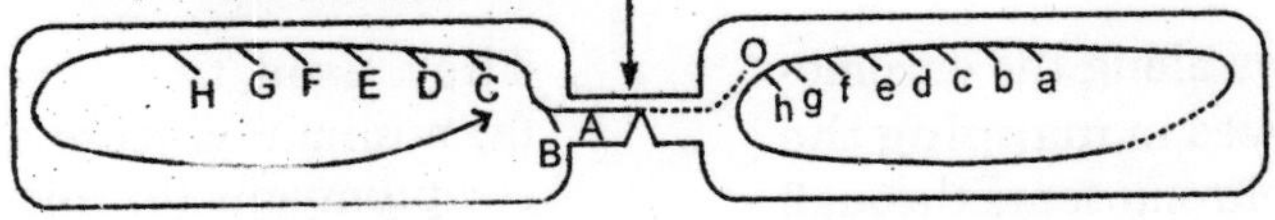

(ii) The Hfr chromosome breaks at the site of incorporation of the F factor. Part of the B factor of the origin O⁺ leads donor DNA through the conjugation tube into the recipient cell. The genes A-H follow in sequence

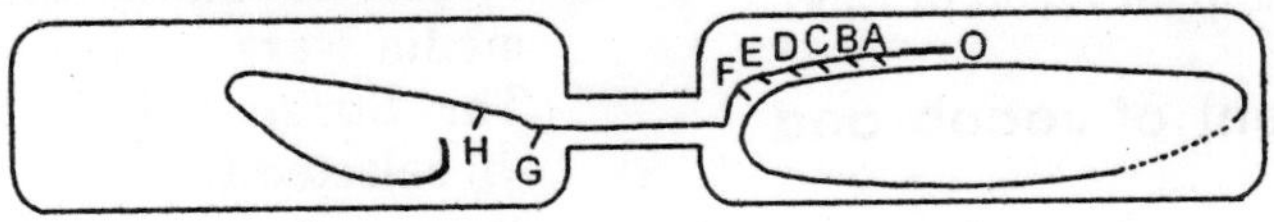

(iii) At some stage the conjugation process is interrupted and the transferring DNA strand breaks this may occur naturally or can be brought about by blender

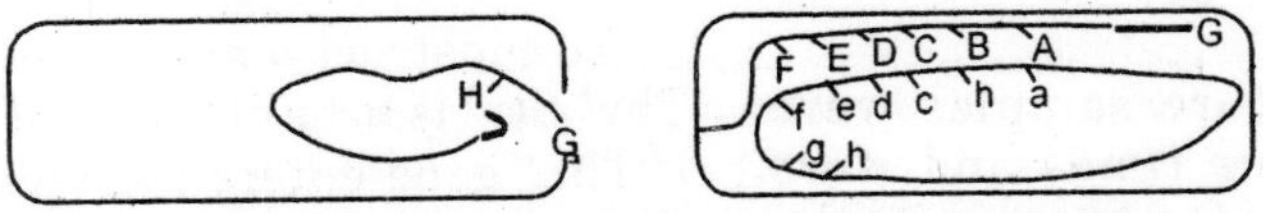

(iv) The cells separate. The recipient now has its own full complement of the genes plus some genes from the donor. Note that the donated DNA and part of the recipient chromosome are homologous and that the genes G and H have failed to enter the recipient.

FIGURE 14.6

The conjugation process in *E.coli*.

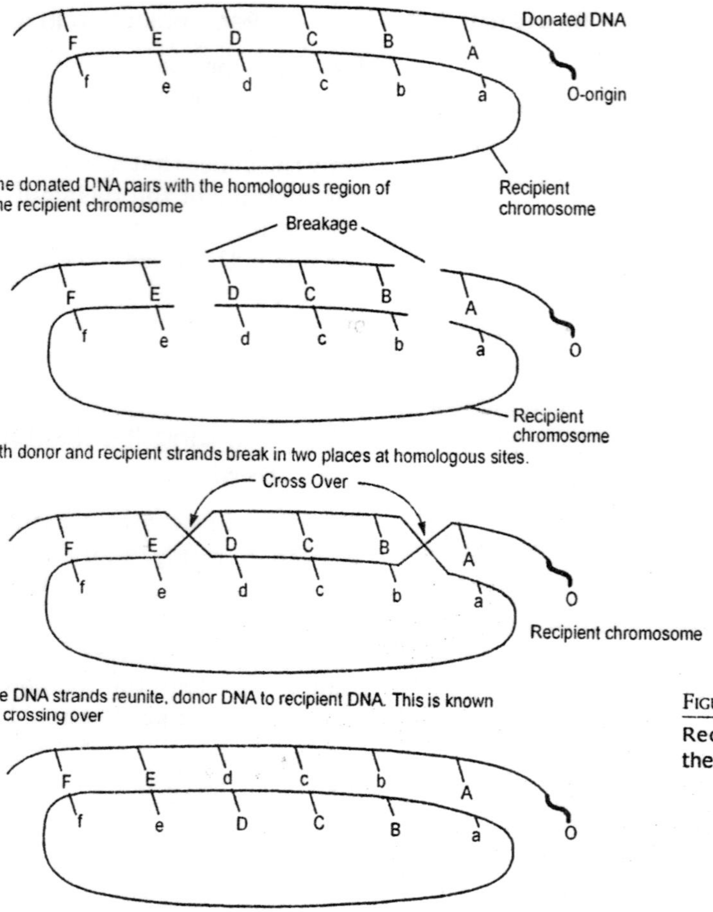

(i) The donated DNA pairs with the homologous region of the recipient chromosome

(ii) Both donor and recipient strands break in two places at homologous sites.

(iii) The DNA strands reunite, donor DNA to recipient DNA. This is known as crossing over

(iv) The intact chromosome has now exchanged the genes B, C, D though genes A, E and F are unaffected. The piece of DNA with the genes AbcdEF will be lost. Progeny of the recipient cell will all be of recombiant type: aBCDef

FIGURE 14.7 Recombination in the merozygote.

the actual order of genes along the chromosome. This principle is used in **mapping the genes** on the bacterial chromosome through the conjugation experiments.

Francis Jacob and **Elie Wolman** produced the first genetic map for *E.coli* in 1956.

14.5.1 The Experiment of Jacob and Wolman

1. They crossed the two strains of bacteria—Hfr Thr^+ Gal^+ Str^s × F^- Thr^- Gal^- Str^r, by mixing them together.
2. Resultant mixture samples were taken at various times and were agitated in a blender, to separate the conjugating bacteria and were then plated on two selective media.
3. They used two types of media—(a) minimal glucose agar containing streptomycin (b) minimal galactose agar containing streptomycin and threonine.
4. The recombinants selected by the two media were— Medium (a) selected Thr^+ Str^r recombinants, and medium (b) selected Gal^+ Str^r recombinants.
5. The number of recombinants growing on the two separate media were recorded at various times (Fig. 14.8).

To construct a map, a recombinant type Thr^+ Str^+ is to be obtained. This can happen if Thr^+ gene enters the recipient cell from the donor. This happens at 8 minutes after mixing the Hfr and F^- strains. On the other hand, the Gal^+ Str^+ recombinants are not

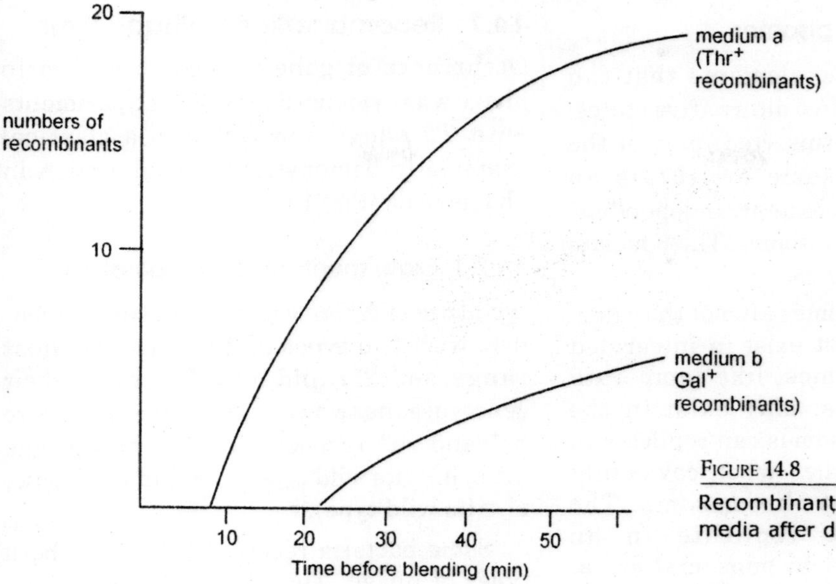

FIGURE 14.8

Recombinants growing on selective media after different incubation times.

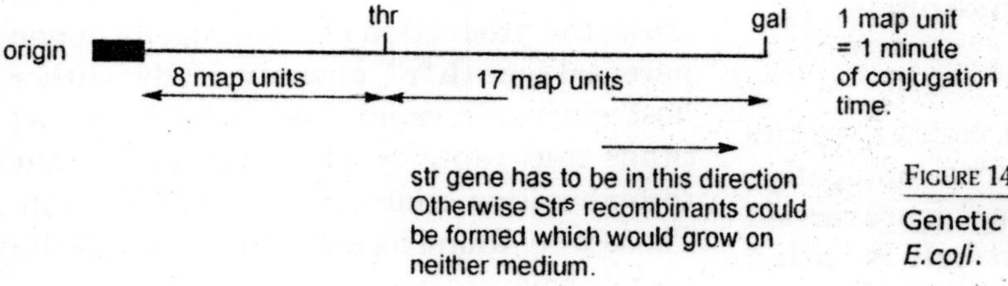

FIGURE 14.9

Genetic map for thr and gal genes of *E.coli*.

produced until conjugation continues for 25 minutes. It means whereas the Thr^+ gene enters the recipient after 8 minutes, the gal^+ gene has to wait for another 17 mins. After getting the recombinants, now, we can draw the map (Fig. 14.9)—

14.6 Plasmids and Episomes

These are two special genetic materials of bacteria.

14.6.1 Definition of Plasmids

A plasmid is a replicon that is stably inherited in an extrachromosomal state in bacteria. It is capable of independent replication. The plasmids are important to study in antibiotic and drug-resistance phenomena in pathogenic bacteria and the instability of industrially important microorganisms.

14.6.2 Types of Plasmids

Three types of bacterial plasmids are found—

1. F and F′ plasmid.
2. R plasmids.
3. CoL plasmids.

They may be also of (1) Conjugative or Transmissible plasmids (F^+, F' etc.) and (2) Non Conjugative or non-transmissible plasmids (many R and Col plasmids).

F and F′ Plasmid:

These are conjugation oriented fertility factors (See 14.4.1).

R Plasmid:

It is also known as RTF. They carry genes for resistance to antibiotics.

Col plasmid:

These code for *colcins,* which are proteins that kill sensitive *E.coli* cells.

14.6.3 Definition of Episomes

Episomes are genetic elements that can replicate in either of two alternative states: (a) as an integrated (inserted) part of the main host chromosome or, (b) as an autonomous genetic element, independent of the main host chromosomes. The F factors are one kind of episomes.

Plasmids and episomes are not the same. Many plasmids do not exist in integrated state; but the episomes, like temperate phage chromosomes, may exist in the integrated state. Episomes can replicate in an integrated state, i.e., when covalently inserted in the host chromosome. The episomes can also replicate in an autonomous or extrachromosomal state. Examples of episomes are the *E.coli* **K 12F factor** and **phage λ chromosome**.

14.6.4 Sexduction

Some recombinant F factors, called F′ factors carry chromosomal genes. The conjugative transfer of donor chromosomal genes carried by F′ factors to recipient cell is called **sexduction**: It is a kind of **transduction**.

14.7 Recombination in Phage Virus

Occurrence of genetic recombination in phage was evidenced in 1940s. Experiments with *T2 phage* and *E.coli* is a classical example to demonstrate recombination in phage virus (Fig. 14.10).

14.7.1 Experiment on Recombination

A culture of *E.coli* was infected simultaneously with two types of T2 phage: (1) **host range**, and (2) **rapid lysis**. In terms of their genotypes the two 'parents' in the cross were hr^+ and rh^+ respectively (h = host range gene, h^+ = its wild type, r = rapid lysis gene, r^+ = its wild type).

Some bacteria received DNA from both types of phage. The potential cross was— hr^+ (**host range**) × rh^+ (**rapid lysis**). In this cross, the production of any progeny of non parental type (**h^+r^+**, phenotypically neither host range, nor rapid lysis, or hr, both host range and rapid lysis) would show that recombination occurs in a mixed infection. These were found as recombinants. Results of the above cross were—

Progeny Type	% of all plaques	Type
h^+r	34	Parental: rapid lysis
hr^+	42	Parental: host range
h^+r^+	12	Recombinant: neither type
hr	12	Recombinant: both types

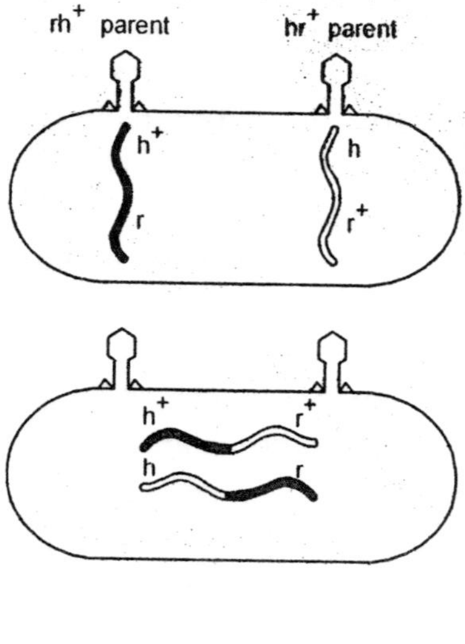

(i) Mixed infection by two different genotypes of phage results in the presence of two different but homologous pieces of phage DNA in a single bacterial cell.

(ii) Recombinant DNA–the genes now united in a different way.

FIGURE 14.10

Recombination in phage

The phenomenon of recombination in T2 phage has been illustrated in (Fig. 14.10).

IMPORTANT NOTES:

1. Genetic materials of bacteria are chromosome, episome and plasmid.
2. Recombination of genetic materials in bacteria occurs by transformation, transduction and conjugation.
3. Bacteriophages (phage viruses) are of two types—lytic and lysogenic.
4. Transduction is the process of transporting or carriage of bacterial genes from the donor to the recipient by a bacteriophage.
5. F-factor is a plasmid (or episome). It is the fertility or sex factor for bacteria, without which a bacterium cannot act as a donor of its genome.
6. There are 3 kinds of plasmids—F and F', R and Col.
7. Genetic mapping of bacterial chromosome is constructed by observing recombination. The experiment was first done by Jacob and Wolman in *E.coli*.
8. The conjugative transfer of donor chromosomal genes carried by F' factors to recipient cell is known as sexduction.
9. The lysogeny of a phage virus may be reverted to lytic cycle spontaneously after some generations, or can be induced by UV light induction.

EXERCISE

1. Give elementary idea on the genetic materials of bacteria.
2. What are the principal methods for genetic recombination in bacteria?
3. What do you understand by transformation?
4. Give a detail account on transduction.
5. Describe the method of conjugation in *E.coli*.
6. What is F-factor? What is its role in conjugation in bacteria? What is Hfr?
7. How would you map the bacterial chromosome?
8. What is merozygote? How is it formed?
9. Write notes on plasmids and episomes. What is a prophage?
10. How does recombination occur in phage virus? Describe it with suitable example.

CHAPTER 15

EXTRACHROMOSOMAL (EXTRA NUCLEAR, CYTOLOGICAL) INHERITANCE

Inheritance other than by chromosome, Criteria of extrachromosomal inheritance, Cytoplasmic organelles, Mitochondria, Plastids and symbiont bacteria, Kappa-particles, Plasmids, Cytoplasmic male sterility in plants, Maternal inheritance, Maternal effect, Snail-shell-coiling, Fertility and survival of *Drosophila*.

So far, we have discussed the mode of inheritance through the chromosome or DNA residing in the nucleus. However, many experiments suggested that the egg-cytoplasm contains certain determinants of development. It is also known that there is little genetic function during the time of cleavage of a zygote. Soon it came to light that some inheritance patterns are non-Mendelian in their behaviour, and in them the determinants lie in the cytoplasm having autonomy from the nucleus. These are called cytoplasmic determinants.

The cytoplasmic determinants have no link with the nuclear chromosome. Thus, **the inheritance which is independent of the role of chromosome is called extrachromosomal inheritance**.

In extrachromosomal inheritance, male and female parents do not make equal contribution to the genetic constitution of the progeny. It is an exception of the Mendelian principle of inheritance. Thus, it is also known as non-Mendelian inheritance.

15.1 Extrachromosomal Inheritance: Criteria

The extrachromosomal inheritance usually involves DNA of mitochondria and plastids; some bacteria and viruses may also play role as agents in extrachromosomal inheritance. There are a few criteria that can distinguish between traits that are controlled by **nuclear** (chromosomal) genes and traits controlled by **extranuclear** genes. Such as—

1. The results of reciprocal crosses indicate that one parent (usually maternal) exerts greater influence than the other on a particular trait.
2. Organelles in the cytoplasm of maternal reproductive cell containing

extranuclear DNA are associated with the extrachromosomal or cytoplasmic inheritance. It is an evidence of maternal transmission of inheritance.

3. Absence of linkage to known chromosomal genes in the product is an evidence of extrachromosomal inheritance.
4. Absence of Mendelian segregation and Mendelian ratios in the product is an evidence of extrachromosomal inheritance.
5. Transmission of traits in enucleated cells also suggests extranuclear inheritance.

15.2 Cytoplasmic Organelles

Mitochondria, plastids (chloroplasts) etc. have DNA (not nuclear DNA) and can reproduce of their own. Similarly, a symbiont bacteria, *Caedobacter taeniospiralis* can reproduce only in the presence of a particular host *Paramoecium*.

15.2.1 Mitochondria

It is now strongly believed that the mitochondria were once free-living bacteria. Later they became hereditary symbionts in eukaryotic cells, and finally they became essential organelle in the eukaryotic cells. They still persist as the **DNA** (which was present in free living life) for genetic purpose.

15.2.1.1 Mitochondrial DNA

Mitochondria contain a small amount of DNA which is very unique. It is an autonomous structure outside the nuclear genome. Mitochondria contain a separate **protein-synthesizing-machinery** (with mRNA, rRNA, tRNA, ribosomes and specific enzymes) for protein synthesis. The ribosomes of this machinery are similar to that of bacteria. The rRNAs are of same size to that of bacteria. Mitochondrial DNA can undergo mutations. Heritable changes in mitochondrial DNA may lead to heritable alterations in the mitochondrial structure. It has been evidenced that the sterility in hybrid *Culex* is caused due to an interaction in mitochondrial DNA. The DNA-genome of mitochondria is called **mtDNA**, which usually exists as small circular molecule. It ranges in size from about 16Kb in mammals to several hundred Kb in certain plants.

15.2.2 Plastids

Plastids are self-reproducing bodies in plant cells. There are two main kinds of it: the **chloroplast** and the **leukoplast**. The chlorophyll of chloroplasts is involved in the photosynthetic process of plants. Plastids (chloroplasts) possess their own hereditary material (DNA). It has been subsequently demonstrated that chloroplasts contain DNA, mRNA, tRNA and ribosomes.

In 1908, C. Correns described cytoplasmic inheritance of colour in plastids in the four-O'clock plant, *Mirabilis jalapa*. There may be three kinds of leaves (and branches) in the plant; **green** (containing chlorophyll), **white** (lacking chlorophyll), and **variegated** (with green and white patches). Corrnes found that (1) flowers on green branches produced only green offspring (regardless of the phenotype of the pollen-parent), while, (2) flowers from the white branches produced only white seedlings (regardless to the type of pollen parent), and the plants developing from these die, because they lack chlorophyll and cannot carry photosynthesis. On the otherhand, (3) flowers in variegated branches produced a mixed progeny of green, white and variegated plants in non-Mendelian ratios. It explains a pattern of **extrachromosomal inheritance**, because the offspring resemble the phenotype of the **female parent**, and the phenotype is related with a cytoplasmic par-

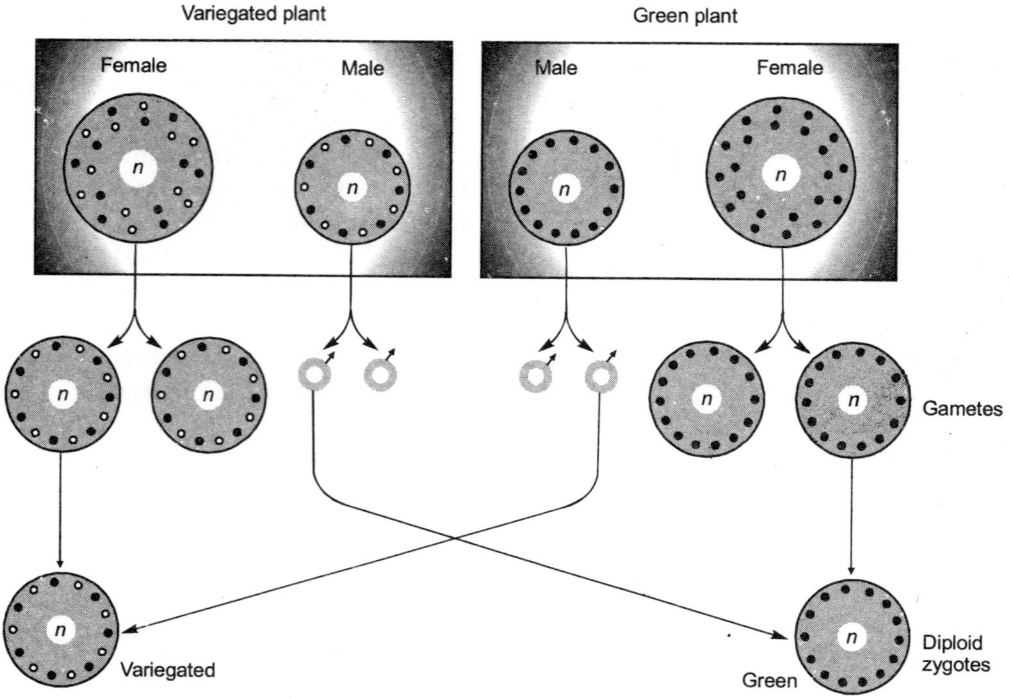

FIGURE 15.1

Diagrammatic illustration of maternal plastid inheritance in the diploid plant *Mirabilis,* which has abundant cytoplasm in the seed-parent gametes but little or no cytoplasm in pollen-parent gametes. *n*, nucleus; *c*, cytoplasm.

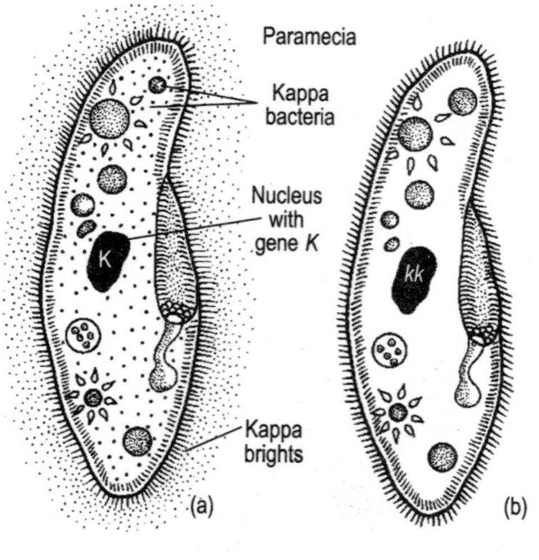

FIGURE 15.2

Kappa particles in *paramoecium aurelia*. (a) Killer with kappa particles inside the paramoecium and paramecin in the liquid medium outside the organism. Gene *K* is present in the nucleus. (b) Sensitive paramecia with no kappa particles, no paramecin, and genes *kk* in the nucleus (After T.M. Sonneborn)

ticle. This is called **maternal inheritance** in those species, where distinct male and female sexes are prevalent (Fig. 15.1).

15.2.3 Symbiont Bacteria in *Paramoecium* (Kappa particle)

T.M. Sonneborn found a persistant extranuclear effect in *Paramoecium aurelia.* Certain strains of it produce a toxic substance that has a lethal effect on other strains of *P. aurelia*. The strains that produce such toxic substance are called '**killers**'. There are about 400 'particles' or 'Kappa' in a 'killer' paramoecium. These 'particles' are called kappa bacteria which are **symbiotic** and named *Caedobacter taeniospiralis.* The killer bacteria or kappa particles produce the 'toxic substance' called **paramecin**, which is diffusible in fluid medium. Paramecin has no effect on killers;

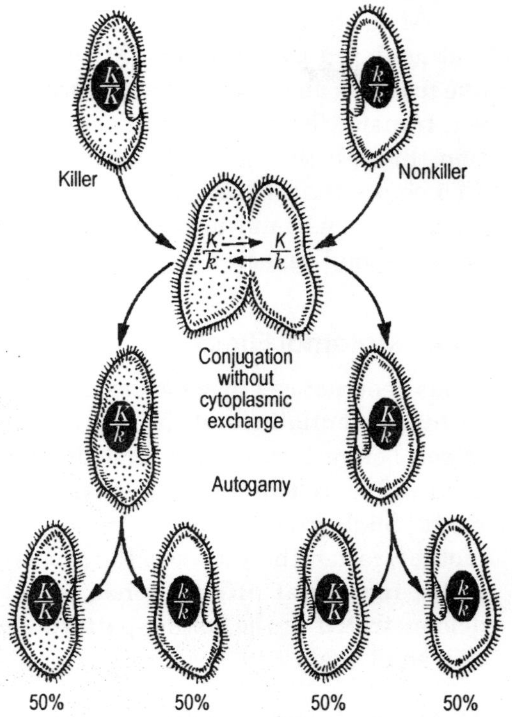

FIGURE 15.3

Conjugation between a killer paramoecium (stippled) and a sensitive paramoecium without cytoplasmic exchange but with an exchange of *K* and *k* genes. The autogamy that follows results in *KK* and *kk* homozygotes, which may give rise to *KK* and *kk* clones. (After T.M. Sonneborn.)

it can kill only the sensitive paramecia. (Figs. 15.2 and 15.3).

Kappa bacteria can reproduce only in those paramecia which contain **dominant nuclear allele K**. When the killer *Paramoecium* conjugates with sensitive *Paramoecium* (in an appropriate condition, so that it does not kill the mate) **without cytoplasmic exchange**, two types of clones are produced. (Fig. 15.3) (1) Paramecia containing allele K (Kk) and kappa bacteria and (2) Paramecia containing allele K (Kk) but no Kappa bacteria. If they undergo autogamy (Fig. 15.3), 50% of the progeny of the killers will be killers and 50% will be sensitives; but all progeny of the sensitives will be sensitives. There was no transfer of cytoplasm in the conjugation, and thus, only the cells from original killers receive that kappa. But the kappa fail to reproduce in the cells where the K allele is not present.

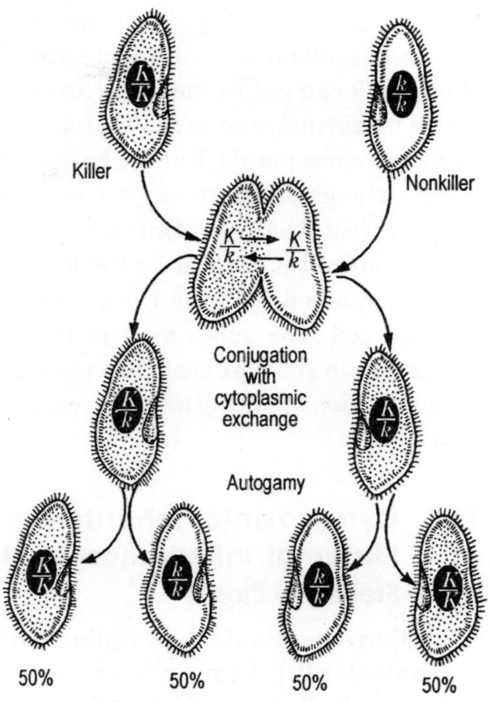

FIGURE 15.4

Conjugation with cytoplasmic exchange followed by autogamy. (After T.M. Sonneborn.)

But when the killer conjugates with sensitive for a longer period facilitating a **cytoplasmic exchange**, the result becomes different (Fig. 15.4). When the conjugants are KK and kk, the alleles are exchanged, and both exconjugants become Kk, as well as, due to **cytoplasmic exchange**, the kappa particles from the killer have been transferred to the nonkillers. Their autogamy produces homozygotes KK and kk in equal ratio. The KK produces clones of killers and the kk produces clones of non-killers. It is a clear demonstration of cytoplasmic inheritance.

15.2.4 Plasmids

Plasmids are almost similar to mitochondrial and plastid-DNA. Because of their ability for self replication, for capacity to combine with other DNA, and for capacity for carrying DNA to cell-synthesis machinery, they are useful for genetic engineering. The plasmid **Ti** can induce tumor in association with a bacterium, and cause 'gall' disease in many economic plants. But the bacteria that initiate the gall disease are not needed for the growth of the tumor. Only a fragment of the **Ti** plasmid is carried by the bacterium and is combined with the DNA segment of the host cell. The genes present in the **Ti**, integrated in the plant cell, are responsible for production of uncontrolled growth of the tumor.

15.3 Cytoplasmic Inheritance or Maternal Inheritance in Male Sterility in Plants

Many flowering plants show **pollen-failure** and male sterility. In maize, wheats etc., fertility is controlled in part by cytoplasmic factors. Male sterility in maize was discovered and analysed by **Rhoades**. Male sterility due to pollen abortion in certain corn plants was transmitted from generation to generation through the egg cytoplasm. The Fig. 15.5 shows that inheritance of male sterility is maternal and not controlled by the chromosomal genes.

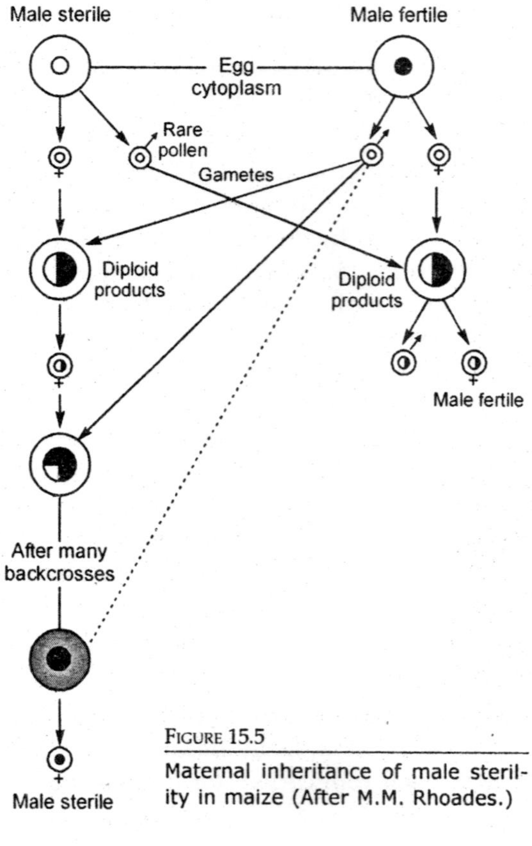

FIGURE 15.5

Maternal inheritance of male sterility in maize (After M.M. Rhoades.)

15.4 Maternal Effect

It has been noticed in some organisms that certain potentialities of the egg are determined before fertilization, and these have been influenced by the maternal environment. Such predetermined genes of the mother, rather than those of the progeny, is called **maternal effect**. It has been well demonstrated in the coiling-pattern of certain snail-shell.

15.4.1 Maternal Effect in Snail-shell-coiling

In *Limnaea peregra* the direction of coiling is determined by a single pair of genes. Snails that coil to the left are called **sinistrals** and that coil to the right are called **dextrals**. The alleles for dextral coiling (R) is dominant over the recessive sinistral coiling (r). It has been surprizingly noticed that in crosses between dextral and sinistral coiling, the direction of coiling of the offspring is always determined by the genotype of the maternal snail, whatever may be the genotypes of the progeny !

The dextrally coiling mother will produce dextrally coiling (Rr dextral) progeny (may their father be sinistral) and a sinistrally coiling mother will produce sinistrally coiling (Rr sinistral) progeny (may their father be dextral) (Fig. 15.6).

A self-fertilized dextrally coiled heterozygote (Rr) snail will always produce eggs that develop into dextrally coiled offspring even though some fertilized eggs are rr (i.e., homozygous recessive) and should be sinistral (Fig. 15.6).

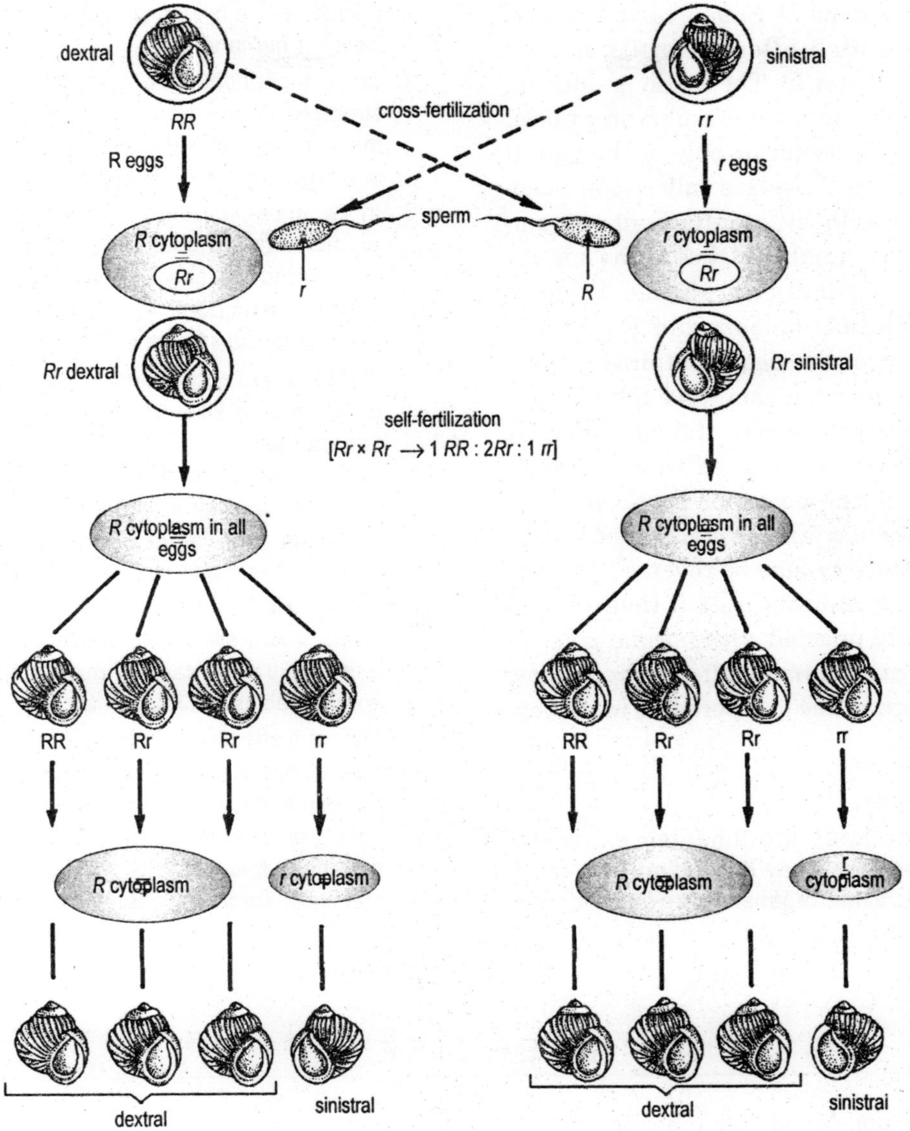

FIGURE 15.6

Inheritance of coiling in some snails, such as *Limnaea peregra*. The egg cytoplasm determines coiling. *RR* and *Rr* maternal genotype produce *R* egg cytoplasm (dextral coiling), and *rr* produces *r* egg cytoplasm (sinistal coiling). Neither the maternal phenotype nor the types of sperm used in fertilizing the eggs affects the direction of coiling.

In the same way, self-fertilized sinistral heterozygotes (Rr) produce dextrally coiled progeny even though some eggs are rr. This phenomenon clearly indicates the maternal effects over the progeny.

However, these maternal effects last for only one generation, because in next generation, the sinistral offspring are produced by the homozygous rr mothers, who are dextrals (Fig. 15.6)!

15.4.2 Maternal Effects in the Fertility and Survival of *Drosophila*

In *Drosophila subobscura*, it was noticed by **Spurway** that the fertility and survival are influenced by maternal effect. He found that although a female *D. Subobscura*, homozygous for **grandchildless** is fertile, her offspring are all sterile. The cause is—during development, a number of embryonic nuclei migrate to the posterior pole of the egg in *D. subobscura*. These are called 'polar cap-cells'. These cells differentiate into an embryonic or 'imaginal disc' that forms mature gonad during metamorphosis. When a **grand-childless** homozygote *Drosophila* produces eggs, the eggs contain modified posterior pole-cytoplasm, and their nuclei are unable to produce functional gonads. It has also been established that **sterility** results when the posterior pole-cytoplasm of the *Drosophila* is made inactivated by UV-rays. The sterility may be reversed and fertility may be re-introduced if these irradiated eggs are injected with normal pole-cytoplasm. In maternal effect phenotypic changes appear due to differences in the egg-cytoplasm.

Important Notes:

1. Some traits are inherited extrachromosomally. They depend on DNA of the cytoplasmic organelles like mitochondria, plastids etc. It is also known as non-Mendelian inheritance.
2. The mitochondrial DNA is an autonomous structure outside the nuclear genome, and the mitochondria have a separate protein-synthesizing machinery.
3. In *Paramoecium aurelia*, kappa particles or bacteria may reside as symbionts, and can reproduce only in those paramecia which contain dominant nuclear allele K. These are called killer paramecia.

   If cytoplasmic exchange occurs during the conjugation of killer and non-killer strains, the kappa from the killer are transferred to the non-killers. Their autogamy produces KK and kk in equal ratio. The KK produces killer– and kk non-killer clones. This is a case of cytoplasmic inheritance.
4. Male sterility in maize and some other crops is controlled by cytoplasmic factors. It is also true in case of male sterility in some species of *Drosophila*.
5. Maternal effect is a phenomenon where a trait is controlled by a gene of the mother but expressed in the progeny. The shell-coiling of *Limnaea* is controlled by maternal effect. It is also observed in the male sterility in some species of *Drosophila*.

EXERCISE

1. Explain non-Mendelian inheritance with example.
2. What do you understand by extranuclear inheritance?
3. Mention the characteristics of extrachromosomal inheritance.
4. Describe the structure and role of mitochondrial DNA.
5. Name one cytoplasmic organelle in plant and explain its role in extrachromosomal inheritance with example.
6. What are cytoplasmic inheritance and maternal effect?
7. Describe the cytoplasmic inheritance observed in a ciliate you have studied.
8. How shell-coiling of *Limnaea* is controlled? Cite example.

CHAPTER 16

POPULATION GENETICS AND EVOLUTION

Population genetics, Gene pool, Gene frequencies and allele frequencies. The Hardy-Weinberg equation, Estimation of equilibrium frequencies in natural population, Changes in gene frequency and allele frequency, Change of frequency due to small population, Non-random mating, Mutation, Random genetic drift, Bottleneck and Founder effects, Natural selection and change in allelic frequency, Darwinian fitness and coefficient of selection, Gametic and zygotic selection, Heterozygote superiority, Migration, Assortative mating, Inbreeding speciation, Barriers to gene-flow (reproductive isolation), Species isolation.

The genetic basis for evolution can be understood by studying population genetics. **Population genetics is the study of the consequences of Mendelian inheritance on the population level, including the mathematical description of a population's genetic composition and how it changes over time**.

A *population* is a group of organisms of the same species found in a defined geographical area.

Modern explanation of **variations** between organisms, mode of their **selection**, **mutation** if any, **migration** etc. show that these phenomena have a collective genetic influence on population. The natural selection operates on *individual* organism of a species, but *the collective genetic responses of the whole population* determine the trend of evolution as well as origin of new species. Thus, a long term effect of *natural selection* spreads over the gene and population through **interbreeding** among the members of the population. The interbreeding brings **exchange of genes**, and **pass on the genes** to the next generation. Thus, there is a **gene flow** between members of population, and the 'fittest' genes are selected. These are all studied under **population genetics**.

16.1 Gene Pool

A gene pool is the total variety of genes and alleles present in a sexually reproducing population. In other words, the genes shared by the individuals of a Mendelian population are called gene pool. A population undergoes evolutionary changes if its gene pool shows consistent change from generation to generation.

16.2 Gene Frequencies and Allele Frequencies

A frequency is a *proportion*, and it always ranges between 0 and 1. If 30 per cent individuals of a population have blue eyes, the frequency of the gene for blue eye in that population is 0.30 and the frequency of the gene of non-blue eye is 1–0.30 = 0.70 = 70%.

The general formula is: **frequency of a gene = frequency of homozygotes for that gene + ½ frequency of heterozygotes.**

Alleles are the alternative forms of a gene. So usually, we use **frequency of alleles** to describe how the gene pool changes over time. Allele frequencies are used because, in sexually reproducing organisms, during gametogenesis, the genotypes break down to *alleles* and the alleles pass in the next generation (as for example, AA × aa = AA, Aa and aa). Thus, only the alleles have continuity over time, and the gene pool undergoes evolution when allele frequencies change.

The formula to calculate an allele frequency by **gene counting** method is:

Allele frequency =

$$\frac{\textbf{Number of copies of a given allele}}{\textbf{Sum of counts of all alleles in the population}}$$

Let us imagine that, there is a population of 2000 diploid individuals, of which 706 are AA, 988 are Aa and 306 are aa. Each AA individual has two A alleles, whereas each Aa has only a single A allele. Therefore, the number of A alleles in that population will be—

(2 × number of AA homozygotes) + (the number of Aa heterozygotes). It means (2 × 706) + 988 = 1412 + 988 = 2400. In all population, the total number of allele is **twice** the number of individuals (because each diploid individual has 2 alleles), i.e., in this imaginary population, the total number of allele is, 2 × 2000 = 4000. Now, using the formula, the *allele frequency* is 2400/4000 = 0.60. When two alleles are present in a locus, the following formula is used to calculate allele frequency—

$$P = f(A)$$

$$= \frac{(2 \times \text{Count of AA}) + (\text{Count of Aa})}{2 \times \text{total number of individuals}}$$

$$= \frac{(2 \times 706) + 988}{2 \times 2400}$$

$$= \frac{1412 + 988}{4000} = \frac{2400}{4000} = 0.60$$

where, f(A) = frequency of A, allele = P.

16.3 The Hardy-Weinberg Equation

The method used to calculate the expected proportions of different genotypes in a population was published in 1908 by the British mathematician **G.H. Hardy** and, independently by the German physician **W. Weinberg**. They developed a mathematical relationship between the frequencies of alleles and genotypes in population. The relationship is called **Hardy-Weinberg equilibrium.** It is based upon a principle which states that "**the frequency of dominant and recessive alleles in a population will remain constant from generation to generation provided certain conditions exist.**" The conditions are:

1. The population must be large.
2. Mating between its members must be random.
3. There occur no mutations.
4. All genotypes are equally fertile, and there is no selection.
5. Generations do not overlap.
6. There is no emigration or immigration from or into the population (meaning, no gene flow between populations).

If there is a violation of any of the above conditions, then changes in

TABLE 16.1: Outcome of a cross where $\frac{3}{4}$ of gametes carry A and $\frac{1}{4}$ carry a

		Gametes from males			
		$\frac{1}{4}$A	$\frac{1}{4}$A	$\frac{1}{4}$A	$\frac{1}{4}$a
Gametes from females	$\frac{1}{4}$A, $\frac{1}{4}$A, $\frac{1}{4}$A	$\frac{9}{16}$ = 0.5625 AA			$\frac{3}{16}$ = 0.1875 Aa
	$\frac{1}{4}$a	$\frac{3}{16}$ = 0.1875 Aa			$\frac{1}{16}$ = 0.0625 aa

allele and genotype frequencies must occur. When changes occur, the Hardy-Weinberg equation helps to study the change and to measure the rate of change.

❑ **EXAMPLE:** If we cross two heterozygous (Aa) parents (♂ Aa × ♀ Aa), the two types of gametes (A and a) occur in equal proportions.

Now, let us consider a situation where two males and two females breed *at* **random** with each other. The females are with genotypes AA and Aa. The males are also AA and Aa. Both genotypes produce the same number of gametes but while Aa produces half of each kind, AA produces all A. Therefore, $\frac{3}{4}$ of all gametes are A and $\frac{1}{4}$ are a. In random mating, the probabilities

TABLE 16.2: Outcome of a cross where frequency of A is 0.75 and frequency of a is 0.25

		Gametes from males	
		0.75 A	0.25a
Gametes from females	0.75A	0.75^2 = 0.5625 AA	0.75 × 0.25 = 0.1875 Aa
	0.25a	0.75 × 0.25 = 0.1875 Aa	0.25^2 = 0.0625 aa

TABLE 16.3: The probabilities of zygotes from union of gametes with frequencies p and q

		Gametes from males	
		p (A)	q (a)
Gametes from females	p (A)	p^2 (AA)	pq (Aa)
	q (a)	pq (Aa)	q^2 (aa)

of the three possible genotypes of the progeny and the proportions of the gametes are shown in Table 16.1.

Next we consider an intermediate number of males and females mating at random. We need to know the proportion or frequency of the two types of gametes. As in table 16.1, the allele frequency of A is 0.75 and that of allele a is 0.25. The probabilities of three types of zygotes are shown in Table 16.2. If the frequencies of alleles are represented by p and q the result will be as that of Table 16.3.

If the whole population is denoted by 1, then the probabilities of the zygotes are $(p + q)^2 = 1$. By expansion the equation, we get $p^2 + 2pq + q^2 = 1$. This equation is called *Hardy-Weinberg Equation*.

Now, the genes in the gametes are important to consider. The sum total of genes in the gametes of a population is known as the **gene pool**. The gene pool consists of all the genes at every locus which can be passed to the next generation. The individuals in a population share the gene pool in common. In population genetics only two alleles of one gene locus is usually considered. In the calculation of allele frequency, p and q must be expressed either as fractions or as decimals and their addition must be 1.

In Hardy-Weinberg equation, p^2 = frequency of one homozygote (AA) [genotype frequency], 2pq = frequency of the heterozygote (Aa) [genotype frequencies], q^2 = frequency of the other homozygote (aa) [genotype frequency], where p = frequency of one of the two alleles (A), q = frequency of the other allele (a); as there are only two alleles, so, $p + q = 1$.

16.4 Estimation of Equilibrium Frequencies in Natural Population

If the frequency of one of the alleles in a gene pool is known, we can use the Hardy-Weinberg equation to estimate the expected proportions of genotypes in the population.

❑ EXAMPLE: **Estimation of genotype frequencies given allele frequencies**

Suppose, a breeding population of *Drosophila* is maintained in a large container in a laboratory. The frequency of the allele for black body (b) in this population is 0.3 and that of normal allele (B) is 0.7. What are the expected proportions of the genotypes—BB, Bb and bb?

From the data we get—
The frequency of b = 0.3 = q
The frequency of B = 0.7 = p.

Using the Hardy-Weinberg equation,

$$p^2 + 2pq + q^2 = 1,$$

We get,

frequency of genotype
$BB = p^2 = 0.7^2 = 0.49$
frequency of genotype
$Bb = 2pq = 2 \times 0.7 \times 0.3 = 0.42$
frequency of genotype
$bb = q^2 = 0.3^2 = 0.09$

Total 1.00

In natural population, allele frequencies are normally calculated from observed genotype frequencies. Again, the genotype frequencies are calculated by using the allele frequencies (according to the question). Calculation of allele frequencies is easier when the alleles are *codominant*.

❑ EXAMPLE: *Calculation of allele frequencies given genotype frequencies.*

In humans, the MN blood group system is controlled by two codominant alleles L^M and L^N.

The blood group M is of genotype $L^M L^M$, group MN is of genotype $L^M L^N$ and group N is of $L^N L^N$.

In a study of 747 people, 233 are with blood group M, 385 are MN and 129 are N. What are the frequencies of alleles L^M and L^N?

To solve the problem our first step is:

Each L^ML^M person contributes two L^M alleles while L^ML^N person contributes one L^M allele. Each L^NL^N person also contributes two L^N alleles.

Thus,

total number of L^M alleles

$= 385 + (2 \times 233) = 851$

total number of L^N alleles

$= 385 + (2 \times 129) = 643$

total number of both L^M and L^N alleles

$= 851 + 643 = 1494$

Therefore,

$$\text{proportion of } L^M = \frac{851}{1494} = 0.57$$

$$\text{proportion of } L^N = \frac{643}{1494} = 0.43$$

16.5 Changes in Gene-frequency or Allele Frequency

Populations do not remain always in equilibrium. There may be several reasons for this; the conditions for Hardy-Weinberg equation may not be maintained. As for example, (1) the population may be a small one (2) mating is not at random (3) mutations may occur (4) all generations are not equally fertile, i.e., selection may take place (5) generations may overlap, (6) there may be regular emigration of immigration allowing gene flow from different populations (see 16.3).

The other reason for loss of equilibrium in population is allele or gene frequencies are not constant from one generation to the next due to any other factor.

16.5.1 Change of Allele Frequency due to Small Population

In **small population**, allele frequencies are likely to fluctuate because of the relatively large contribution of each individual to the gene pool. The fluctuations may result in **random genetic drift** (see 16.5.4).

16.5.2 Non-random Mating and Change in Allele Frequency

Random mating does not occur frequently in nature; rather **non-random mating** among individuals in a population is common. There are specific choices in the searching for mating partners in many species; some species of plants are habitually self-fertilizing. The choice and selection of mating partner in higher animals are predominant in females. These non-random mating alters genotype frequencies.

16.5.3 Mutation and Change in Gene or Allele Frequency

An important factor affecting gene or allele frequency is the incidence of **mutations** that introduce a particular gene in a population. As for example, if, in a population, gene **A** mutates to **a**, and a reverse mutation does not take place, the chances of **a** to increase in frequency will enhance in each generation, and after a considerable period of time, the **a** can totally replace **A**. This is called **forward mutation**, and its rate is designated as **U**. The rate of reverse mutation is designated as **V**. The frequency of **A** is called **p** and that of **a** is **q**.

Gene mutations consist of heritable changes in the DNA that occur within a locus. The rate of occurring of mutation is usually *low* but varies in species. Mutation is the source of all genetic variation, and is the **raw genetic material for evolution**.

The change of allelic frequency can be calculated by a formula— $\Delta p = Vq - Up$, where, p = frequency of A, V = rate at which **a** mutates to **A** $(A \leftarrow a)$, q = frequency of a, U = rate at which **A** mutates to **a** $(A \rightarrow a)$.

16.5.4 Random Genetic Drift

In small populations, *chance factors* may produce large changes in allelic frequencies. Random change in allelic frequency due to chance is called **random genetic drift** or

genetic drift. **Ronald Fisher** and **Sewall Wright** described how genetic drift affects the evolution in population. Thus, the effect of genetic drift is also referred to as **Sewall Wright effect**.

Variation in gene frequencies within population can occur by chance rather than by natural selection. The Sewall Wright effect may be effective in evolutionary change in small populations.

As for example, let us assume, a small group of humans inhabit a very small island; the total number of reproductive individual is 10, and 5 of them have green eyes (bb) and the other 5 have brown eyes (BB, Bb), and the frequency of allele of green eyes is 0.6 in the island. Let us assume that in this state, 50% of the population is killed by a typhoon in that small area, all of whom are brown eyed. The survival of the green-eyed persons is only due to chance. After the typhoon, the allelic frequency for green eyes is 1.0. Thus, the change of allelic frequency becomes from 0.6 to 1.0, simply due to a chance factor. But if the number of persons in the island were higher, the chance of death of all brown-eyed people where not possible; as a result the allelic frequency would not change in this ratio. It illustrates the characteristic of genetic drift. Chance factors produce drastic changes in allelic frequencies only in small populations.

It is possible for an allele to disappear from a small population, or it is also possible for it to appear in higher frequency, simply by chance, due to genetic drift. The genetic drift is random, and so it is unpredictable. It may lead to the extinction of one species, or may lead to result in the development of a better adapted species, which may lead to the **origin of new species** in due course of time.

16.5.4.1 Bottlenecks and Founder Effects

There are several ways in which the genetic drift may arise. One of them is **founder-effect**. When a population is initially established by a small number of breeding individuals, the **founder-effect** is seen to occur. Later the population may grow in size and consist of a large number of individuals but, originally, the gene pool of the population is derived from the genes present in the **initial breeding individuals** (called 'founders'); and chance factors may determine which genes were present among the 'founders'. This exerts effects on gene pool of subsequent generations. The founder effect or founder principle was first postulated by **Mayr** in 1965.

Bottleneck effect is another form of genetic drift. It means, the changes in allelic frequency due to chance factors may arise in a population, which is at first, drastically reduced in size (as if in the narrow neck of a bottle), and then, suddenly expanded (as if in the expanded body of a bottle). The former is called **population crash** (in the bottle-neck) and the latter is called **population flush** (in the body of the bottle). The expanded population that results may have distinctly different gene frequencies from that of the bottleneck (population crash). Thus, bottleneck effect may encounter the effect of previous selection for a short time, when some genes may be lost from the gene pool. In our previous example (16.5.4) of small population with brown eye and green eye, we got only the green-eyed people due to a typhoon (chance factor), and the frequency of green eyes changed from 0.6 to 1.0. This is an example of bottleneck effect. It is historically important to note that this effect was found to occur in the population of **Tristan da Cunha**.

16.5.5 Natural Selection and Change in Allelic Frequency

Natural Selection produces organisms that are finely **adapted** to their environment. Although, genetic drift, migration, mutation etc. may influence the pattern of adaptation, the Natural Selection is the chief force for producing adaptation. Nowadays, Natural Selection is defined as 'differential reproduction of genotypes'. Therefore, those genes increase in frequency in a population which produce more better adapted offspring than the others.

The best example of selection was found in peppered moth, *Biston betularia* in England. The typical form of this moth is **grayish-white** with black mottling over body and wings. Before 1848 (in pre-industrial period) the moths showed no other phenotype, but in 1848, a single **black** moth was found; this new phenotype was called **carbonaria**. By 1900, they achieved a frequency of more than 90% in industrial areas, while the typical grayish white phenotype remained common in non-industrial areas. **Kettlewell** found that this **melanism** is an effect of industrial progress in England. He demonstrated that the increase of **carbonaria** is due to selection against the typical grayish-white phenotype. The cause behind it is: these typical peppered moths are nocturnal and in daytime they rest on the lichen-covered barks of trees. Their predator birds could not detect them in daytime due to camouflaging of moth-colour and the tree-barks. In 1850s, in industrial areas, the trunks of trees were covered with **black soot** instead of lichens. The **carbonaria** survived better in this black background, so more carbonaria—genes were flowed to the next generation and the *carbonaria* phenotype increased in frequency. This phenomenon is called **industrial mela-nism** that strongly suggests the role of **selection** changing allelic frequency.

16.5.5.1 Darwinian Fitness and Coefficient of Selection

Natural selection is measured in terms of **Darwinian fitness**. The Darwinian fitness

TABLE 16.4: Relative fitness and selection coefficients calculated from data taken in the same generation of population

Data : Number of individuals in the population according to the genotype

	A/A	A/a	a/a
Before selection	4100	5000	2200
After Selection (same generation)	3900	4000	1200

Calculations:

(a) Survival rate

$$A/A = \frac{3900}{4100} = 0.95$$

$$A/a = \frac{4000}{5000} = 0.80$$

$$a/a = \frac{1200}{2200} = 0.55$$

(b) Relative fitness W (compared with A/A's maximum survival rate)

$$W_{AA} = \frac{0.95}{0.95} = 1.00$$

$$W_{Aa} = \frac{0.80}{0.95} = 0.84$$

$$W_{aa} = \frac{0.55}{0.95} = 0.58$$

(c) Selection coefficients (s)

$$SAA = 1 - W_{AA} = 0$$

$$SAa = 1 - W_{Aa} = 0.16$$

$$Saa = 1 - W_{aa} = 0.42$$

is the relative reproductive ability of a genotype, and it is symbolized as W.

Natural Selection includes three parameters: **survival rate, relative fitness, selection coefficient**. The genotype with the maximum survival rate is used as the standard for **relative fitness (W)** of all other genotypes.

The table 16.4 furnishes an example of calculation of survival rate, relative fitness and Selection Coefficients. Genotypes are AA, Aa and aa, and the numbers of individuals before and after Selection are arbitary.

The **Selection Coefficient** is represented by **S**. The **W** reflects the chances of reproductive success of an organism while the S reflects the chances of reproductive failure of an organism. Thus, S can be calculated as 1 – W. It is a measure of relative intensity of selection against a genotype.

16.5.5.2 Gametic Selection and Zygotic Selection

Selection may act on the genotype of gametes or zygotes or both. When selection occurs in gametes, there is no difference between dominant and recessive genes. But in most diploid higher organisms selection takes place in zygotes. There are 3 possible genotypes for a single gene difference (AA, Aa, aa) in the diploids. In zygotic selection, the effectiveness depends upon that degree of dominance. If complete dominance exists and selection occurs only against the recessive aa, the heterozygote is guarded against selection, and harmful recessive genes may persist for many generations.

16.5.5.3 Heterozygote Superiority or Heterozygote Advance

Some forms of selection maintain balanced model of genetic variation. One of the balancing selection is **heterosis**. It is also known as **overdominance** or **heterozygote superiority** or **heterozygote advance**. Balanced model of genetic variation may arise due to **stabilizing selection**. One of its kinds is observed in the variants of human hemoglobin. One of the variants, **Hemoglobin-S** (HbS) produces **sickle-cell** (when RBC are sickle-shaped) **anemia**, which is a severe and fatal disease. The HbS is a recessive allele to normal hemoglobin (Hb^+). This fatal anemia is manifested only in homozygous (HbS/HbS) condition. In heterozygote people (Hb^+/HbS), the RBC is only slightly distorted, and O_2-concentration in blood is low. But these heterozygote people can live their normal lives.

Sickle-cell anemia is overspread in USA, India, Indo-China, Middle East and Africa. In Africa, the proportion of heterozygotes (Hb^+/HbS) is about 40%. Interestingly, it has been noticed that in some districts of Africa, people with normal hemoglobin with homozygous genotype (Hb^+/Hb^+) are susciptible to malignant malaria, caused by *Plasmodium falciparum*. But heterozy-gotes (Hb^+/Hbs) for sickle cell anemia are much resistant to this fatal malaria. This incidence categorically supports that the heterozygotes are adaptively superior to the normal homozygotes. It was conclusively demonstrated that sickle-cell trait carriers (heterozygotes) have a **selective advantages** over the other genotypes in areas with malignant falciparum malaria. It is called **heterozygote superiority**.

16.5.6 Migration

In population genetics, 'migration' is related to the movement of genes from one gene pool to the other (gene flow). Movement of genes takes place when organisms or gametes migrate and contribute their genes to the gene pool of the recipient population.

The gene flow through migration may (1) introduce new alleles to the recipient population, and (2) change the allelic frequencies within the recipient population (if the genes carried by the migrants differ from those of the recipient population).

Migration tends to reduce genetic divergence and increases the effective size of the population.

16.5.7 Assortative Mating, Inbreeding

In many cases, mating between members of population is **non-random** which is an anti-conditions of Hardy-Weinberg **equilibrium.** If non-random mating occurs, the genotypes do not exist in the proportions of the Hardy-Weinberg equilibrium. Non-random mating is of two types: **Positive assortative mating** and **Negative assortative mating**.

Mating of individuals by preference of similar phenotypes is called **positive assortative mating**, while non-random mating of phenotypically dissimilar individuals is called **negative assortative mating**. Marriage of tall men with tall women, black men with black women by preference is example of positive assortative mating. It does not affect allelic frequencies of a population but may influence genotype frequencies. Marriage of tall men with short women, black men with fair women by preference, on the other hand, is example of negative assortative mating. It may affect both the allelic and genotypic frequencies of a population.

Another type of non-random mating is observed in **inbreeding**. Inbreeding involves preferential mating between close relatives. The extreme case of inbreeding is **self-fertilization**. Inbreeding has very similar effects to genetic drift in small population.

Continuous inbreeding in a population decreases heterozygosity and increases homozygosity, which may reduce fitness. The reduction in fitness due to continued inbreeding bringing increasing homozygosity in a population is called **inbreeding depression**.

16.6 Speciation (Formation of new species)

The general principle for **speciation** are (1) Barriers to gene flow through **reproductive isolation**, and (2) Genetic basis for **species isolation**.

16.6.1 Barriers to Gene Flow (Reproductive isolation)

The barriers to gene flow may be studied under two heads: (1) **Post-zygotic barriers** and (2) **Pre-zygotic barriers**.

The **post-zygotic barriers** result in poor fitness of hybrid offspring. Post-zygotic isolations may occur through—**hybrid sterility, hybrid inviability** and **hybrid breakdown.**

- **Hybrid sterility**: Mating of two species may give rise to a hybrid, but the hybrids are sterile. This is called hybrid sterility, as is examplified by the **mules**, which are products of the mating between a male ass and a female horse.

- **Hybrid inviability**: After mating between two species, fertilization may occur but the zygote or hybrid does not survive. This is called hybrid inviability, as seen in the mating of goat and sheep, where fertilization occurs but the hybrid embryo dies in an early phase of development.

- **Hybrid breakdown:** Sometimes the hybrids are partly or wholely fertile, but give birth to abnormal or sterile progeny. The hybrid females of the cross between *Drosophila pseudoobscura* and *D. Persimilis* are fertile, but cannot produce second generation if backcrossed with either of the parents. This is called hybrid breakdown.

The **prezygotic** or **premating** isolation provides mechanisms that prevent **intraspecific mating**, fertilization or zygote formation. It includes the following types: (1) Ecological Isolation, (2) Temporal Isolation, (3) Ethological Isolation, (4) Mechanical Isolation and (5) Gametic Isolation.

1. **Ecological isolation** is also called **habitat isolation**. When population live in the same region but in different habitats, the potential mates do not get opportunity for mating. This is called habitat or ecological isolation.
2. **Temporal or seasonal isolation**: The males and females of a species may become sexually mature at different seasons, as a result mating remains unsuccessful.
3. **Ethological or behavioral isolation**: Potential mates reject each other due to specific behaviour or some other reasons called behavioral incompatibility. Thus, mating remains prevented.
4. **Mechanical isolation**: The reproductive organs of potential mates (genitalia of animals and flowers of plants) may be of different structure and may prevent mating or pollination.
5. **Gametic isolation**: If by any means, mating is performed, there still remains a highly complex process of gametic fusion (fertilization), that can fail, and zygote is not formed. It is called gametic isolation.

16.6.2 Genetic Basis for Species Isolation

In nature, it is often observed that the male hybrids are sterile, but female hybrids are fertile. This widespread pattern of male hybrid sterility was noticed by Haldane and the phenomenon is called **Haldane's Rule**. For this type of male hybrid sterility, and many other post-zygotic isolation, it is now known that **many genes** are involved in the fertility of the male hybrids.

Thus, by keeping potential males **isolated**, and **preventing reproduction**, genetic diversities are achieved in separate populations.

In the long run, these **genetic diversities**, when selected, give rise to new species.

Important Notes:

1. The sumtotal of all genes and alleles in a sexually reproducing population is called gene pool.
2. The genetic structure of a population can be described in terms of allelic and genotypic frequencies.
3. The Hardy-Weinberg law describes the equilibrium state of allelic and genotypic frequency in a population, which is large, where mating is random, and mutation migration and selection are absent.
4. Mutation, genetic drift, migration and natural selection can alter allelic frequencies in a population.
5. Most efficient force to bring change in Hardy-Weinberg equation is provided by mutation.
6. Genetic drift may have important evolutionary role in small population. It leads to loss of genetic variation within populations, genetic divergence among populations, and random fluctuation in the allelic frequencies of a population over time.
7. Migration of gene flow tends to reduce genetic divergence among populations and increases the effective size of the population.
8. Natural selection involves differential reproduction of genotypes. It is measured in terms of Darwinian fitness (W).
9. Heterozygote superiority or heterozygote advance or heterosis within a population is a result of balancing selection.
10. Continuous inbreeding increases homozygosity within a population and results in reduced fitness. It is called inbreeding depression.
11. Speciation results due to genetic interference like mutation, genetic drift, migration as well as prolong reproductive isolation, selection and adaptation in a large population.

EXERCISE

1. What is Hardy-Weinberg equilibrium? What are the conditions necessary for the maintenance of the Hardy-Weinberg equilibrium?
2. Distinguish between allelic frequency and genotypic frequency.
3. Define Hardy-Weinberg Law with reference to the Synthetic Theory of Evolution.
4. How would you estimate equilibrium frequencies in natural population?
5. Phenylketonuria is a hereditary disease which, if untreated may result in severe mental retardation. PKU are homozygous recessive disorders and occur at a frequency of one in 10,000 births.
   (a) In a population of 1000, how many people would be expected to be heterozygous?
   (b) A normal couple have already had one PKU child. What is the probability that their second child will be heterozygous for the condition?
6. The ability to taste phenylthiocarbamide (PTC) is dominant to inability to taste. Some people were tested and the following results were found—

	Tasters	Non-tasters	Total
Male	98	31	129
Female	104	19	123
Total	202	50	252

   (a) Calculate the frequency of the recessive allele amongst (i) males (ii) females.
   (b) The gene is not sex-linked. What is the most probable explanation of the sex-difference in this small sample?
7. Sickle-cell anemia is a serious hereditary disease and suffers a homozygous for an allele HbS, which codes for abnormal hemoglobin. Normal Hb is coded by the allele HbA. The heterozygote HbA/HbS are healthy but can be recognized by the shape of their RBC. They have sickle-cell trait.
   (a) In a District of Central Africa 2.5% of newborn babies have sickle-cell anemia. What is the frequency of the HbS allele in this population?
   (b) How many of 500 babies are expected to be heterozygous?
8. Malaria is caused by *Plasmodium*, and causes death in West Africa. Where malaria is prevalent, the sickle-cell allele HbS is maintained at a high frequency because heterozygote HbA/HbS are resistant to malaria and are more likely to reproduce than are normal homozygotes. In one population HbS occurs in the frequency of 0.1.
   (a) Out of 1000 new born babies in this population, how many would there be of each of the three genotypes?
   (b) In this population of 1000, all the sickle-cell homozygotes and 20% of the normal homozygotes die before reproducing. What is the frequency of HbS allele in the remaining population?
9. What is gene pool? What is genetic drift? What are its effects in a small population?
10. Define bottleneck and founder effect.
11. How does mutation can change allele frequency?
12. What are Darwinian fitness and coefficient selection.
13. What is heterozygous superiority? Explain with suitable example.
14. What is role of migration of gene flow in a small population?

15. Describe assortative mating and inbreeding.
16. What do you mean by speciation? Describe the steps leading to speciation.
17. What are the barriers to gene flow? Explain them.
18. What is Haldane's rule? What is the genetic basis for species isolation?
19. What do you understand by reproductive isolation? What role does it play in speciation?
20. Describe the pre-zygotic and post-zygotic barriers to gene flow.

CHAPTER 17

HUMAN GENETICS

Human chromosome, Chromosomal aneuploidy and Syndromes, Trisomy-21 and Down Syndrome, Trisomy-13 and Patau Syndrome, Trisomy-18 and Edwards Syndrome, Aneuploidy and Turner Syndrome, Aneuploidy and Klinefelter Syndrome, Cri-du-Chat, Chronic myelocytic leukemia, Burkitt's lymphoma, Embryonic tumor retinoblastoma, Sex-linkage, X-linked recessive inheritance, Hemophilia A, Green and Red colour-blindness, Y-linked inheritance, Mutational disorders, Metabolic disorders, Genetics of Huntington's Disease, Cystic fibrosis, ABO blood group genes, Hemoglobin genes, Recessive and Dominant mutant traits, Genetic diseases, Blood-group genetics, Mendelian principles in human genetics, Pedigree analysis, Eugenics and Euthenics.

This chapter deals with the genetics related to humans. Many of its topics are also common to other organisms, and have been discussed in previous chapters. Here, we will discuss certain exclusively human-related genetic aspects.

17.1 Human Chromosome

There was much doubt about the exact number of human chromosome upto 1956. In 1956, however, Tijo and Levan first correctly demonstrated that there were 23 pairs or 46 chromosomes in humans. Out of 23 pairs, 22 pairs are **autosomes** and one pair is **sex-chromosome**.

Human chromosomes are classified into 7 groups according to their size and location of centromeres (Table 17.1).

Group A (1-3) and **Group B** (4-5) are large, **Group C** (6-12 and X) (6-12 and X) and **Group D** (13-15) are medium sized, **Group E** (16-18) is short-medium sized, **Group F** (19-20) is short and **Group G** (21-22 and Y) is very short sized chromosome (Fig.17.1).

17.2 Chromosomal Aneuploidy and Syndromes

For **aneuploidy**, please refer to chapter 8.

Various types of human syndromes (a group of concurrent symptoms of a disease) like **Down**, **Patau**, **Edwards**, **Turner** and-develop due to aneuploid abnormalities. In human, aneuploidy may occur both in autosomes and sex-chromosomes. **Autosomal aneuploidies** are seen in Down (Trisomy-21), Patau (Trisomy-13) and Edwards (Trisomy-18) syndromes.

17.2.1 Trisomy-21 and Down Syndrome

Trisomy involves a single extra chromosome, i.e., a trisomic cell's chromosome

TABLE 17.1: Human chromosomes grouped according to size and centromere location

Chromosome Group	Group Designation	Size	Centromere Location
1-3	A	large	approximately median
4-5	B	large	submedian
6-12 and X	C	medium	submedian
13-15	D	medium	acrocentric
16-18	E	short-medium	median or submedian
19-20	F	short	approximately median
21-22 and Y	G	very short	acrocentric

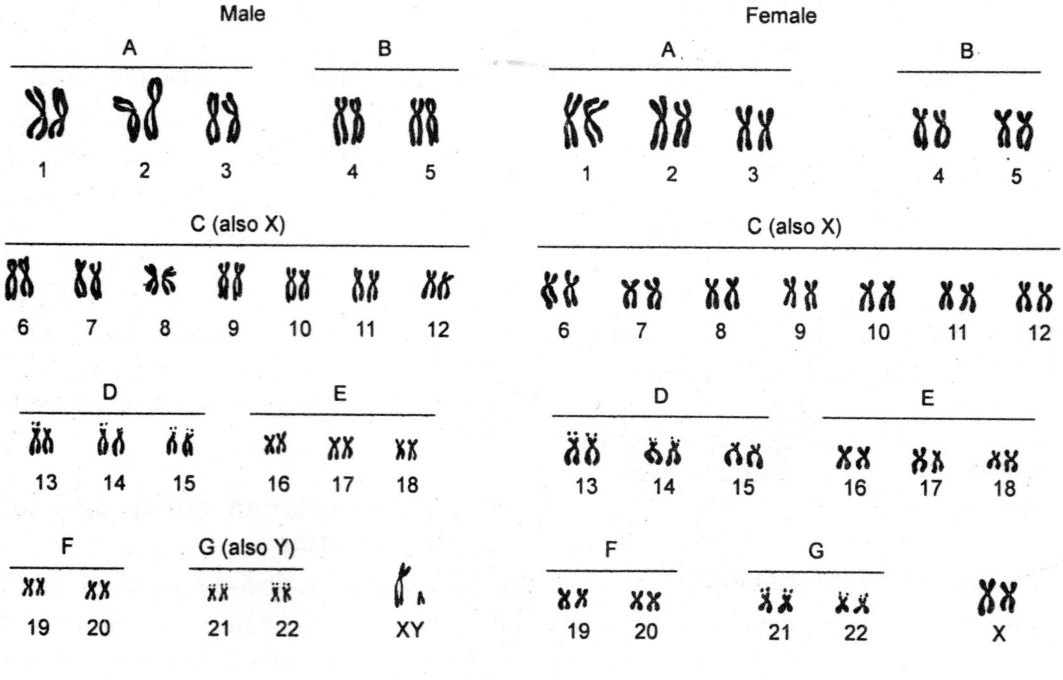

FIGURE 17.1

Metaphase chromosomes of mitosis in a human male and female arranged according to homologous pairs.

number is 2n + 1 (See chapter 8.5.2). When 3 copies of chromosome-21 occur, a **trisomy-21** or **Down syndrome** develops. The syndrome was first discovered by **John Langdon Down** in 1866, and it is named after him.

17.2.1.1 Characteristics of Down Syndrome (Fig. 17.2)

1. Occurs in about 1430 per 1 million live births.
2. Low I.Q.; mentally retarded.

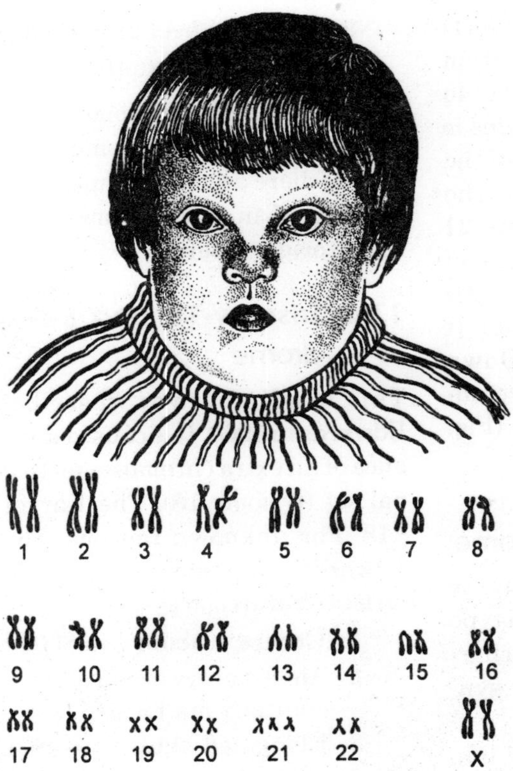

FIGURE 17.2
Appearance and karyotype of a female with Down syndrome showing three 21 chromosomes.

3. Epicanthal folds over eyes (thus, formerly known as 'mongolism').
4. Short stature (about 4ft in adult).
5. Short and broad hands.
6. Short and broad skull.
7. Wide nostrils.
8. Large, furrowed tongue, protruded through open mouth.
9. Simian crease on the palm and single crease on the fifth digit in hands.
10. Loose jointedness in ankles (hyperflexibility of joints).
11. Average life span 16-20 years.

There is a relationship between the age of the mother and risk of having a Down-syndromebaby. Mothers of the age group between 27 and 34 are less-risked.

17.2.1.2 Karyotype of Down Syndrome

Down syndrome individuals usually have 47 chromosomes, of which the chromosome-21 has a trisomy, i.e., three copies of chromosome-21 are present in a cell (47, +21).

In some cases, a Down syndrome individual may have 46 chromosome instead of 47, but in that case, a *translocation* joins the long arm of chromosome-21 with another chromosome (mostly with the chromosome-14) in the same complement.

17.2.1.3 Prediction of Down Syndrome

When an inherited translocation is noticed in an infant, the risk of the second occurrence in the same family can be calculated. Karyotypes of parent who have already produced a baby with inherited translocation Down syndrome are the mother with 45, XXt (14q 21q) and the normal father 46, XY. The gametes from the mother (first parent) will be of 6 types: (1) Normal 14, 21 (2) balanced translocation (14q 21q) (3) t (14q 21q), +21 (4) 14 (5) t (14q 21q), +14 (6) 21. If each of these gametes is fertilized by a normal gamete (14, 21) of the father (second parent), the ratio of the children will be **2 normal : 1 trisomy-21** (Down syndrome). Out of the 6 types of zygotes, only those with the normal (14, 21/ 14, 21), normal translocation carrier [t (14q21q)/14,21] and trisomy-21 Down syndrome [t(14q21q),+21/14,21] chromosomes, will survive. The rest will die (Fig. 17.3).

17.3 Trisomy-13 and Patau Syndrome

Trisomy-13 causes Patau syndrome. In 1960, K. Patau described that the Karyotype 47, +13 produces abnormal babies. The characteristics together are called Patau syndrome. The symptoms are—

1. Occurs in about 2 in 10,000 live births.
2. Cleft lip and palate.
3. Small eyes, small brain.
4. Polydactyly is noticed.
5. Mental retardation.
6. Deafness.
7. Growth retardation.
8. Cardiac anomalies.
9. Posterior heel prominence.
10. Most die before the age of 3 months.

3 copies of chromosome no. 13 are associated in a Patau syndrome individual; as a result, there are 47 chromosomes due to the addition of an extra chromosome (the 13th) in the original set (47, +13).

17.4 Trisomy-18 and Edwards Syndrome

This syndrome was described by J.H. Edwards in 1960. It is caused due to the presence of an extra chromosome-18 in the original set (trisomy-18). The Karyotype is 47, +18. For unknown reasons, about 80% of Edwards syndrome are female. The characteristic features are—

1. Occurs in about 1 in 8000 live births.
2. Almost every organ of the body is congenitally malformed.
3. Elongated skull, low-set malformed ears.
4. Receding mandible.
5. Small mouth.
6. Short sternum.
7. Horse-shoe shaped kidney.
8. Mentally and developmentally retarded.
9. 90% of the babies die within six months.

Cross: 45, XX, t(14*q*21*q*) × 46, XY

GAMETES OF SECOND PARENT	GAMETES OF FIRST PARENT					
	14,21	*t*(14*q*21*q*)	*t*(14*q*21*q*), + 21	14	t(14*q*21*q*), + 14	21
14,21	14,21 14,21	*t*(14*q*21*q*) 14,21	*t*(14*q*21*q*), + 21 14,21	14 14,21	t(14*q*21*q*), + 14 14,21	21 14,21
PHENOTYPES OF ZYGOTES	Normal	Normal translocation carrier	**Trisomy 21, Down syndrome**	Monosomic 21, lethal	Trisomic 14, lethal	Monosomic-14, lethal

FIGURE 17.3

Diagram of a cross between an individual with an inherited translocation, carrying chromosome 21, 45, XX, t(14*q*21*q*) and a normal person, 46, XY.

17.5 Aneuploidy and Turner syndrome (45, XO)

Turner syndrome people are **monosomic** (see 8.5.2) for one X chromosome. There is a loss of one X chromosome and the total number of chromosome is 45. The Karyotype is 45, XO. (44 autosomes and one X chr.). This syndrome was first described by **H. H. Turner** in 1938. 45, XO individuals are phenotypically female and **lacks a Barr body** (i.e., sex-chromatin negative) (Figure 17.4).

17.5.1 Characteristics of Turner Syndrome

The characteristic features of the syndrome are—

1. Occurs in about 1 per 2500 live female births.
2. More than 90% are spontaneously aborted.
3. Adults almost without ovaries.
4. Secondary sexual characters are poorly developed.
5. Short stature (140-145 cm).
6. Low-set ears.
7. Webbed neck.
8. Shield-like chest (no breast development), with widely spaced nipples.

Turner phenotype is mostly controlled by genes present in the short arm of X-chromosome. In some instances, a Y chromosome may also be present. These individuals are **mosaic** for 45, X/46, XY. There are instances that Turner syndrome may occur in some **males** even. The male Turner syndrome is characterized by defective testis development, sterility, scanty male secondary sexual characters and all other usual somatic features of Turner syndrome. Their Karyotype is 46, XY, but the X lacks the short arm or the Y lacks the long arm.

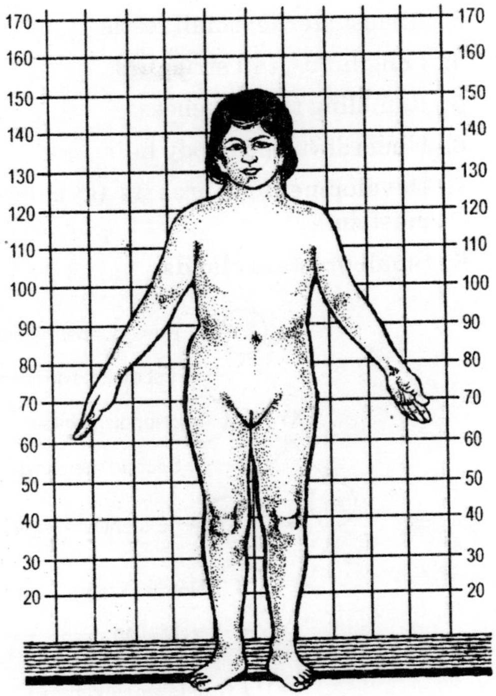

FIGURE 17.4

Female with Turner syndrome (45, X).

17.6 Aneuploidy and Klinefelter Syndrome (47, XXY)

Certain males with trisomy of sex chromosomes, XXY, or double trisomy, XXYY may exhibit **Klinefelter syndrome**. In 1942, **H.F. Klinefelter** described this abnormal male syndrome which was due to the presence of an extra X chromosome to the usual male XY chromosome complement, forming a Karyotype 47, XXY. It means, these males have 47 chomosome, of which 44 are autosomes and 3 are sex chromosomes. Out of the sex chromosomes, the normal male complement (XY) is associated with an extra X.

17.6.1 Characteristics of Klinefelter Syndrome (Fig. 17.5)

1. The syndrome occurs 1 per 500 live male births.
2. They are phenotypically male showing a tendency toward femaleness.

3. Almost sterile; small testis.
4. Long limbs; tall structure.
5. Rambling talkativeness.
6. Underdeveloped body hair.
7. Development of breasts (gynaecomastism).
8. Small prostate glands.

17.6.1.1 Origin of Klinefelter Syndrome

The development of XXY condition may result either by—(1) fertilization of an exceptional XX-egg (formed due to nondisjunction of X-chromosome) by a Y-sperm, or (2) fertilization of an X-egg by an exceptional XY-sperm (formed due to nondisjunction).

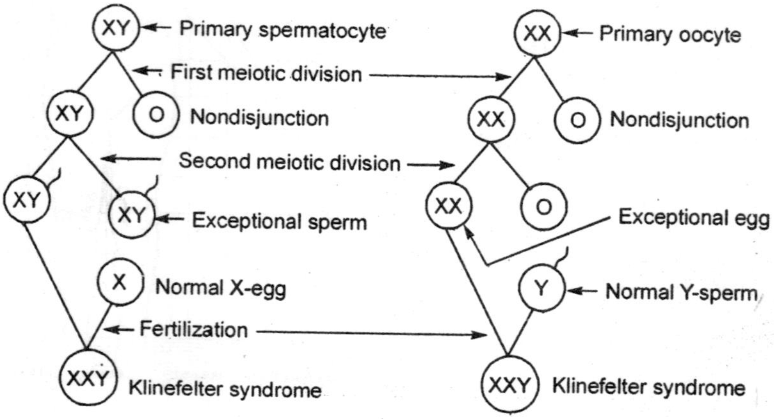

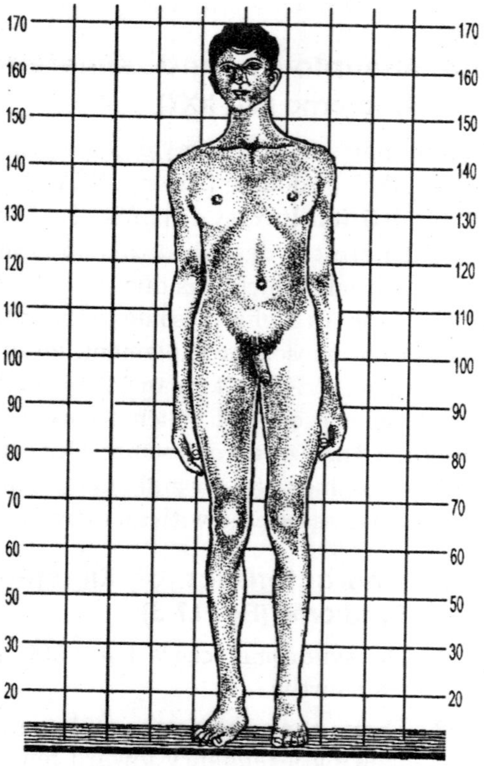

FIGURE 17.5

Male with Klinefelter syndrome.

17.6.1.2 Karyotype of Klinefelter Syndrome

About 75% of Klinefelter Karyotype is 47, XXY. But when number of X chromosome increases the syndrome persists. As for example, 46, XY is a normal male (disomic); 47, XXY is a trisomic with Klinefelter syndrome; 48, XXYY is a double trisomic with Klinefelter syndrome; while 48, XXXY is a tetrasomic, and 48, XXXXY is a pentasomic, both showing extreme Klinefelter syndrome with mental retardation. The condition of XXXYY, XXXXYY and XXXXY also exhibits Klinefelter syndrome.

17.7 Deletion (chromosomal aberration) and cri-du-chat

J. Lejèune *et al* discovered a syndrome in human called *cri-du-chat* that developed due to chromosome **deficiency**.

The characteristics of cri-du-chat are—

1. Small and weak infants with cat-like mewing cry (cri = cry, chat = cat).
2. Small head.

3. Broad face and saddle nose
4. Widely spaced eyes.
5. Epicanthic folds.
6. Both physical and mental retardation.
7. I.Q. range 20-40.
8. Death at very early stage.

This syndrome develops due to the chromosome deficiency or deletion in the short arm of chromosome 5 (designated as $5p^-$). The Karyotype is written as 46, XX, $5p^-$.

17.8 Translocation (chromosomal aberration) and Chronic Myelocytic Leukemia

A deletion in chromosome 22 was first discovered in persons in Philadelphia city; this chromosome is known as **'Philadelphia'** **(Ph')** chromosome, and is always found associated with the chronic myeloid leukemia patients. Later, **J Rowley** discovered that actually it was a case of translocation. It was demonstrated that a part of the long arm (q) of chromosome 22 was translocated to the long arm of another chromosome (usually chromosome 9) forming a deletion in the long arm in chromosome 22. This type of **translocation** (and deletion) causes the chronic myelocytic leukemia, and the karyotype of the patient is designated as (46, XX $9q^+$, $22q^-$).

17.9 Burkitt's Lymphoma and Embryonic Tumor Retinoblastoma

A **translocation** between chromosome 14 and chromosome 8, forming a Karyotype [t ($8q^-$, $14q^+$)] results in the malignancy, **Burkitt's lymphoma**.

A case of **deletion**, is associated with human embryonic tumor, **retinoblastoma (RB)**. Here, a deletion in the long arm of chromosome 13 is found to occur.

17.10 Sex-linkage and Inheritance in Human

Inheritance through sex-chromosome is called **sex linked inheritance**, and the traits associated with sex-chromosome is called **sex linkage**.

17.10.1 X-linked Recessive Inheritance

The trait resulting from a recessive mutant allele carried on the X chromosome is called an **X-linked recessive trait**. A few of the X-linked recessive traits in humans are the red-green colour blindness, hemophilia A, Duchenne muscular dystrophy etc.

EXAMPLE: **Hemophilia A (Fig. 17.6)**

There are two types of hemophila—hemophilia A and hemophilia B. Hemophilia A is called "Royal hemophilia" or "Bleeder's disease" and the hemophilia B is called "Christmas disease". Both types are X-linked traits. They are caused by mutations of two different genes in the X-chromosome. Hemophilic persons cannot produce a factor required for blood-clotting. As a result, any wound or cut causes continuous bleeding and finally death. Persons affected by hemophilia are nearly always male.

Queen Victoria was a carrier (heterozygous) for the gene (Fig. 17.6) and passed the mutant allele to some of her children. Her daughters became carriers, and one of her sons was affected with hemophilia (hemizygous).

Since hemophilia is caused by recessive X-linked gene, a woman may be the carrier of the disease (but not affected), and would transmit it to 50% of her sons, if her husband is normal (Fig.17.7).

On the other hand, if the mother is normal and the father is hemophiliac the result will be as follows: **father's gene is inherited to grandson through his daughter, called a criss-cross pattern of inheritance (Fig. 17.8).**

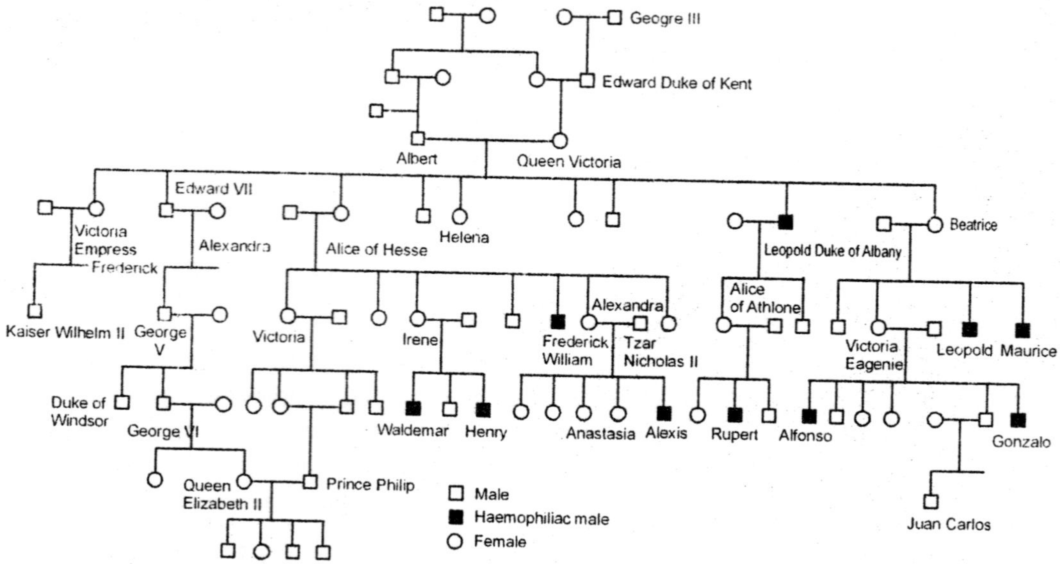

FIGURE 17.6

Pedigree of the Royal Family showing the inheritance of hemophilia.

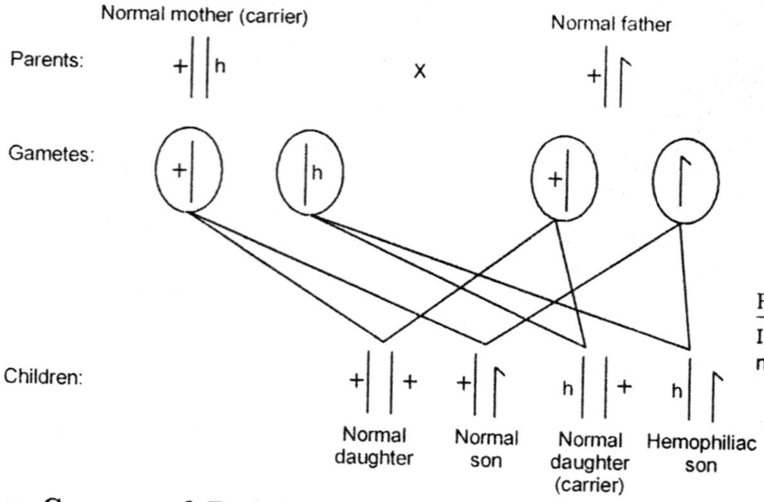

FIGURE 17.7

Inheritance of hemophilia A.

EXAMPLE: Green and Red Colourblindness

The genes that encode the green-and red colour-receptor proteins are located on the X-chromosome. Defect in **green** colour vision is called **deuteranopia** or deutan colour blindness. On the other hand, defect in **red** colour vision is called **protanopia** or protan colour blindness. Both the visions are controlled by a recessive gene located on X-chromosome.

A man defective in green colour vision (deuteranopia) has a single **recessive** allele (g) on his ***X* chromosome**. Since the Y chromosome carries no colour-vision allele, the single allele on X is capable to cause colour blindness in male (hemizygosity). If the man's wife is homozygous for the dominant allele (g^+) for normal colour-vision, all of their daughters will be heterozygous (receiving X with g^+ from mother and X with g from father) (Fig. 17.9). Sons, with only

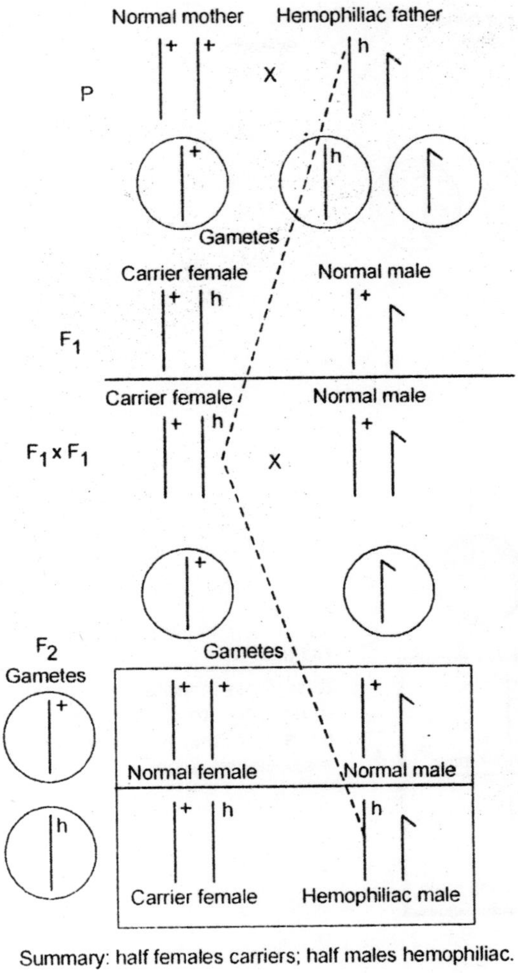

FIGURE 17.8

Diagram with genes on the X chromosome. The dotted line through the center illustrates the **criss-cross pattern of inheritance**; from father through daughter to grandson. The symbol "h" here represents the sex-linked recessive gene for hemophilia.

one X chromosome will have only one allele (g^+) from the mother and will show normal colour vision (Y with no such allele will come from the father). In the next generation, about half of the sons of the carrier (heterozygous) females will be normal and half will have defective colour vision. If the fathers are normal (g^+), half of the daughters of carrier mothers will be normal in vision but carrier in genotype.

17.10.2 Criteria for Detecting X-linked Recessive Trait in Human

1. The trait will be exhibited by many more males than in females.
2. All sons of an affected mother should exhibit the trait.
3. The sons of the carrier mothers should show an approximately 1 : 1 ratio of normal to affected.
4. The trait is transmitted from an affected man through his daughters to half his grandsons (criss-cross).
5. The X-linked trait is never transmitted directly from father to son.
6. All affected females have an affected father and a carrier or affected mother.

17.10.3 Y-linked Inheritance

Traits due to mutant genes located on Y chromosome having no counterparts on the X chromosome are called **Y-linked** or

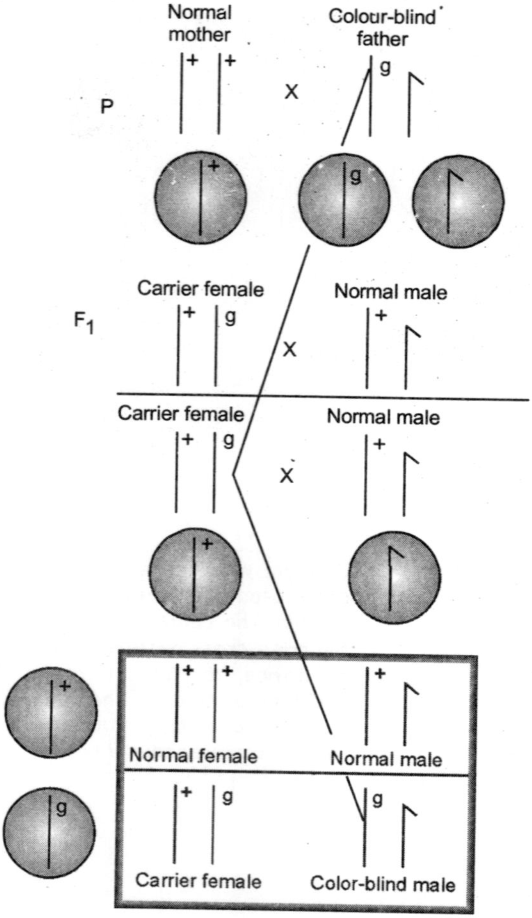

FIGURE 17.9

Criss-cross inheritance from father to daughter to grandson (of original father) that is characteristic of a sex-linked gene.

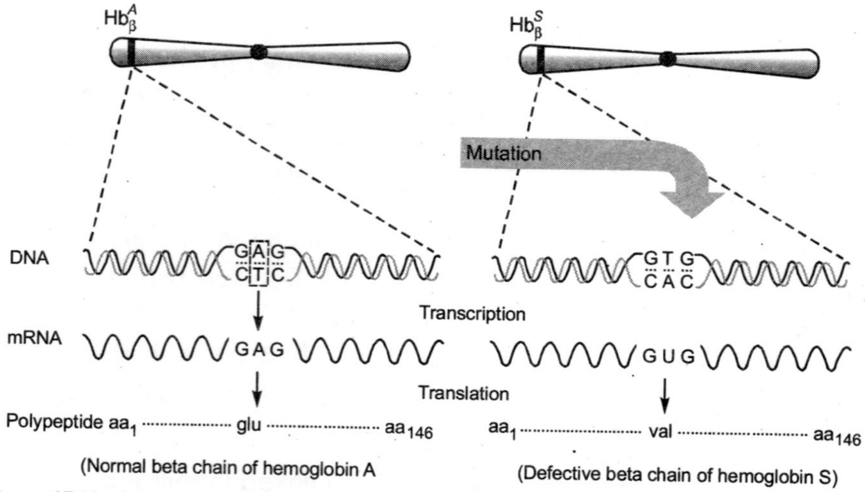

FIGURE 17.10

Mutational origin of sickle-cell hemoglobin (hemoglobin S). The amino acid sequences of the "normal" (hemoglobin A) beta polypeptide and the sickle-cell beta polypeptide have been determined by direct chemical analysis (amino acid sequencing).

holandric traits. A number of genes located on Y chromosome have been discovered (See 10.3). A common but not doubtless example of Y-linked inheritance is that of hairs on pinna. Although the trait is transmitted from father to son through Y-chromosome (and never to daughter), there are scopes to consider that the characters may also develop due to testosterone hormone or due to an autosomal inheritance.

17.11 Mutational Disorders in Some Human Genes

1. The major form of hemoglobin in adults (HbA) contains two identical α-chains and two identical β-chains of globin protein. The α-chain consists of 141 amino acids and the β-chain consists of 146 amino acids. There arise some **hemoglobin variants** in human due to mutation. The mutated Hb can cause severe diseases. One of them is **sickle-cell anemia**. In most cases, interestingly, there is no reduction in the hemoglobin content in the RBC. The RBC of people with sickle-cell anemia undergo a structural change, from biconcave to *crescentic sickle* (having Hb_{β}^{S}-chain), when the O_2 tension around them is reduced.

   The sickle-cell trait is the heterozygous manifestation of a gene which, when homozygous, results in sickle-cell anemia.

   The variant sickle-cell hemoglobin or **Hemoglobin S** (Hb_{β}^{S}) is produced from Hb_{β}^{A} by mutation. The sixth amino acid of the β-chain of HbA is **glutamic acid**, while the sixth amino acid of β-chain of HbS is **valine**. The α-chains of both hemoglobins remains same. Thus, the change of a single amino acid in one polypeptide chain of HbA can produce HbS, causing a severe disease, the **sickle-cell anemia** (Fig. 17.10).

- **Causal Events**

The trait is autosomal recessive. It is a classic example of **point mutation** causing the replacement of the *glutamic* acid residue at position 6 in the beta amino acid chain of hemoglobin by *valine*. The consequences are as follows—

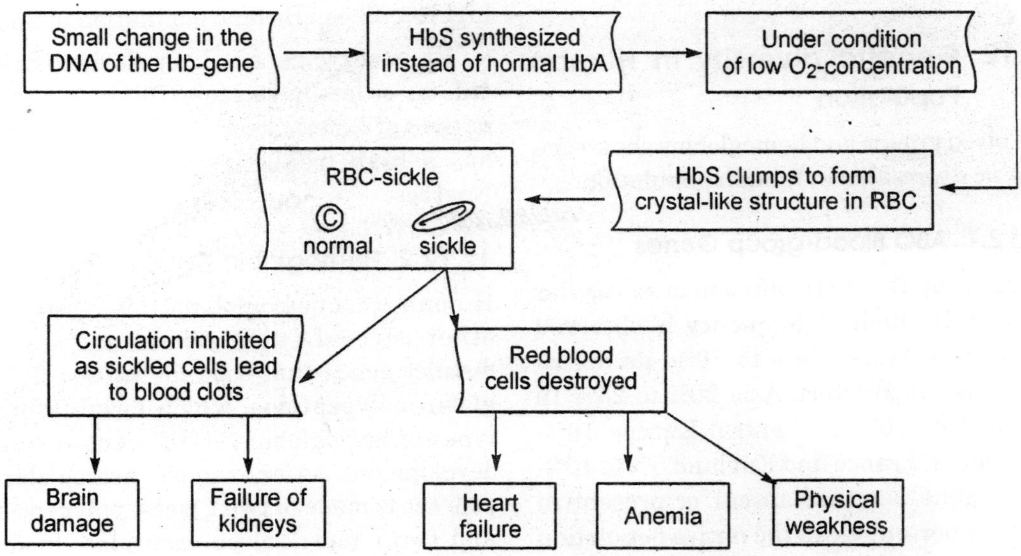

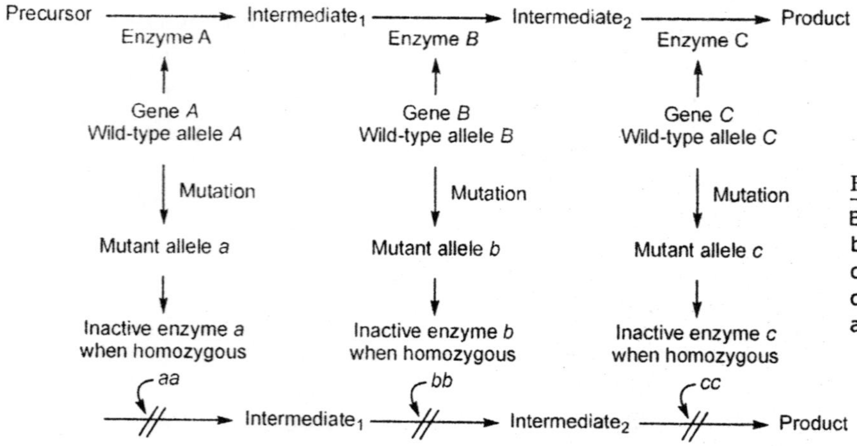

FIGURE 17.11 Blocks in metabolic pathways caused by recessive mutant alleles.

A minor change in a single gene can cause a large number of phenotypic effects (pleiotropism).

2. **Metabolic disorders**: In inborn metabolic disorders (see chapter 9), it has been established that the 'errors' in metabolism results from **mutation**. Metabolism occurs by sequences of chemical reactions, each step of which is catalyzed by a specific **enzyme** encoded by **one or more genes**. Mutation in these genes produce blocks in metabolic pathways. (Fig. 17.11).

17.12 Genetic diversity in Human Population

The blood groups and hemoglobins show wide genetic diversities in human population.

17.12.1 ABO Blood-group Genes

Blood group B is distributed in most regular pattern. Its highest frequency is observed in Central Asia— 25% to 30%, its distribution is—in Western Asia 20% to 25%, in Russia 15%-20%, in Central Europe 10%-15%, and in France and England 5% to 10%. The B gene is either absent or present in very low percentage in the native population of Australia and South America.

Blood group gene A is more universal but less regular in distribution. Blood group gene O is the most frequent allele and is found in high frequency all over the world.

For **Rh factor**, three closely linked genes, **cde**, each with a dominant and recessive allele are responsible for the phenotype. There is only one Rh-negative genotype, which is denoted as **rh/rh** or **cde/cde**. The Rh-positive genotypes are characterized by at least a single **Rh** or **D** gene. The cde is 40% in Europe, 0-5% in China and Japan, 23% in Africa, absent from the natives of Australia and America, and highest, 55% in Basque population in Spain.

Rh-positive gene is irregularly distributed, 5% or less in Europe, China, and in the natives of America and Australia. This gene has a high frequency (60%) in African Negroes.

17.12.2 Hemoglobin Genes

Human adult hemoglobin, HbA is one of the other several types. There are several distinct genes that code for α- and β- like globin polypeptides, which form different types of hemoglobins at different stages of development. As for example, hemoglobin of yolk sac is made of two ζ (zeta) polypeptides and two ε (epsilon) polypeptides. In fetal hemoglobin or HbF (made in liver and spleen) the polypeptide chains of the globin are made of two α- and two beta-like γ (gamma)-chains. After birth, the γ-chain synthesis stops and hemoglobin synthesis takes place in bone-marrow, where α- and β- polypeptides along with some β-like δ- (delta) polypeptides are produced. This is what we call HbA, which is usually made up of $\alpha_2\beta_2$ tetramer of polypeptides.

Chromosome 16 contains the α-like genes (2α gene and 1 γ gene) while **chromosome 11** contains the β- like genes (ε, γ A, γ G, δ, and β).

The mutated Hb genes, the 'sickle cell' genes are found in highest frequency in Africa, Mediterranean region, the Middle East, Southern India and Indo-China. The 'sickle-cell' gene, when homozygous, is lethal gene; thus, the *heterozygotes* of them are *adaptively superior* to the normal homozygotes (See 16.5.5.3).

Another Hb-gene, 'thalassemia', is normally homozygous lethal. It is believed that the thalassemia heterozygote has some resistance to the malaria caused by *Plasmodium vivax* and *P. malariae.*

17.13 Blood Group Genetics (Multiple allele system)

According to the presence of antigen in RBC and antibody in serum the human blood groups are—A, B, AB and O. The table 17.2 will help us to understand the distribution of antigen and antibody in the RBC and serum of blood respectively. When only the A antigen is present on the red cells, the blood is group **A**, when only the B antigen is present the blood group is **B**; when both antigens (A and B) are present, the blood group is **AB**, and when neither antigen is present, the blood group is **0** (Table 17.2).

The gene responsible for producing the A and B antigens is denoted by the letter I. It has three alleles—I^A, I^B and i (thus called multiple alleles). The **I^A allele** produces the A antigen, and the **I^B allele** produces the B antigen. The **i allele** does not produce any antigen. There are 6 possible genotypes for the 4 blood groups. (Table 17.3). In this system, the I^A and I^B alleles are **codominant** (each is expressed equally in AB) and the **i** allele is recessive to both of them. Thus I gene is **polymorphic**.

The genetics of the **ABO blood group** system follows Mendelian Principles. A person of O blood group is of ii genotype. Both parents of this person could be O (ii × ii), or A (I^A i × I^A i), or B (I^B i × I^B i), or one parent could be A and the other could be B ($I^A{}_i$ × $I^B{}_i$). Each parent must have i as one of its two alleles.

In case of dispute of paternity or maternity genetic analysis of the blood group can be done, only to know that an individual is *not* the parent of the baby in question. For example, a baby of blood group AB (genotype I^AI^B) could not be the child of a parent of blood group O (genotype ii). In recent days DNA typing is used to detect actual parent (See chapter 18).

17.13.1 H Antigen, ABO Blood Group and Bombay Blood-group

The ABO locus encodes an enzyme called **glycosyltransferase**. It adds sugar groups

TABLE 17.2

Blood Group	Antigen in Cell (RBC)	Antibody in Serum
A	A	β
B	B	α
AB	AB	None (0)
O	None (0)	αβ

TABLE 17.3

Genotype	Blood Group (Phenotype)	A antigen	B antigen
I^AI^A or $I^A i$	A	+	–
I^BI^B or $I^B i$	B	–	+
I^AI^B	AB	+	+
ii	O	–	–

to a preexisting polysaccharide that combines with lipids to form glycolipids. This glycolipid associates with red cell membrane and form blood group **antigen** (Fig. 17.12).

People usually produces a glycolipid called the **H antigen**. The I^A allele produces an enzyme that adds the sugar α-N-acetyl galactosamine to the pre existing polysaccharide to produce **A antigen**. Similarly, the I^B allele produces an enzyme that adds galactose to pre existing polysaccharide to produce **B Antigen**. In I^AI^B heterozygotes both antigens (AB) are thus produced. People, homozygous with i allele produce no enzyme, thus no antigen (O) is produced.

The H antigen is a product of a dominant H allele, located separately from the ABO locus. People, who are homozygous for the recessive mutant allele h, (hh), cannot produce H antigen; as a result, neither A nor B antigens are produced in them, regardless the presence of I^A or I^B at their ABO locus. These rare hh people lack A and B antigen and are like O group, although they have I^A or I^B alleles. They are called **Bombay blood group persons**.

17.14 Recessive and Dominant Mutant Traits and Genetic Disease

Homozygosity of some *recessive* mutant alleles may produce diseases in human. **Albinism** is an autosomal recessive trait; these people (albino) cannot produce melanin pigment, which protects the skin from UV radiation. Their skin and eyes become very sensitive to sun-light.

There are also some autosomal dominant traits, like **PTC**, **achondroplasia**, **brachydactyly** etc. An affected person with autosomal dominant trait is likely to be heterozygote, and he or she must have at least one affected parent. The autosomal dominant trait usually does not skip any generation in a pedigree (See 17.15). Usually the affected heterozygous person will transmit the mutant gene to half of his or her progeny.

PTC (Phenyl Thio Carbamide) is bitter in taste. The person who can taste it, is normal, and who cannot is non-taster. The gene for tasting bitterness of PTC is an autosomal dominant and is inherited in simple Mendelian Principle.

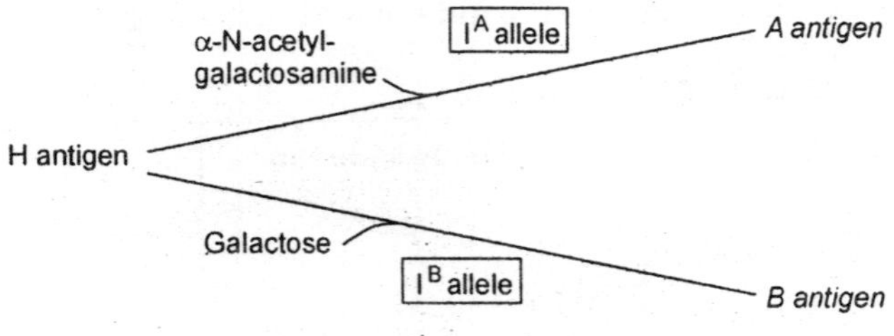

FIGURE 17.12

Transformation of H antigen (glycolipid) to A or B antigen by adding α-N-acetylgal actosamine (sugar) or galactose (sugar) respectively.

Huntington's Disease (HD) is a neural disorder that affects a person in middle age. The affected person gradually lose motor control and mental function. They show involuntary writhing and swaying called 'chorea' (= dance); thus, the disease is also known as **Huntington's chorea**; the symptoms worsen with age, eventually causing death. The disease was discovered by George Huntington. **HD** is inherited in a family as an *autosomal dominant trait*.

- **Cause of HD**: Autosomal dominant gene is located on the tip of the short arm of chromosome 4.

- **Frequency**: 5 per 1,00,000 of the population are HD. Sufferers can be homozygous or heterozygous for the defective gene. HD is a neuropsychiatric disorder resulting the involuntary muscle movement and progressive mental deterioration. Age of attack is usually in 40s and 50s, so people can have children before they realise they have the disorder. There is no known treatment. Detection of carriers of the gene is now possible using DNA probes and restriction fragment length polymorphism (**RFLP**). Isolation and basic molecular characterization of the HD gene was possible by James Gusella, Nancy Wexler *et al* (See chapter 18).

Cystic Fibrosis (CF) is an example of *autosomal recessive trait*. It is caused due to the mutant gene probably located on the long arm of chromosome 7. Its frequency is common in Europe; 1 is affected in every 1600 live births.

- **Symptoms and detection**: This is a disorder of mucus-secreting glands, particularly of pancreas, intestine and lungs. The mucus becomes very thick and sticky so that ducts are choked leading to intestinal blockages and pneumonia. The food is not digested and absorbed properly due to blockage of the pancreatic duct. Many patients die at an early age. Detection of sufferers prenatally may be possible by using recombinant technology and DNA probes.

17.14.1 Hereditary Diseases due to Enzyme Deficiency

Hereditary disease is due to a more or less rare genotype that produce ill adapted phenotypes in environments in which more common genotypes give rise to healthy phenotypes. Many of them are caused by deficiencies in enzyme activities. Most of them are inherited as recessive traits, and some are inherited as dominant traits.

In general, an enzyme deficiency caused by a mutation may produce disease. The consequence of the mutation may be simple or **pleiotropic** (showing multiple phenotypic effect). Some of the genetic disorders due to enzyme deficiency or block as a result of mutation in one or multiple allele have been discussed in chapter 9. Some of the genetic disorders and their enzyme deficiencies are lisled below (Table 17.4).

17.15 Mendelian Principles in Human Genetics

The genetic analysis of human heredity depends on family records, which may not be always sufficient. The number of progeny in a family is not very large; moreover humans are not laboratory animals, cannot be a subject of experiments. Manifestation of a particular trait may take much time. For such reasons human genetic analysis is very difficult and often is not error-free. Despite all these obstacles, progress in analysis of human inheritance has achieved a good position. Some thousands of genes control dominant or recessive traits following Mendelian Principles. **Pedigree analysis** has become an important tool to identify dominant and recessive traits in a particular family. The analysis may help genetic counsellors to assess the risk of any harmful trait that may be inherited to a person in that family.

TABLE 17.4

Genetic Disorder	Enzyme Deficiency
1. Alkaptonuria	Homogentisic acid oxidase
2. Cystic fibrosis	Cystic fibrosis transmembrane conductance regulator (CFTR)
3. Cataract	Galactokinase
4. G6PD deficiency (Favism)	Glucose 6-phosphate dehydrogenase
5. Hypoglycemia and acidosis	Fructose-1, 6 diphosphatase
6. Immunodeficiency	Uridine monophosphate kinase
7. Intestinal lactase deficiency (adult)	Lactase
8. Kidney tubular acidosis with deafness,	Carbonic anhydrase B
9. Male pseudohermaphroditism	Testicular 17, 20 desmolase
10. Phenylketonuria	Phenylalanine hydroxylase
11. Tay-Sachs disease	Hexoseaminidase A
12. Defect in thyroid hormone synthesis	Iodide peroxidase or deiodinase.

17.15.1 Pedigree Analysis

Using pedigree charts we can identify the patterns of inheritance of a particular trait. Before we go into details, some **symbols** and **signs** used in pedigree are to be studied (Fig. 17.13).

17.15.1.1 Autosomal Dominant

The Fig. 17.14 is a sample pedigree showing the pattern of inheritance of an *autosomal dominant* trait. We say this is a trait determined by **dominant allele** since:

- It appears in every generation.
- Every affected individual has an affected parent.
- When one generation does not express the trait, the trait is lost and does not reappear in future generation (look at children, grand-children and great grand-children of II-3 in Fig. 17.14.

To confirm that it is an autosomal allele, and not sex-linked, there are a number of clues. *Firstly,* equal numbers of males and females are affected. In a X-linked inheritance, usually the males are more affected. *Secondly,* the male III-1 passes the trait on to his sons. Since a man only passes Y chromosome to his sons, the gene cannot be an X-chromosome. *Thirdly,* the affected IV-1 has got two unaffected daughters as

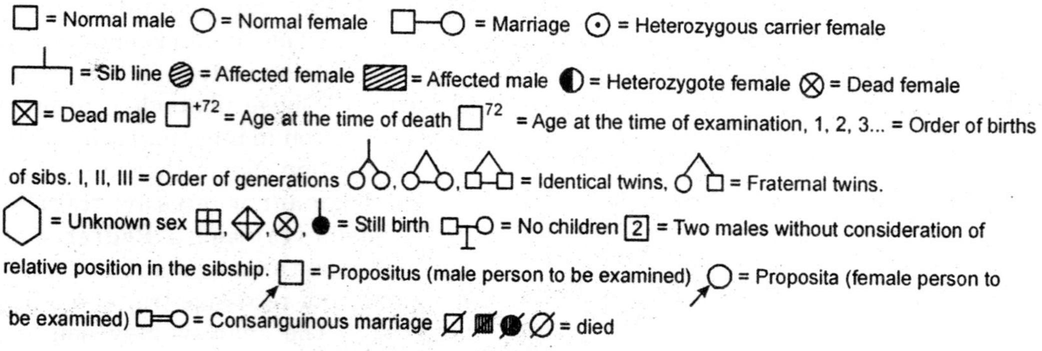

FIGURE 17.13

Signs and Symbols used in Pedigree Chart.

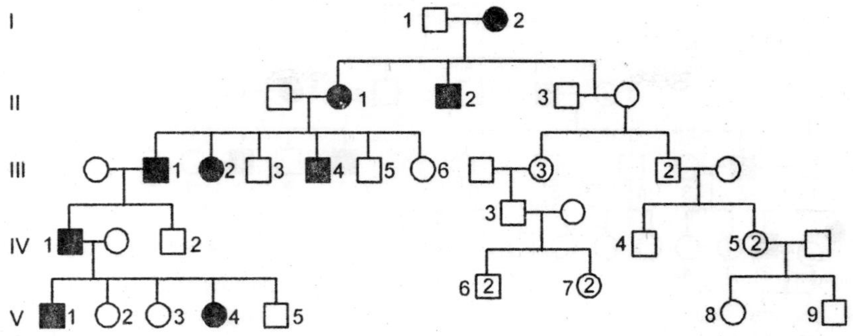

FIGURE 17.14

A hypothetical example of inheritance of autosomal dominant trait.

well as one affected daughter. If the gene were an X-linked dominant then all his daughters would be affected, since he transmits his X-chromosome to each of his daughters. The trait is not Y-linked since the affected male III-1 transmits the trait only to one of his sons, while the other son remains unaffected. If the trait were an X-linked then all the sons would be affected.

Now we look at Fig. 17.15. This shows a pedigree for muscular dystrophy (an autosomal form, not the X-linked Duchenne Muscular dystrophy, see 17.10.1). In this pedigree, the illness appears in every generation except for the first and the last. It looks like a dominant trait but the original parents (I-1, 2) and the children (V-1, 2) do not show the disease. The effect of a gene which causes a disease can vary because people vary. In particular, the age at which a disease develops will be affected by other genetic and environmental factors. When an individual carrying the **relevant gene** is so mildly affected that the disease seems unnoticed, it is called the **incomplete penetrance** of the gene. In Fig. 17.15, one of the two original parents (I) must have had the disease, but they were both dead at the time of examination of a particular person in the pedigree. If one of them only developed the disease very mildly then their children might not have noticed. So this pedigree chart shows a pattern of inheritance consistant with a dominant allele showing incomplete penetrance. In the figure, there is an arrow mark below the person III-7. It means, he first drew the counsellor's attention to this trait in his family. He is called **propositus**. The propositus is the male individual who first draws an attention to geneticist for a particular trait. (In case of female person, it is called **proposita**. **Proband** is another term for an affected person with whom the study of a trait in a family begins). The children of the last generation do not show the trait, may it be, that they will exhibit the disease in their later life.

17.15.1.2 Autosomal Recessive

Fig. 17.16 shows the pattern of inheritance indicating an **autosomal recessive trait**. The recessive nature of the trait is indicated by—

- the rarity of the trait
- skipping of generations
- the observation that IV-1 and IV-2 are the children of a **consanguinous marriage** (here, between cousins III-1 and III-2).

This pedigree chart is not a recessive X-linked trait since the trait effects almost equal numbers of males and females (whilst X-linked trait affects more males than females). The appearance of the trait in IV-

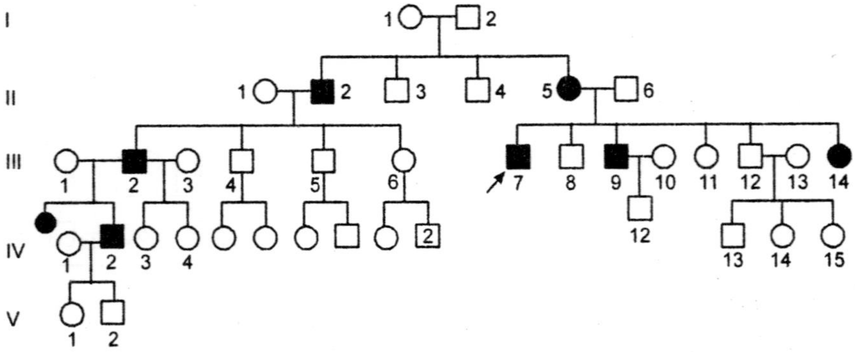

FIGURE 17.15

Pedigree illustrating autosomal inheritance of muscular dystrophy which shows incomplete penetrance.

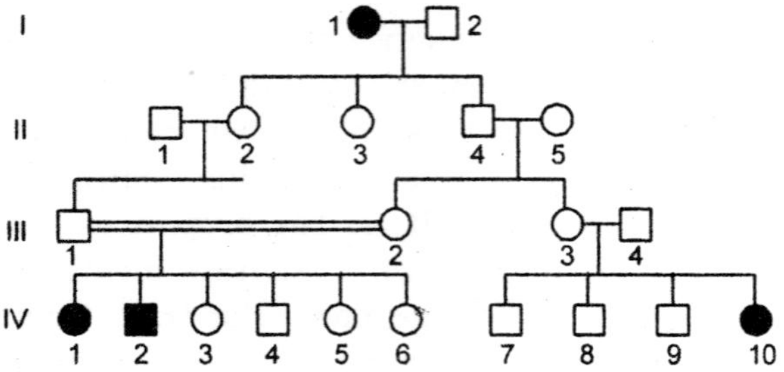

FIGURE 17.16

Pedigree showing inheritance pattern of autosomal recessive trait.

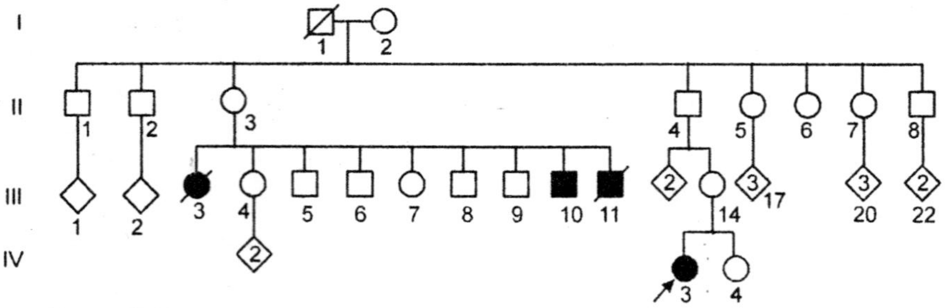

FIGURE 17.17

The pedigree showing inheritance pattern of cystic fibrosis, an autosomal recessive trait in a family.

10 is only possible if III-3 and III-4 are the carriers of the recessive allele. Here it has happened.

Let us now look at Fig. 17.17. It is a pedigree chart of inheritance of **cystic fibrosis** (See sub-chapter 17.13). It shows a **recessive autosomal** pattern of inheritance—since

- the trait is rarely exhibited.
- it affects equal number of males and females.
- the trait skips generations.

17.15.1.3 Sex-linked (X-linked) Traits

Fig. 17.18 shows the characteristic pattern of inheritance for an **X-linked recessive** trait. The important features of this inheritance are—

- Many more males than females are affected.
- If the recessive gene is very rare almost all observed cases will be male.

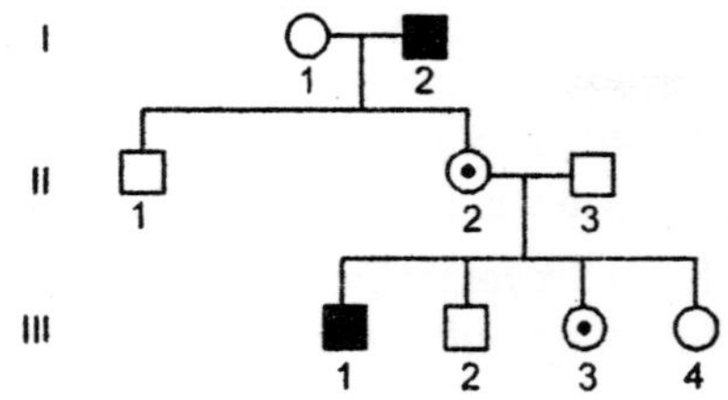

FIGURE 17.18

X-linked recessive trait inheritance pattern in a pedigree.

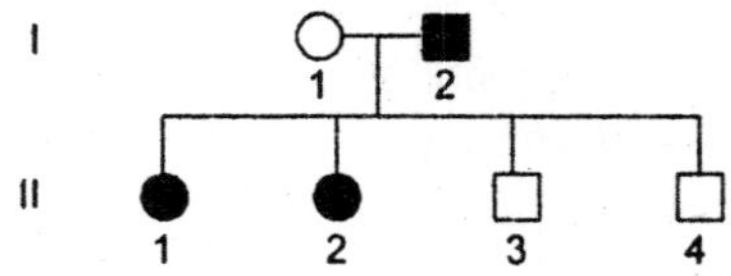

FIGURE 17.19

Pedigree pattern of X-linked dominant trait.

- usually none of the offspring of an affected male will be affected, but all his daughters will carry the allele, so half of their sons, on average, should be affected.
- none of the sons of an affected male will be affected.
- the inheritance follows a **criss-cross** pattern, i.e., from father to daughter (carrier) and again from daughter (carrier) to grandson.

In Fig. 17.18, since males are XY they only need one copy of the allele to show the trait (hemizygous) whilst females will have to be homozygous for the allele to show the trait. A son gets the Y chromosome from his father whilst the daughters will inherit the X chromosome carrying the trait making themselves *carriers*. A daughter will only be affected if her mother is a carrier and her father has the trait. Fig. 17.19 shows the pattern of inheritance for an X-linked dominant trait. In this case, only the daughters show the trait because the trait is due to an X-linked dominant allele. Wherever the X-chromosome will go (all daughters), it will show the trait due to its dominant nature.

17.16 Eugenics and Euthenics

Eugenics is related to the betterment of human race. The Greek word **eugenes** means 'well born'; thus eugenics is a science of well born, in which benefits of genetic knowledge are applied to improve the qualities of human race.

There are two branches of the science: the improvement of present-day humans called **euthenics**, and the improvement of future human race called **eugenics**. The former gives stress to the betterment of environmental factors so that the existing humans can achieve better life and the latter deals with **germplasm** of present-day humans so that future human race can be designed and improvised by selected artificial breeding.

Several steps or measures can be adopted for the betterment of human race through application of the knowledge from genetics. Reproduction and selection are two chief instruments to provide positive eugenics. The desirable traits of humans are selected through—

1. Advising early marriage of the persons having desirable traits.
2. Advising and forming laws for artificial insemination for very resourceful and outstanding persons. Preservation of their germplasm through **spermbank** and **eggbank** is encouraged to get desired future human race.
3. Proper **social education** and **genetic counselling** are two important steps to improvise human race and to protect them from germinal wastes. Counselling before marriage is one of the important measures to prevent genetic disorders.
4. Better pollution-free environment, nutrition and education are to be provided for people. Eugenics is concerned with genetic health of a per-

son, as such, field of genetic research is to be enhanced. H. J. Muller, one of the founders of eugenics, is of strong opinion for utilization of improved and selected **germplasms** through artificial insemination so that mother earth can have only genetically healthy and quality-people in future.

5. Rejection of germplasm of unhealthy and genetically disordered people can be achieved through their sterilization and sexual separation after genetic counselling. Culmination of defective germplasm is aimed for a future society free from genetic disorders.

Genetic counselling plays a great role in eradication of genetic disorders. Further, genetic engineering (see chapter 18) and gene therapy may provide a bright future for the human race.

IMPORTANT NOTES:

1. Human chromosomes are 23 pairs in number; 22 pairs of them are autosomes and 1 pair is sex-chromosome. The autosome pairs are all alike in both male and female but the sex chromosomes of male are XY and that of female is XX.
2. Both autosomal and sex chromosomal aneuploidy may occur in human. A few of the autosomal aneuploidy are Down syndrome or Trisomy-21, Patau syndrome or Trisomy-13, and Edwards syndrome or Trisomy-18.
3. When 3 copies of chromosome-21 occurs, a Down syndrome develops. Its Karyotype is (47, +21).
4. In sex-chromosomal aneuploidy, monosomy for one X chromosome is called Turner syndrome. Its Karyotype is (45, XO). People with Turner syndrome are phenotypically female with no Barr body.
5. Abnormal male syndrome due to presence of an extra X chromosome (47, XXY) is called Klinefelter syndrome. These abnormal males possess a Barr body.
6. Chronic myelocytic leukemia develops when a part of the long arm of chromosome 22 is translocated to the long arm of another chromosome (usually 9) causing a deletion in chromosome 22 (called Philadelphia chromosome).
7. A few of the X-linked recessive trait in humans are the red-green colour blindness, hemophilia A and Duchenne muscular dystrophy.
8. X-linked recessive trait follows a criss-cross transmission, i.e., the trait is inherited from an affected man to his grandsons (through his daughters).
9. An example of autosomal dominant trait is the Huntington's disease (HD). An example of the autosomal recessive trait is the cystic fibrosis (CF).
10. In general, an enzyme deficiency caused by a mutation may produce genetic disorders, the consequences of which may be simple or pleiotropic.
11. Some genetic disorders due to mutation and enzyme deficiencies are alkaptonuria, cystic fibrosis, phenylketonuria, albinism, cataract, etc.
12. Blood group A is produced by the genotype I^AI^A or I^Ai Blood group B is produced by the genotype I^BI^B or I^Bi. Blood group AB is produced by I^AI^B. Group O is produced by ii genotype.
13. Pedigree analysis is an important tool to identify dominant and recessive traits and their mode of transmission (autosomal or sex-chromosomal) in a family.
14. Eugenics is related to the betterment of human race. Genetic counselling and genetic engineering may play great roles for a bright future of the human race.

EXERCISE

1. How are the human chromosomes classified?
2. What is aneuploidy? What are its types? Mention two syndromes due to autosomal and sex chromosomal aneuploidy.
3. Mention cause, origin and characteristics of Down syndrome.
4. Mention cause, origin and characteristics of Turner syndrome.
5. Mention cause, origin and features of Klinefelter's syndrome.
6. What do you know about cri-du-chat?
7. How does chronic myelocytic leukemia develop?
8. What is sex-linked inheritance? Describe two instances of human sex-linked inheritance.
9. Write notes on hemophilia A, green and red colour blindness and Duchenne muscular dystrophy.
10. Mention criteria of a pedigree by which X-linked recessive trait can be detected.
11. Discuss the genetics of sickle-cell-anemia.
12. How does Huntington's disease develop?
13. Describe the genetics of cystic fibrosis.
14. Discuss the genetics of ABO blood group.
15. What do you mean by genetic disease? Mention and describe one human genetic disease with its cause of development.
16. Write notes on H-antigen, ABO blood group and Bombay blood group.
17. What do you understand by pedigree analysis?
18. Mention criteria for detection of autosomal dominant, autosomal recessive, and sex-linked recessive inheritance pattern in a human pedigree.
19. Draw two pedigree charts to explain the inheritance of an autosomal recessive trait and an autosomal dominant trait.
20. Draw a pedigree chart to explain an X-linked recessive trait.

CHAPTER 18

GENE MANIPULATION OR GENETIC ENGINEERING OR RECOMBINANT TECHNOLOGY

Genetic engineering, Cloning, Restriction enzyme, Vectors, DNA ligase, Isolation and cloning of a gene, cDNA, λ-phage as a vector, Genomic libraries, cDNA libraries, PCR, Analysis of DNA and RNA, RFLP, Application of recombinant technology, DNA Polymorphism, DNA molecular testing, DNA fingerprinting, Paternity test, DNA forensics, Gene therapy, Economic production of proteins, Synthesis of hGH, Transgenic organisms, Human genome project.

This chapter deals with the study of *gene manipulation* or *genetic engineering* which includes some methodologies forming new-combinations (*recombinants*) of genetic materials by manipulation. It is also called **recombinant DNA technology**. The gene manipulation includes the following procedures:

1. DNA fragments are produced artificially.
2. Recombinant DNA is produced by splicing the DNA fragments into a composite molecule.
3. This recombinant DNA is used as a *vector* or incorporated into a vector.
4. The vector is transferred into a cell where it replicates.
5. The cells with recombinant molecules of DNA are selected and *cloned*.

- **Cloning**

Cloning is the process of production of large amount of identical pure DNA, which are manipulated in various ways. There are several steps for cloning of DNA of which **isolation of DNA, cutting of it by restriction enzyme, insertion of DNA piece into a cloning vector producing recombinant DNA, introduction of the recombinant DNA into a host-cell and its replication inside the host-cell producing many identical copies (clones)** are most relevant. By this way, innumerable copies of desired piece of DNA can be isolated and sequenced which can be transcribed into mRNA and translated into proteins.

Genetic engineering allows geneticists to isolate, purify and study individual genes. Application of cloned DNA is a part of genetic engineering. The basic tools of genetic engineering are:

1. Restriction enzymes
2. Use of vectors
3. DNA ligase
4. Host systems and
5. Radioactive DNA probes

18.1 Restriction Enzymes

In 1970 it was discovered that the defence mechanism of some strains of bacteria against the attack of phage was due to a group of enzymes. These bacterial enzymes restrict the growth of foreign DNA in its body; thus these enzymes are called *restriction enzymes* or *restriction endonucleases*. These are endonucleases that recognize a specific sequence of DNA and *cleaves* the DNA molecule at that specific site.

The site of cleavage by the restriction enzyme is called *restriction site*. These sites are determined by the sequence of bases in the DNA. Inside the bacterial cell, the restriction enzyme acts as a molecular scissors or chopper cutting up DNA molecules into *restriction fragments*.

A restriction enzyme of the bacteria *Haemophilus influenzae* is called **Hin DII** (Hindee 2). It cuts the DNA of Phage T7 into 40 pieces. This is very specific because the enzyme cuts the DNA only at specific sequence. The Fig. 18.1 shows the point of cut (indicated by an arrow). In the figure, Py is a pyrimidine and Pu is a purine. The arrangement of bases is very similar on both sides. The sequence around the dot (point of cut) on both sides remain unchanged. This specificity of sequence allows the **Hin DII** to cut in into 40 pieces.

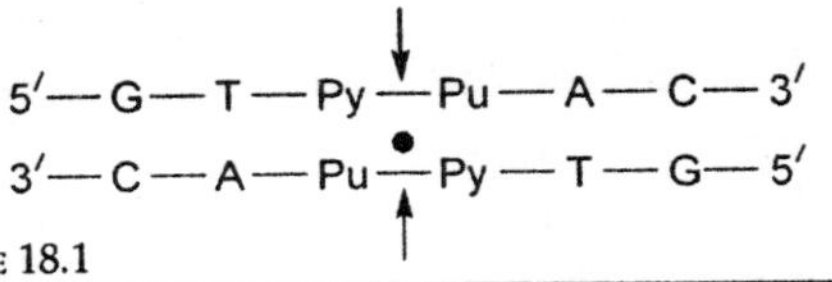

FIGURE 18.1

Restriction site for the restriction enzyme Hin DII.

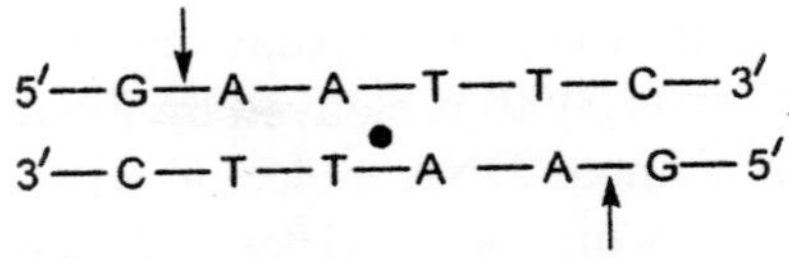

FIGURE 18.2

The EcoRI cuts the DNA to produce staggered cuts. The ends of the staggered cuts are "sticky".

The Hin DII enzyme produces *blunt ended* cuts of DNA.

However, there are some other restriction enzymes which behave in a different manner and may produce *staggered cuts* (instead of blunt ended cuts) in DNA. The restriction enzyme **Eco RI** is produced in an R plasmid of *E.coli.* This enzyme is specific for the site GAATTCC, i.e., it cuts DNA where the base sequences are present in the order GAATTCC and, are palindromic. Fig.18.2 shows the restriction site for **Eco RI** restriction endonuclease. In the figure, the arrows indicate the cleavage site. The DNA sequences on both sides of the central dot are the same, i.e., palindromic. The cuts produced by the **Eco RI** are staggered, which plays important role to open a circular DNA.

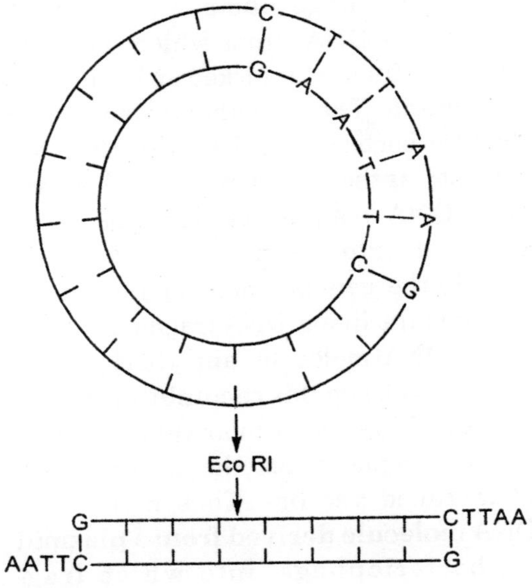

FIGURE 18.3

Action of Eco RI on the circular DNA of SV40 virus to cut it and to produce a linear DNA with staggered ("sticky") ends.

The Fig. 18.3 shows that the circular DNA of virus SV40 is cleaved by **Eco RI** to produce a linear DNA molecule with staggered ends which are "sticky". The ends of each strand bear a series of unpaired bases, AATT on the left and TTAA on the right. Such strings of unpaired bases will 'want' to pair up with a complementary base string; this 'desire' of complementary base pairing makes the two end-sequences "sticky".

Thus, a restriction enzyme provides; a way of cutting DNA from an organism at a specific sequence. It produces a collection of DNA fragments of variable length. These fragments of DNA may have sticky ends. The restriction sites vary for different enzymes in different organisms.

18.2 Vectors

In recombinant DNA technology, after cutting DNA by restriction enzyme vectors are used to carry the DNA segments into bacterial cells. So vectors are referred to as 'molecular vehicles'. After digestion by restriction enzyme, a mixture is produced that contains huge number of different fragments of DNA, from which only one fragment of interest is picked up by the help of a bacteria. The recombinant engineers induce the bacteria, which take up the DNA fragments, as they stand; but since these are linear DNA fragments, they must be *recombined* into the circular molecule of DNA. So the genetic engineer's aim is to recombine the linear DNA fragment with a circular DNA molecule, and then get the bacteria to take up this **recombinant DNA molecule**. These 'molecular vehicles' carry the DNA fragment into the bacterial cell, and are called '**vectors**'. Thus, **a vector is a DNA molecule derived from a plasmid or a bacteriophage, into which fragments of DNA may be inserted**.

There are two common vectors used in, recombinant technology. One is the bacterial plasmid pBR 322, and the other is λ-phage (lambda-phage). The λ-phage has been designed to produce specialized vectors called λ-charon phage (See 18.6).

18.3 DNA Ligases

DNA ligases are used for splicing of DNA. The **Eco RI** cuts both vector molecule and DNA from a target organism. The staggered 'sticky' ends produced on the vector and restriction fragments are complementary. These molecules 'want' to pair up together, by a process called **annealing**. After annealing they produce a **recombinant DNA molecule**.

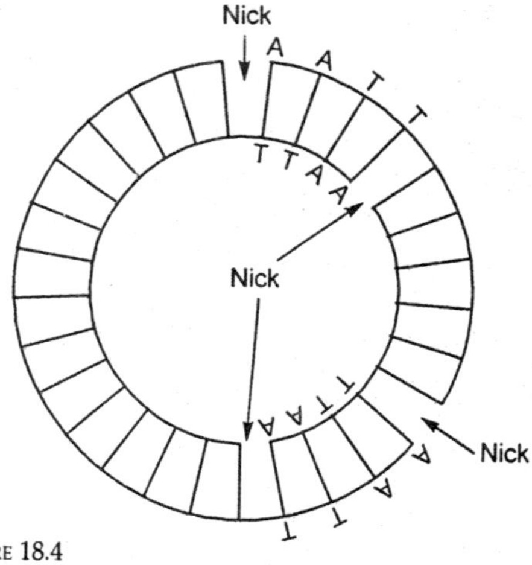

FIGURE 18.4

An unspliced recombinant molecule showing 'nicks'.

The Fig. 18.4 shows a recombinant DNA molecule that *lacks* the **covalent bonds** (which holds the sugar-phosphate backbone together) between the two sticky complementary DNA strands, represented by 'gaps'. These gaps are called 'nicks'. To produce a stable recombinant DNA molecule these nicks are to be sealed by a viral enzyme called *DNA ligase*. The sealing together is called **splicing or ligation**. Thus, it is possible to cut and recombine any two DNA molecules by using restriction enzymes and DNA ligase to produce recombinant DNA molecule *in vitro*, provided they possess complementary sticky ends.

There are three common splicing methods:

1. Sticky ends formed by a particular restriction enzyme (*Eco RI*), allowing different DNA fragments to pair at these ends, and then are spliced together by DNA ligase.
2. *Phage T4 ligase* join 'blunt ended' DNA fragments together.
3. Terminal transferase enzymes can add sequences of the same nucleotide to the 3′ end of a DNA chain. Thus, complementary homopolymer 3′ tails are generated in two types of DNA. These ends anneal and ligate together.

18.4 Host Systems

Vector DNA molecules do not replicate *in vitro*. Usually, the host cell *E.coli* are used to take up the vector within it, where the 'vector-molecules' can replicate. Other host systems like yeast and mammalian cells are also used for some specific purposes like production of vaccines.

18.5 Isolation and Cloning of a Gene

For isolation and cloning of a gene (e.g. HGH), the following procedure is adopted. The process of **gene cloning** which involves the production of multiple copies of the same gene includes the following stages:

1. A means for producing DNA fragments which carry the HGH gene.
2. A method for inserting these DNA fragments into a suitable vector (e.g. pBR 322) to produce recombinant DNA molecules.
3. Inducing host cells to pick up the recombinant DNA molecules.
4. Finding the cell which contains the required DNA-fragment.

18.5.1 Shotgunning Procedure

The common gene-cloning procedure is called **shotgunning**. It includes the following steps:

1. Pure simple DNA taken.
2. Digested with restriction enzyme.
3. Restricted fragments inserted into a suitable vector (e.g. pBR 322) to produce recombinant molecules.
4. Transformation of host-bacteria with the recombinant DNA molecule.
5. Multiplication of bacteria and increasing the number of each recombinant plasmid (DNA), facilitating cloning.

18.5.2 Reverse Transcriptase and Production of cDNA

The central dogma of biochemistry (i.e., DNA → RNA → Protein) is not maintained

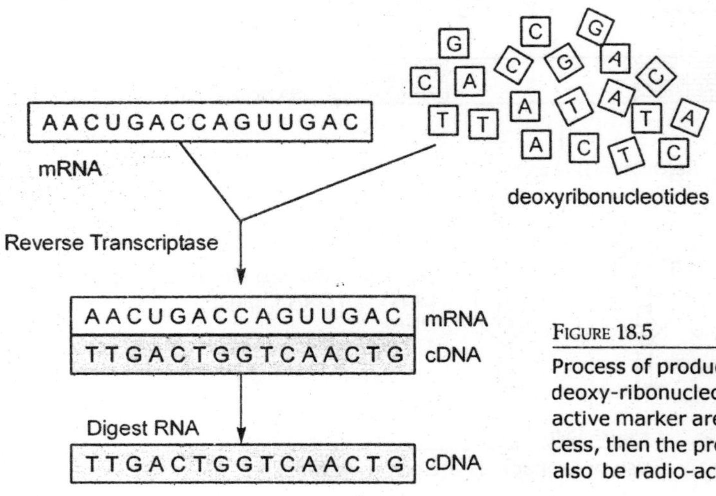

FIGURE 18.5

Process of production of cDNA. (If deoxy-ribonucleotides with radio-active marker are used in the process, then the produced cDNA will also be radio-active).

by the retroviruses. They possess *reverse transcriptase* and produce protein through the RNA → DNA → RNA → protein pathway. Genetic engineers use this enzyme to make DNA that complements any RNA molecule; these DNA are called **complementary DNA (cDNA)**. The Fig. 18.5 shows the process of production of complementary DNA or cDNA.

Each cell of an organism contains the same genes but different genes are transcribed into mRNA in the cells of different tissues. These cells contain only two copies of gene encoding a particular protein but may produce innumerable copies of mRNA transcript directing the translation of a specific protein. These mRNA can be isolated and used to produce that particular protein. If reverse transcription is allowed with these isolated mRNA by using reverse transcriptase, then cDNA molecules can be produced to these mRNA. These cDNA can be made double stranded, recombined into a vector and used to transform a bacteria.

18.6 Use of λ-page as a Cloning Vector

Lambda phage (λ-page) is used to clone larger fragments of DNA (between 15-20 kb). They are used to produce chimeric DNA molecules in which the central portion of the DNA is replaced by a target (desired) DNA fragment (Fig. 18.6).

The replaced part of phage-genome controls the recombination of the prophage into the bacterial DNA during lysogenization. The stored parts of the phage-genome control replication of the phage in the host *E.coli*. These specially designed λ-phages are called **λ-Charon** vectors.

18.7 Genomic Libraries

The rapid replication of these recombinant λ-phage, which carry large desired (target)

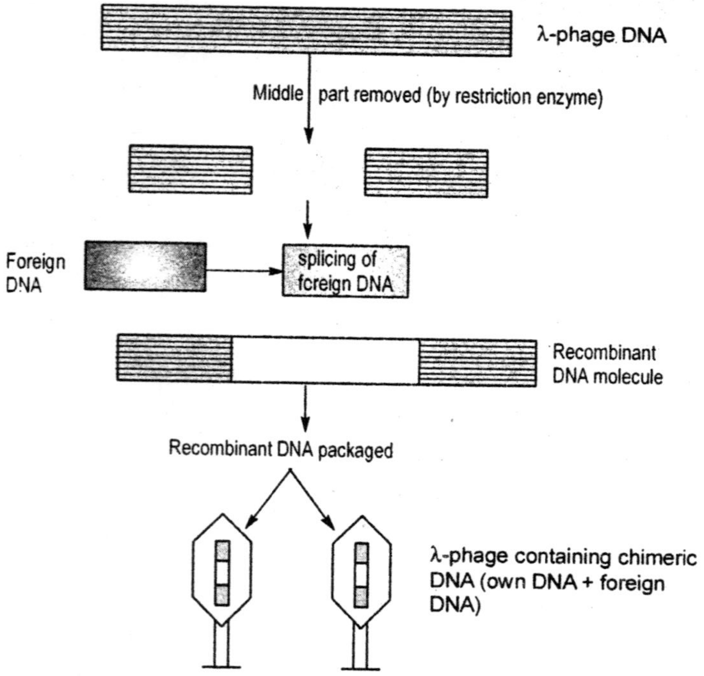

FIGURE 18.6

Use of λ-phage as a cloning vector.

DNA fragments, makes it very useful to create **recombinant-gene-libraries** by shotgunning. Gene-libraries are collection of innumerable clones of DNA fragments. The method of making of gene-libraries with λ-charon vector is to clone enough DNA fragments so that the collection of recombinant λ-phages should contain at least one copy of every DNA sequence in the target organism's genome.

18.7.1 Recombinant DNA Libraries (for human genome)

Genomic library is a collection of cloned DNA fragments that contain at least one copy of every DNA sequence in the genome. For this, genomic DNA is isolated from an organism and cut with a restriction enzyme, and the produced DNA fragments are cloned in a cloning vector.

To make **a human genome library**, we can follow the steps as shown in Fig. 18.7.

1. Human genome is collected and digested with **Eco RI**.
2. On the other hand, λ-Charon DNA is digested with **Eco RI** to remove the middle part.
3. Both types of fragmented DNA are annealed and ligated.
4. These recombinant DNAs are packaged into λ-phage coats (recombinant phages).

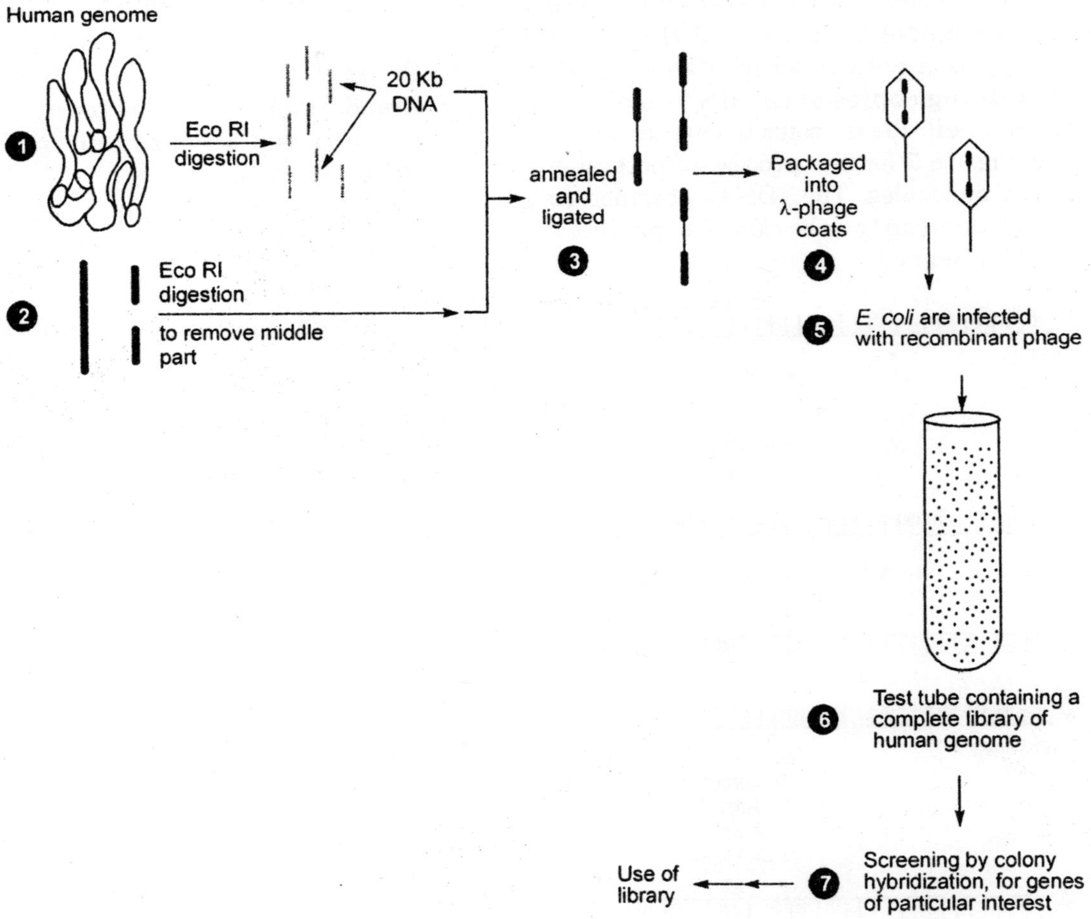

FIGURE 18.7
Making of a human genome library using λ-Charon phage.

5. Host *E.coli* are made infected with these recombinant phages for replication and cloning.
6. The test tube contains a complete library of human genome (each phage clone is the equivalent of a book in a library).
7. Such a library are then screened using colony-hybridization, for genes of particular interest. (In the same way as the books in a library can be checked to see which one is of our interest).

18.7.2 cDNA Libraries (Complementary DNA libraries)

cDNAs are molecules synthesized **in vitro** from an mRNA template, and **the cDNA library is a collection of cDNA clones containing copies of the RNAs isolated from a cell**. Please refer to chapter 18.5.2 and Fig. 18.5 for the process of obtaining cDNA molecules. The cDNAs obtained in this process can then be *cloned* to produce a **cDNA library**.

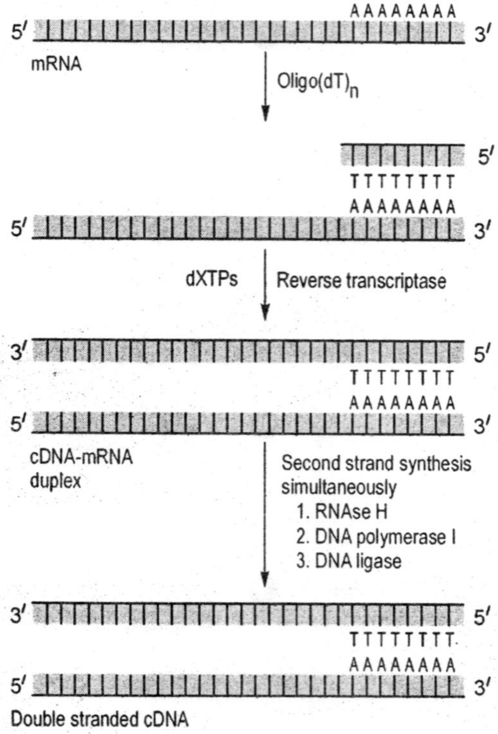

FIGURE 18.8

The synthesis of double-stranded cDNAs from mRNA molecules.

18.7.2.1 Synthesis of Double-stranded cDNA from a poly (A) mRNA

Process of synthesis of double-stranded cDNA molecule is shown in Fig. 18.8.

1. The mRNA from which cDNA is produced contains 3′ poly (A) tail.
2. A short oligo (dT) primer is annealed to the poly (A) tail.
3. The primer is extended by reverse transcriptase to make a DNA copy of the mRNA strand.
4. A DNA-RNA double stranded molecule is produced.
5. Next, a ribonuclease called *RNase H*, *DNA polymerase I* and *DNA ligase* are used.
6. *RNase H* degrades the RNA template strand, and short RNA fragments are produced that serve as primers for DNA synthesis.
7. *DNA polymerase I* synthesizes the second DNA strand and replaces the RNA primers.
8. *DNA ligase* seals the remaining single-strand-breaks in the double stranded DNA molecule.
9. These double stranded cDNAs are cloned by inserting in a plasmid or λ-phage.

18.7.2.2 Features of cDNA Library

1. Intron sequences are absent in cDNA clones.
2. The molecules in the cDNA library represent the mature mRNAs found in the cell, and are smaller than the equivalent gene clone.
3. cDNAs are useful for isolating the gene that codes for that mRNA.

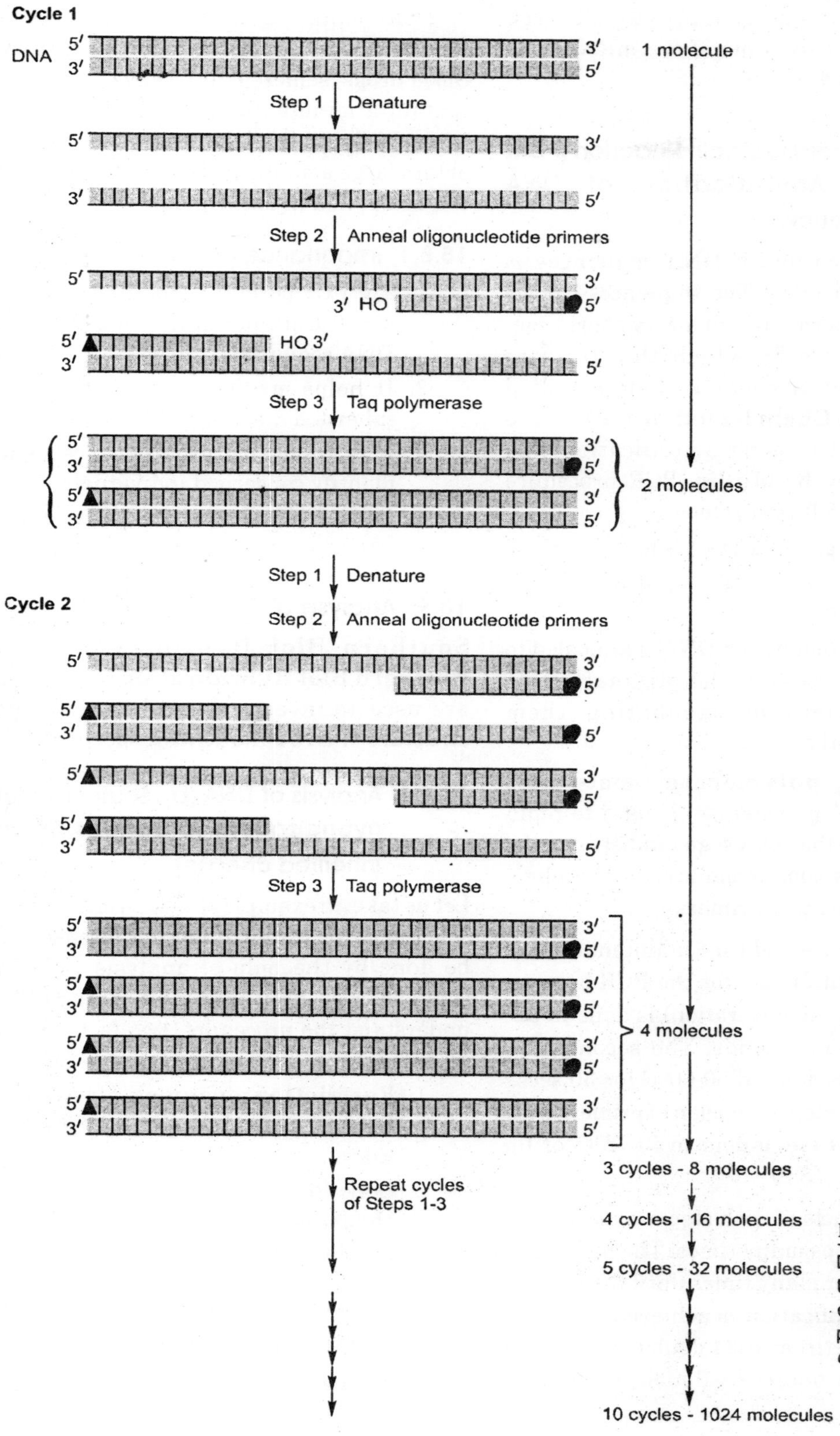

FIGURE 18.9
Diagram illustrating the use of PCR to amplify DNA molecules *in vitro*.

4. It reflects the gene activity of the cell type at the time of the mRNAs are isolated.

18.8 Polymerase Chain Reaction (PCR) and Amplification of DNA Sequence

The amplification of DNA sequences is performed *in vitro*. The sequences can be multiplied enormously in a very short time. The procedure by which this *in vitro* amplification of genes is done is called **Polymerase Chain Reaction** or **PCR**. The PCR method of gene amplification was discovered by **K. Mullis**. PCR procedure includes the following steps:

Step I. The genomic DNA consisting of the sequence to be amplified is denatured.

Step II. The denatured DNA is annealed to an excess of the oligonucleotide primers by incubating them together.

Step III. **Taq polymerase** (heat-stable DNA polymerase) is used to replicate the DNA segment between the sites complementary to the oligonucleotide primers.

Temperature and time maintenance are two important factors for the PCR process. The first step (denaturation) is done in 92-95°C for 15-30 seconds. The second step (annealing) is done in 50-60°C for 30 secs., and the third step (extension of primers with **Taq polymerase**) is done in 70-72°C for 1.5 minutes (Fig. 18.9).

The products of each step are allowed to follow the steps many times. The procedure is repeated for many times until the desired level of amplification is achieved. After 10 cycles of repetition 1024 double helices are yielded from one DNA double helix (Fig. 18.9).

PCR-amplification allows to obtain DNA sequences from a very little amount of DNA. Blood drops, semen, hair or any tissue may contain a minute amount of DNA, but by using PCR-amplification technique we can obtain large amount of DNA sequences that can be applied in many fields.

18.8.1 Importance of PCR

1. The PCR permits us to obtain definite structural information of genes and DNA sequences.
2. It helps medical men to diagnose inherited diseases.
3. It is applicable in forensic cases to identify concerned individuals.
4. **PCR-DNA-fingerprinting** plays important role in legal cases.

18.9 Analysis of DNA and RNA

Southern Blot Hybridization and **Northern Blot Hybridization** techniques are used to investigate and analyse the structure of DNA and RNA respectively.

18.9.1 Analysis of DNA (by Southern blot hybridization) [Diagnosis of an inherited disease]

Let us take an example of sickle-cell anemia. The diagnosis of this inherited disease can be done by the help of analysis of DNA sequences by Southern Blot method. To understand the procedure, two facts are to be noted:

1. **Restriction enzymes** always cut DNA at a specific sequence of nucleotides.
2. Genes (or the stretches of DNA) in a chromosome are linked to one another during inheritance.

18.9.1.1 Restriction Fragment Length Polymorphism (RFLP)

To diagnose the inherited disease, sickle-cell anemia, please refer to Fig. 18.10(a) and

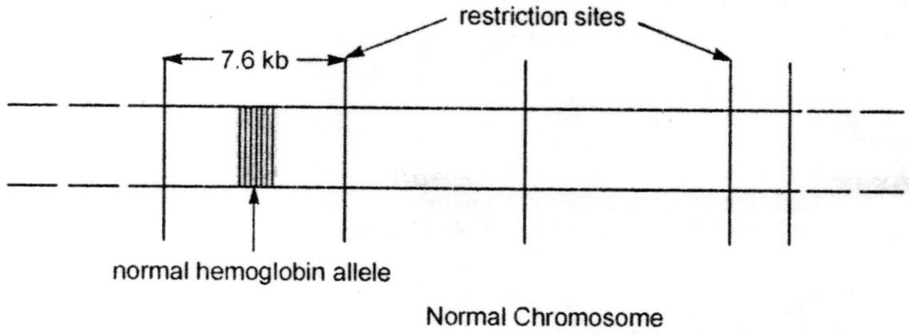

FIGURE 18.10(a)

RFLP of human chromosome containing normal hemoglobin allele.

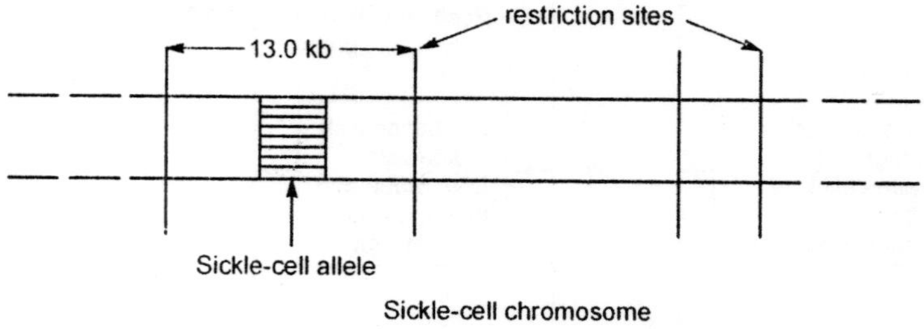

FIGURE 18.10(b)

RFLP of human chromosome containing sickle-cell allele.

18.10(b). The Fig. 18.10(a) shows the restriction sites for the enzyme *Hpa I* on a chromosome which contains the normal hemoglobin allele. Hpa I cuts the DNA, producing several pieces with a characteristic length. The *normal* hemo-globin allele is present in a piece of DNA which is 7.6 kb long.

Now, refer to Fig. 18.10(b). It shows the restriction sites for a chromosome which contains the sickle-cell allele. It is important to note that the pattern of restriction sites in the two chromosomes is different. The **restriction fragment** (applying Hpa I) with the sickle-cell allele is 13.0 kb long (instead of 7.6 kb). This variation in size of restriction fragment is called **restriction fragment length polymorphism (RFLP)**.

Analysis of the lengths of the fragments is the basis for the detection of sickle-cell allele, and consequently, is the tool for diagnosis of sickle-cell anemia.

18.9.1.2 The Technique for Analysis of DNA

Figs. 18.11 and 18.12 show the technique for **detection of RFLP** and **Southern blotting** respectively. (The name **Southern** blot is due to its inventor E.M. Southern) Fig.18.11.

For detection of RFLP—(1) Purified DNA of a suspected carrier is digested with restriction enzyme. (2) The resulting mixture of restriction fragments is then separated by agarose gel electrophoresis. The fragments are separated on the basis of size. (3) The separated fragments are now transferred to a nitrocellulose filter by **Southern blotting** (Fig. 18.12). (4) Particular restriction fragments are then located by hybridization with a radioactive gene probe followed by autoradiography (radioactive gene probe or DNA probe is a cDNA which is made with nucleotides containing radioactive p hosphorus, ^{32}P).

18.9.2 Analysis of RNAs by Northern Blot Hybridization

Likewise DNA molecules, the RNA molecules separated by agarose gel electrophoresis can be transferred to a nylon membrane filter and analysed. The RNA blots are called **northern blots** due to the fact that the procedure almost resembles the Southern blotting technique; the only difference is instead of transferring to a nitrocellulose filter, the separated RNA molecules are transferred to a nylon membrane (Fig. 18.13). After transfer to the membrane the RNA blot is hybridized to RNA or DNA

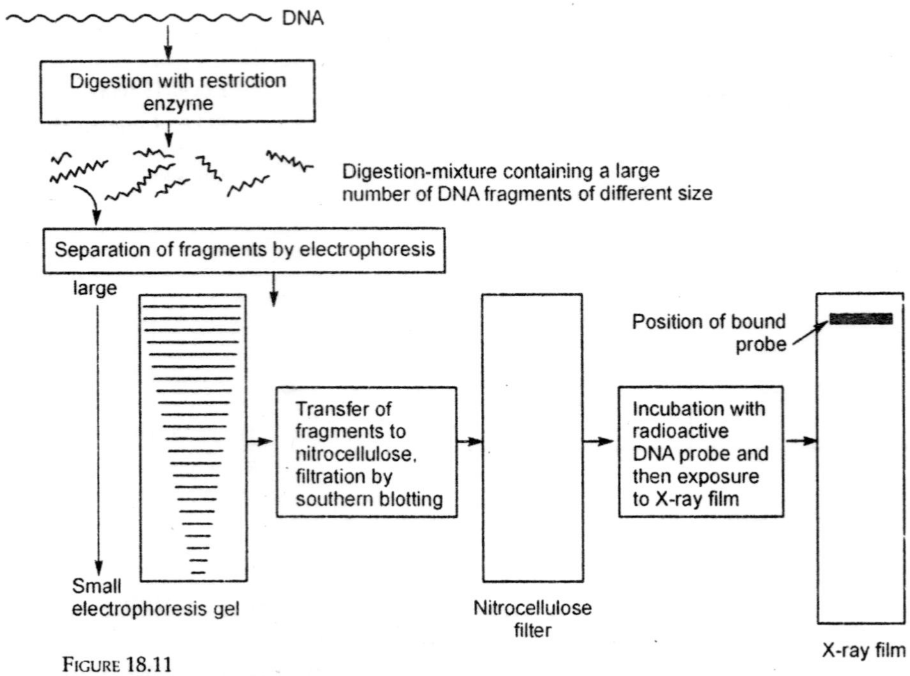

FIGURE 18.11

Detecting RFLP-After digestion with a restriction enzyme, the fragments are separated on the basis of size, by agarose-gel electrophoresis. The fragments are transferred to a nitro-cellulose filter by Southern blotting (Fig. 18.12). Particular restriction-fragments are then located by hybridization with a radioactive gene probe followed by autoradiography.

probe. The northern blot hybridization is useful to determine when and where a particular gene is expressed.

18.10 Application of Recombinant Technology

1. Most common application of recombinant DNA technology (biotechnology) is found in the field of medicine and agriculture.
2. It can be applied in the field of anthropological studies, conservation of ecosystems, industries and forensics.
3. Genes can be studied *in vivo* by causing specific mutation in the clone of them *in vitro* by applying recombinant DNA technology.
4. Human genetic diseases can be detected by using recombinant DNA and PCR technology.
5. Since two individuals (except identical twins) never have the same genome sequence, DNA typing or DNA fingerprinting may be applied in the field of forensics and in paternity testing.
6. Genetic disorders can be treated by gene therapy, which is a method of introducing a normal gene in the body of a patient replacing the defective one.
7. Transgenic or genetically modied crops and animals can be produced with the help of recombinant DNA and PCR technology.

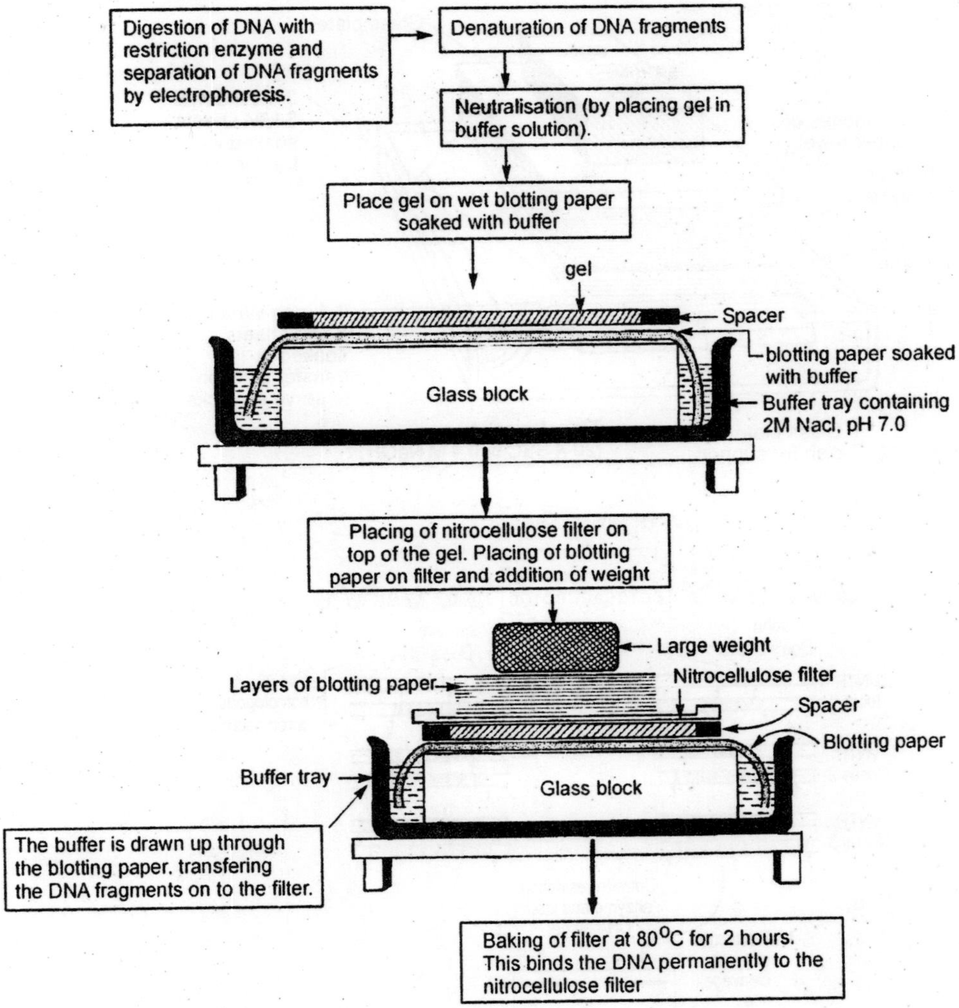

FIGURE 18.12

Southern Blotting. The aim of this technique is to transfer single stranded DNA (denatured) fragments from the agarose gel used for electrophoresis to a nitrocellulose filter.

18.10.1 DNA Polymorphism

DNA polymorphism is defined as one of the alternative forms of a locus that have either a different nucleotide sequence or have variable numbers of tandemly repeated sequences. Polymorphic loci can be used in gene mapping. Examples of DNA polymorphisms are single nucleotide polymorphisms (SNPs), short tandem repeats (STRs) and a variable number of tandem repeats (VNTRs).

18.10.2 DNA Molecular Testing

Please refer to the chapters 18.9.1, 18.9.1.1 and 18.9.1.2. Besides the diagnosis of an inherited disease, the human genetic testing (or DNA molecular testing) can be done for **prenatal diagnosis, newborn screening and heterozygous or carrier** detection. In all cases tests are done by RFLP analysis and Southern blotting or by PCR techniques, followed by allele-specific oligonucleotide (ASO) hybridization.

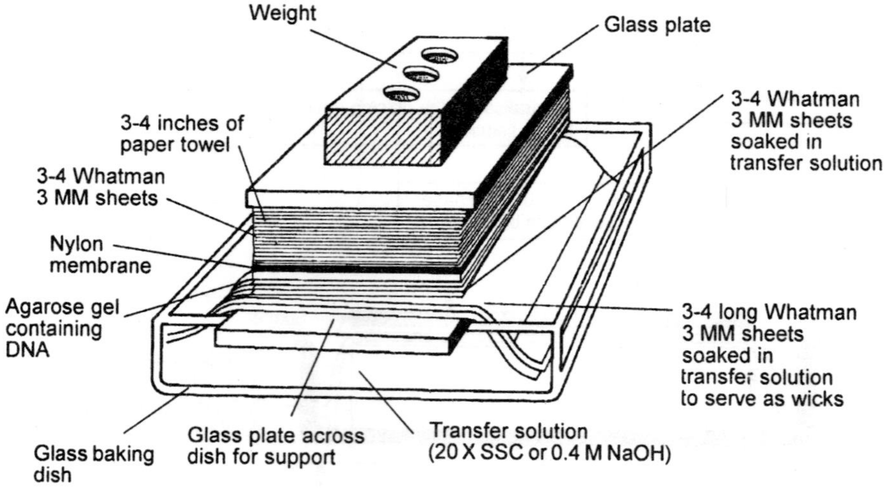

FIGURE 18.13 Gel electrophoresis with Nylon membrane (Northern blotting).

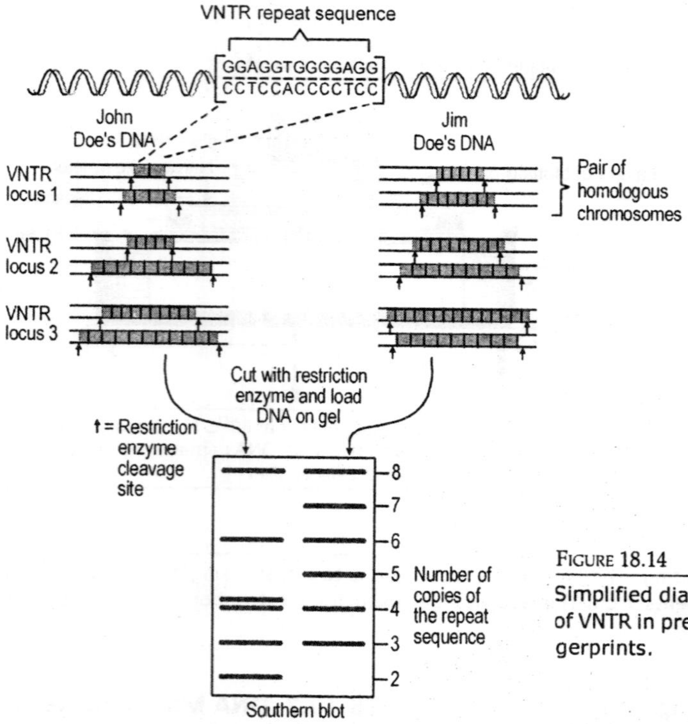

FIGURE 18.14 Simplified diagram of the use of VNTR in preparing DNA fingerprints.

18.10.3 DNA Typing (DNA fingerprinting or DNA profiling)

DNA typing (or, DNA fingerprinting or DNA profiling) is used in paternity or maternity test and in forensic science (to detect a criminal).

18.10.3.1 Paternity Test by DNA Typing (DNA fingerprints)

DNA typing or DNA fingerprints provide a positive identification of the true father. When a dispute arises regarding the paternity of a child, the following is the

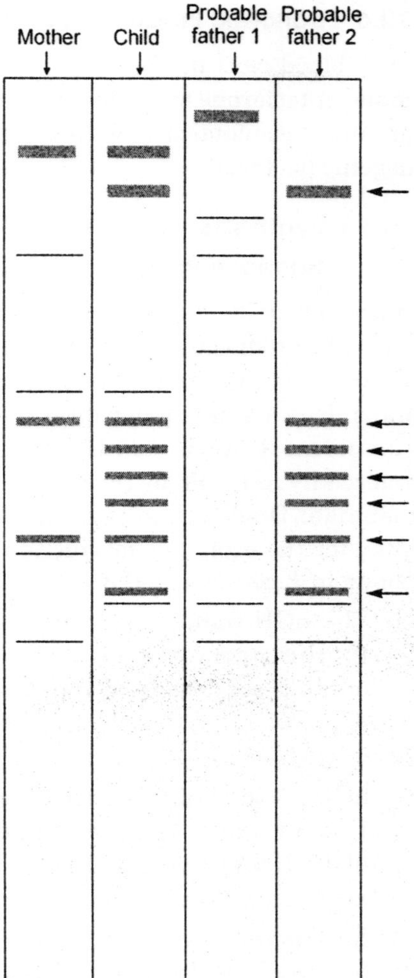

FIGURE 18.15

DNA fingerprinting for paternity test. Arrows mark bands that probable father no. 2 is actual father.

procedure to test the biological father of the child concerned.

1. DNA samples are obtained from cells of the child, the mother and *possible fathers*.
2. Fig. 18.14 shows the method of preparation of DNA fingerprints. DNA is extracted, amplified by PCR, and analysed with DNA probes by Southern blot method. The human genome contains many short DNA sequences present in tandem repeats of varied lengths. These variable number of tandem repeats (VNTRs) are the basis of DNA-fingerprints.
3. The fingerprints of each individual done in the above method are compared.
4. In comparison, all the bands in the child's DNA type are expected to be present in the combined DNA type of the parents.
5. For each pair of homologous chromosomes, the child would have received one from each parent, i.e., half of the bands in the child's DNA type have come from the mother and the other half from the father.
6. Fig. 18.15 shows that, in this case, the DNA types indicate that the second *possible* father is the child's **actual biological father**. The arrows at the side of the types indicate the bands that have inherited to the child from the true parents.

18.10.4 DNA Forensics: Crime Investigation

In crime investigations, DNA samples of suspects are analysed to build a case against or to exclude a suspect. If DNA samples collected from crime spot (blood stains, hairs, semens etc.) match with that of the suspect, then, the person can be convicted. The procedure of matching the VNTR is the same as in case of a paternity test.

18.10.5 Gene Therapy

RFLP (or restriction fragment length polymorphism) and other recombinant DNA techniques allow geneticists to find **defective** alleles in human genome. In this case, recombinant DNA techniques are used to rectify the gene expression or to insert a normal gene replacing the defective one.

Gene therapy involves introduction of a normal copy of gene in the genome of a person carrying a defective gene. For many inherited diseases gene therapy may be applied in a promising way.

A gene that is introduced into an organism is called **transgene**. The organism which carries the introduced gene (transgene) is called **transgenic**. In successful gene therapy the transgene synthesizes normal gene product and cure the disease.

Viruses are ideal organisms to introduce genes into the cell. Modified viruses are most effective "vector" systems in gene therapy. Geneticists try to "design" viruses that give human cells the genes they need to compensate for defective genes causing genetic diseases, such as thalassemia or other disorders. The retroviruses are particularly useful to ferry the genes into the cells. As part of their natural life cycle, retroviruses insert a DNA copy of their genetic material into the chromosome of an infected cell. So, if a gene therapist desires to make some new gene a permanent part of the cell's DNA (as transgene), the first step is to insert the gene into the genome of a retrovirus, and then let the retrovirus insert the gene into the chromosome of the patient.

Gene therapy is of two types: (1) Somatic cell gene therapy and (2) Germline gene therapy.

Human gene therapy is performed under guidelines of NIH (National Institute of Health). The first case of gene therapy was done in 1990, in a 4 yr-old girl suffering from *Adenozine deaminase deficient severe combined immuno-deficiency disease* (ADĀSCID).

The gene therapy is still now under research and a hundred per cent cure has not been achieved. It is hopefully expected that in future, the gene therapy with gene—replacement will become a tool for the treatment of human genetic diseases.

18.10.6 Economic Production of Proteins

Insulin, blood clotting factor VIII, growth hormone, interferons and other proteins can be produced economically with the help of transgenic bacteria.

18.10.6.1 Synthesis of Human Growth Hormone (hGH)

Human growth hormone (hGH) is biologically produced in anterior pituitary gland and is required for normal growth of humans. It is a single polypeptide chain of 191 amino acids. It can now be synthesized commercially in *E.coli*, by the application of recombinant DNA technology. The method adopted for the economic production of the hormone in *E.coli* is as follows:

1. The hGH-coding sequence is joined to the promoter and ribosome-binding sequences of the *E.coli* lac operon.
2. Two parts of the DNA sequence of hGH—one synthetic DNA encoding 1 to 23 amino acids and the other hGH cDNA encoding 24 to 191 amino acids are involved.
3. Restriction enzyme Hae III is used to make a cleavage site in the nucleotide-pair triplet specifying codon 24.
4. The codon 24 (Ala) is used to fuse the two parts of the DNA sequence of hGH.
5. This chimeric DNA is now inserted into an *E.coli* plasmid. The plasmid carries the lac operon, and is introduced into *E.coli* by transformation (Fig. 18.16). This machinery produces hGH in *E.coli*.
6. The hGH thus produced in bacteria consists of **Met** as its initial amino acid at the amino terminus. But the human native hGH contains **Phe** at amino terminus (by removing the original Met enzymatically).

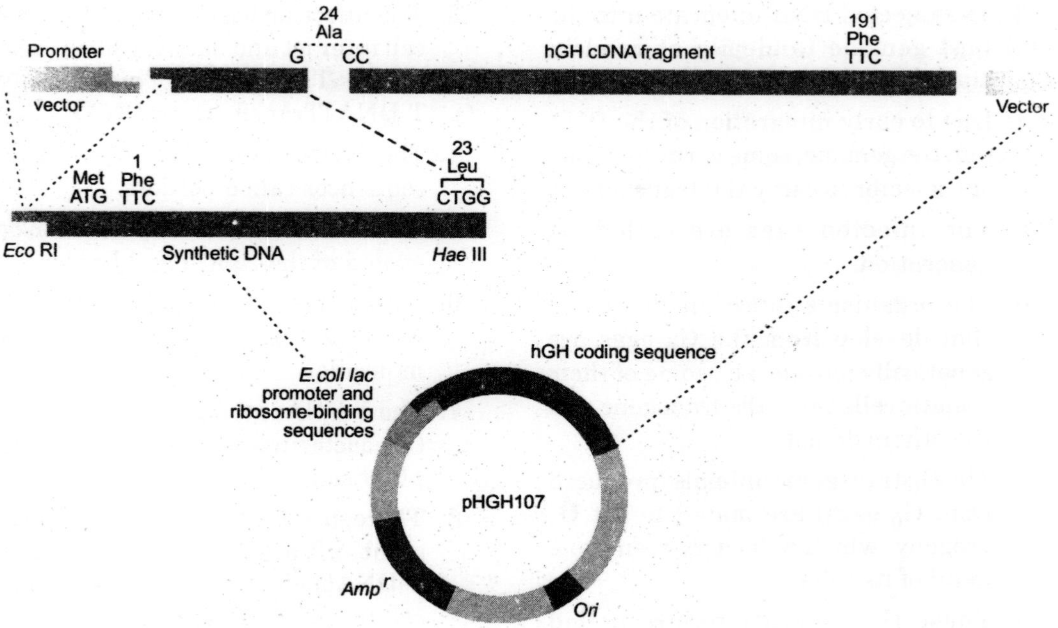

FIGURE 18.16

The synthesis of human growth hormone in *E.coli*. The DNA segment encoding amino acids 1-23 was assembled *in vitro* from 12 synthetic oligonucleotides. The rest of the coding sequence (amino acids 24-191) was derived from a cDNA. The desired cDNA fragment was cleaved from the complete cDNA with *Hae* III.

7. The *E.coli* does not excise the terminal Met from its hGH; this hGH (with an extra amino acid, Met), synthesized in *E.coli*, however, works good in human.
8. However, recently, a DNA sequence encoding a signal peptide is added to the hGH gene, by which the hGH produced in *E.coli* is correctly processed. The signal peptide helps to remove the amino terminal methionine of the hGH in *E.coli*, and the product now becomes identical to the original (native) human growth hormone.
9. In this way, hGH is now commercially produced in *E.coli* and is used in treatments of diseases related to growth in human.

18.11 Transgenic Organisms (Genetically Modified Organism)

Artificial introduction of genes in an organism makes it a transgenic organism. Such genetically engineered (**m**odified) **o**rganisms (GMO) are used to study the function of the genes (which are either modified or foreign or synthetic). The transgenic organisms or GMO are also used to produce medicinal products, food products or for studies of genetic diseases.

18.11.1 Transgenic Animals

Transgenic animals can be produced by two methods—(a) Microinjection of DNA into a fertilized egg and(b) Infection of early embryos with retroviral vectors.

The process for microinjection includes:

1. Eggs are removed from the mature female and are fertilized *in vitro* (in laboratories, usually mice are used as an object).
2. The foreign DNA (of desired nature) is then pushed into the male haploid nucleus through micro glass needle.

3. The injected DNA integrate into the host genome tandemly at random sites.
4. Due to early integration of the DNA into the genome, some germ-line cells of the embryo carry the transgene.
5. The injected eggs are called G_0 generation.
6. The organisms (mice, in this case) that develop from the G_0 eggs are genetically mosaic, i.e., some of their somatic cells carry the transgene and the others do not.
7. The G_0 transgenic animals (produced from G_0 eggs) are mated to get G_1 progeny, which will carry transgene in all of its cells.
8. These G_1 generations are called **transgenic animals** (mice).

In laboratories, the hGH gene has been used as transgene in many experiments and transgenic rats, mice, hamsters, pigs etc. have been developed that show enormous growth. Transgenic animals are excellent tools for the study of gene function. Many different kinds of transgenic animals have been developed (using different kinds of transgenes) that reproduce normally showing normal patterns of inheritance. It indicates that the transgenes have been integrated into the host genome.

18.11.2 Transgenic Plants

To develop transgenic plant is difficult, and laborious technologies are applied to produce a genetically modified plant. One of them is to produce a gall (tumor) genetically. The general features of the gall formation are—

1. *Agrobacterium tumefaciens* bacterium causes crown gall (tumor) in a wound in dicot plants.
2. The transformation of the wounded plant cells is caused by a plasmid of the bacteria, called **Ti** plasmid. It is a big circular DNA.
3. The infecting bacterium and the host cell interact and excise a 30 kb region of the **Ti** plasmid. This is called T-DNA (Transforming DNA).
4. The T-DNA is flanked by two 25 bp sequences called borders.
5. The T-DNA is excised by a nick in one strand of the right-hand border.
6. A second nick in the left hand border releases a single stranded T-DNA molecule.
7. The T-DNA is then transformed from the bacterium to the genome of the plant cell.
8. There, the T-DNA integrates and the plant cell acquires the genes of the T-DNA including the genes for plant cell transformation (gall formation).

Using recombinant DNA technology, the excision, transfer and integration of the T-DNA are carried on. The **Ti plasmid** and the T-DNA constitute a vector that may introduce new DNA sequence into the host-cell genome. Thus, any gene (transgene) placed between the 25 borders of the T-DNA will integrate into the host-plant-cell genome and transform the host. The plant, thus become a transgenic one.

Transgenic plants are used in agriculture to improve crop-quality and quantity.

18.12 Human Genome Project

Human Genome Project was launched in 1990. The aims of the project were—

1. To map all the human genes
2. To determine the nucleotide sequences of all human chromosomes (22 autosomes and X and Y = 24 chromosomes).

First director of this project was James Watson (the DNA-double helix discoverer, Nobel Laureate), but he was replaced in 1993 by Francis Collins. A detailed map of the human genome was published in 1996. The first complete sequence of the human chromosome, the chromosome 22, was published in 1999, and that of the chromosome 21 in 2000.

A Private Company, **Celera Genomics** led by J. Craig Venter, worked independently, after the intervention of the White House, published their work (first draft of sequence of human genome) in the same year (2001), along with the draft of genome sequence made by Human Genome Project.

The discovery of the sequence of the human genome led to the discovery that there are only about 30,000-35,000 genes in human. The project was further advanced by the work of the **International Human Genome Sequencing Consortium** who estimated and published (2004) that there are only 22,287 protein encoding genes and other genes for RNA products.

Another organ **ENCODE** (Encyclopedia of DNA elements) was formed to identify the nongenic functional elements of human genome. Another body of similar nature is **HUPO** (Human Proteome Organisation), which has the aim to determine the structure and function of all the proteins encoded in the human genome.

IMPORTANT NOTES:

1. Genetic engineering deals with some methodologies to produce recombinant DNA artificially, and their application in biological fields. It is also called recombinant DNA technology and gene manipulation.
2. DNA cloning is the process of production of large amounts of identical pure DNA.
3. Restriction enzyme is an endonuclease that recognizes a specific sequence in DNA and cleave the DNA molecule at that site.
4. Vectors are molecular vehicles that carry the DNA fragments into a bacterium. It is derived from a plasmid or bacteriophage.
5. Reverse transcriptase enzyme is used to make DNA that complements any RNA molecule. These DNA molecules are called complementary DNA (cDNA) molecules. They can be made double stranded, recombined into a vector and used to transform a bacteria.
6. Genomic libraries are collection of innumerable clones of DNA fragments that contain at least one copy of every DNA sequence in the genome. By recombinant technology DNA library, cDNA library, human genome library etc. can be produced.
7. The Polymerase Chain Reaction or PCR is a procedure by which *in vitro* amplification of genes can be performed. The PCR was discovered by K. Mullis.
8. Investigation and analysis of the structure of DNA and RNA are done by Southern blot Hybridization and Northern Blot Hybridization.
9. Restriction Fragment Length Polymorphism (RFLP) helps to detect and diagnose inherited disease and other researches in human genetics.
10. Recombinant DNA technology is applied in various fields like medicinal, agricultural, anthropological, genetical, conservational and ecological. It can also be applied in forensics and industries. Human genetic diseases are tested by this technology and PCR. Paternity test and crime investigations can be done by recombinant DNA technology that includes DNA fingerprints.
11. Gene Therapy involves introduction of normal gene into the genome of a patient carrying a defective gene; the normal gene replaces the defective gene and restore normal function, thereby curing the disease.
12. A foreign gene introduced into an organism is called a transgene, and the organism that carries the transgene is called transgenic. Transgenic animals and plants have been developed artificially by applying gene manipulation. These genetically modified organisms (GMO) are useful tools in various researches in the field of gene study.
13. Some human and other eukaryotic proteins like insulin, hGH etc. have been produced in commercial basis by the help of transgenic bacteria.

14. Human genome project was launched in 1990 with the aim of mapping and sequencing all the human genes. In association with such other organizations a nearly correct mapping and sequence of nucleotides have been reported in 2004; human genome contains about 22,287 protein encoding genes and other genes for RNA products.

EXERCISE

1. Define genetic engineering.
2. What do you understand by cloning? How DNA-cloning is done?
3. What is a restriction enzyme? How does it act? Cite a suitable example.
4. Write notes on vectors. What is a λ-phage vector?
5. How is a recombinant DNA constructed?
6. Describe the procedure of isolation and cloning of a gene.
7. How cDNA can be produced? What is its significance? What is a DNA probe?
8. What are the roles of restriction enzyme, vector, ligase and cloning in recombinant DNA technology?
9. What is a genomic library? How can a recombinant DNA library and cDNA library be produced?
10. What are the features of cDNA library?
11. What is PCR? How does it act? What is its role? What is meant by gene-amplification?
12. What do you mean by DNA analysis? How can it be done?
13. What is RFLP? Discuss its role with example.
14. Describe the technique for DNA analysis.
15. What are the fields of application of recombinant DNA technology. Illustrate how can it be applied in hereditary disease diagnosis?
16. What do you understand by DNA polymorphism?
17. How are the DNA molecules tested? Give example.
18. What do you know about DNA fingerprinting? What significant role can it play?
19. What do you know about paternity test by DNA fingerprinting?
20. How can an actual father be confirmed in a paternity dispute case?
21. What is the role of DNA typing in forensics?
22. What is gene therapy? Discuss it in detail.
23. How can an eukaryotic protein be produced commercially? Describe the process in detail.
24. What is transgene? What is a transgenic animal? How can it be developed? Give example.
25. How can a transgenic plant be developed?
26. Write a short essay on Human Genome Project.

INDEX